Biochar in European Soils and Agriculture

T0199525

This user-friendly book introduces biochar to potential users in the professional sphere. It de-mystifies the scientific, engineering and managerial issues surrounding biochar for the benefit of audiences including policy makers, landowners and farmers, land use, agricultural and environmental managers and consultants, industry and lobby groups, and NGOs.

The book reviews state-of-the-art knowledge in an approachable way for the non-scientist, covering all aspects of biochar production, soil science, agriculture, environmental impacts, economics, law and regulation and climate change policy. Chapters provide 'hands-on' practical information, including how to evaluate biochar and understand what it is doing when added to the soil, how to combine biochar with other soil amendments (such as manure and composts) to achieve desired outcomes, and how to ensure safe and effective use.

The authors also present research findings from the first coordinated European biochar field trial and summarise European field trial data. Explanatory boxes, infographics and concise summaries of key concepts are included throughout to make the subject more understandable and approachable.

Simon Shackley is a Lecturer in the School of GeoSciences at the University of Edinburgh, UK.

Greet Ruysschaert is a Senior Researcher at the Institute for Agricultural and Fisheries Research (ILVO), Belgium.

Kor Zwart has recently retired as Senior Researcher at Alterra Wageningen-UR, Netherlands.

Bruno Glaser is Professor of Soil Biogeochemistry at the Martin Luther University Halle-Wittenberg, Germany, and Managing Director of Biochar Europe UG.

Biochar in European Soils and Agriculture

Science and practice

Edited by Simon Shackley, Greet Ruysschaert, Kor Zwart and Bruno Glaser

Routledge
Taylor & Francis Group

LONDON AND NEW YORK

from Routledge

First published 2016
by Routledge

2 Park Square, Milton Park, Abingdon, Oxfordshire OX14 4RN
52 Vanderbilt Avenue, New York, NY 10017

Routledge is an imprint of the Taylor & Francis Group, an informa business

First issued in paperback 2020

British Library Cataloguing-in-Publication Data
A catalogue record for this book is available from the British Library

Library of Congress Cataloging in Publication Data
Names: Shackley, Simon, editor. | Ruysschaert, Greet, editor. | Zwart, Kor,
editor. | Glaser, B. (Bruno), 1966- editor.
Title: Biochar in European soils and agriculture : science and practice /
edited by Simon Shackley, Greet Ruysschaert, Kor Zwart and Bruno Glaser.
Description: London ; New York : Routledge, 2016. | Includes bibliographical
references and index.
Identifiers: LCCN 2015034323| ISBN 9780415711661 (hbk) | ISBN 9781315884462
(ebk)
Subjects: LCSH: Biochar. | Ashes as fertilizer—Europe. | Soil amendments—
Europe. | Carbon dioxide mitigation—Europe. | Carbon sequestration—Europe.
Classification: LCC QH545.S63 B46 2014 | DDC 628.5/32—dc23
LC record available at http://lccn.loc.gov/2015034323

ISBN: 978-0-415-71166-1 (hbk)
ISBN: 978-0-367-60604-6 (pbk)

Typeset in Bembo
by Fish Books Ltd.

Contents

Figures

Tables

Boxes

Acronyms and abbreviations

Å	Angstrom
AD	anaerobic digestion
AD	anno domini (after Christ)
ADD	anaerobic digestion digestate
ADE	anthropogenic dark earths
AEC	anion exchange capacity
ANOVA	analysis of variance
APECS	Anaerobic pathways to renewable energies and carbon sinks
As	arsenic
Atm	atmospheric pressure
BC	Before Christ
BC	black carbon
BET	Brunauer, Emmett and Teller theory for measuring specific surface area
BMCs	biochar-mineral complexes
BQM	Biochar Quality Mandate
BTP	biochar testing protocol
C	carbon
Ca	calcium
CC	carbon content
CCS	CO_2 capture and storage
Cd	cadmium
CDM	Clean Development Mechanism (UNFCCC)
CEC	cation exchange capacity
CEF	carbon emission factor
CERs	certified emission reductions (under the CDM)
CFI	Climate Farming Initiative (Australia)
CH_4	methane
CHP	combined heat and power
Cl	chlorine
C:N	carbon to nitrogen ratio
C:O	carbon to oxygen ratio
Co	cobalt

CO	carbon monoxide
CO_2	carbon dioxide
CO_2e	carbon dioxide equivalent (including methane, nitrous oxide and three other gases)
-COOH	carboxylic acid (the line refers to an additional chemical bond)
COP	Conference of the Parties (to the UNFCCC)
COP-15	the 15th meeting of the COP in Copenhagen in 2009
Cr	chromium
Cu	copper
d.b.	dry basis (when weighing)
DBC	dissolved black carbon
DE	Germany
DM	dry matter
DOC	dissolved organic carbon
DW	dry weight (when weighing)
EBC	European Biochar Certificate
ETS	Emissions Trading Scheme
EU	European Union
EU-25	the 25 member states of the EU in the mid 2000's.
EU-28	the 28 member states of the EU as of 2015.
EU-ETS	European Union Emissions Trading Scheme
Fe	Iron
FTIR	Fourier Transform InfraRed spectroscopy
G77	a group of 77 developing countries that was established in 1964
GHGs	greenhouse gas(es)
GPP	gross primary productivity
GPS	global positioning system
GW	gigawatt (10^9)
GWh	gigawatt hour
GWP	global warming potential
H	hydrogen (atom)
H_2	hydrogen (gas)
H_2O	water
H_2O_2	hydrogen peroxide
Ha	hectare
H:C	hydrogen to carbon ratio
Hg	mercury
HHT	higher heating temperature
HNO_3	nitric acid
HTC	hydrothermal carbonisation
IBI	International Biochar Initiative
ICP-OES	inductively coupled plasma optical emission spectrometry
IP	intellectual property
IPCC	Intergovernmental Panel on Climate Change
IRR	internal rate of return

ISO	International Standards Organisation
K	potassium
KCl	potassium chloride
Kg	kilogramme
KP	Kyoto Protocol (under UNFCCC)
kpa	kilopascal (a measure of pressure)
kW	kilowatt (a measure of energy capacity)
kWh	kilowatt hour (a measure of energy generation)
LCA	life cycle assessment
LCI	life cycle inventory
LHV	lower heating value
M	mole (the amount of a chemical substance that contains as many elementary entities as there are atoms in 12 grammes of pure carbon-12)
Mg	magnesium
Mg	a million grammes, which is the same as one tonne
MJ	megajoule (10^6 joules, which is a measure of energy or work done)
Mn	manganese
Mo	molybdenum
MPa	megapascal (10^6 pascals, a measure of pressure)
MPL	maximum permissible limit
MRT	mean residence time
MS	mass spectroscopy
MT	million tonnes
MW	megawatt (10^6 watts)
MWh	megawatt hours
N	nitrogen
N_2O	nitrous oxide
Na	sodium
NGO	non-governmental organisation
NH_3	ammonia
NH_4^+	ammonium ion
Ni	nickel
NO_2^-	nitrite ion
NO_3^-	nitrate ion
NHC	Nutrient holding capacity
NMR	nuclear magnetic resonance
NPK	nitrogen, phosphorus and potassium containing fertiliser
NPP	net primary productivity
O	oxygen atom
O_2	oxygen (gas)
OC	organic carbon
O:C	oxygen to carbon ratio
OECD	Organisation for Economic Cooperation and Development
-OH	hydroxyl (the line refers to an additional chemical bond)

OPRA	operational risk appraisal
OSRP	oil seed rape pellets
P	phosphorus
PAHs	polycyclic aromatic hydrocarbons
PAS	publicly-available specification
PAW	plant-available water
Pb	lead
PBS	biochar-pyrolysis system
PCBs	polychlorinated biphenyls
PCDDs	polychlorinated dibenzodioxins
PCDFs	polychlorinated dibenzofurans
pF	a measure of the soil moisture tension
PLFA	phospholipid-derived fatty acids
PTEs	potentially toxic elements
P=0.05/ p<0.05	indicates statistical significance at 95% certainty
R&D	research and development
RD&D	research, development and demonstration
REACH	Regulation, Evaluation, Authorisation and Restriction of Chemicals Regulation
RED	Renewable Energy Directive (EU)
ROCs	renewable obligation certificates
RomChar	biochar provided from Romania for the Interreg IVb field ring trial
S	sulphur
Se	selenium
SE	Sweden
SEM	scanning electron microscope
SO_4^{2-}	sulphate ion
SOC	soil organic carbon
SOM	soil organic matter
SSA	specific surface area
SSW	sewage sludge water
SWP	soil water potential
TEM	transmission electron microscopy
UNFCCC	United Nations Framework Convention on Climate Change
USDA	United States Department of Agriculture
USEPA	United States Environmental Protection Agency
w/w	weight by weight
WFD	Waste Framework Directive (EU)
WHC	water holding capacity
WID	Waste Incineration Directive (EU)
XRCT	X-ray computer tomography
Zn	zinc

Notes on authors

David Andersson, CEO, EcoEra AB, Sweden. Email: david.andersson@ecoera.se

Peter Brownsort, research associate, School of GeoSciences, University of Edinburgh, UK. Areas of interest: pyrolysis process technologies for biochar production; carbon capture and storage.

Esben Wilson Bruun, Orbicon A/S, Roskilde, Denmark. Expertise: biochar (effect in soil, on GHG emission, degradation), sludge treatment in reed bed systems. Email: esbr@orbicon.dk

Alice Budai, research fellow, Department of Soil Quality and Climate Change, Norwegian Institute of Bioeconomy Research (NIBIO), Norway. Email: alice.budai@nibio.no

Andrew Cross, ECCI, School of GeoSciences, Edinburgh. Andrew is a research associate at the University of Edinburgh. His background is in carbon cycling in soils, and his research interests include the long-term stability and agronomic potential for biochar and soil carbon storage associated with 'climate smart' agricultural systems. Email: andrew.cross@ed.ac.uk

Dane Dickinson, doctoral researcher, Department of Biosystems Engineering, Ghent University, Belgium and UK Biochar Research Centre, University of Edinburgh, UK.

Achim Gerlach is a veterinarian working for the Schleswig Holstein region in Northern Germany. Email: taiyang@gmx.de

Bruno Glaser is professor of soil biogeochemistry at the Martin Luther University Halle-Wittenberg, Germany. Email: bruno.glaser@landw.uni-halle.de

James Hammond, research fellow, World Agroforestry Centre and Bangor University, UK. James is a social scientist working on system solutions to environmental and agricultural problems. His work focuses on developing metrics and indicators to assess the human, environmental and economic dimensions of decision making. Email: J.Hammond@cgiar.org

Henrik Hauggaard-Nielsen, Department of Environmental, Social and Spatial Change, Roskilde University, Roskilde, Denmark. Expertise: climate change adaptation and mitigation addressed in a circular perspective to optimize future production and consumption. Email: hnie@ruc.dk

Lars D. Hylander, Swedish University of Agricultural Sciences (SLU), Uppsala, Sweden, is an agronomist with research activities in carbon sequestration with biochar and nutrient recycling. Email: Lars.Hylander@slu.se

Rodrigo Ibarrola, consultant, Environmental Resources Management (ERM), Mexico City, Mexico. Email: ibas79@hotmail.com

Claudia Kammann is based at Geisenheim University, Germany. Her main research interests are: (1) in the effects of rising atmospheric CO_2 concentrations on plants and ecosystems; (2) biogeochemical feedback effects of climate change; and (3) the development of strategies to implement biochar for mitigating climate change. Email: claudia.kammann@hs-gm.de

Jürgen Kern, senior scientist and head of the working group Biogeochemistry, Leibniz-Institute for Agricultural Engineering Potsdam. Email: jkern@atb-potsdam.de

Tor Kihlberg is an organic chemist whose research interests include climate mitigation, renewable fuels, urban food production and biochar. He is founder of Carboinventus AB, Uppsala, Sweden. Email: tor.kihlberg@hotmail.com

Kirsi Kuoppamäki, researcher, University of Helsinki, Department of Environmental Sciences, Lahti, Finland. Email: kirsi.kuoppamaki@helsinki.fi

Achim Loewen is professor for energy technology and environmental management at HAWK, University of Applied Science and Arts in Göttingen, Germany. He is director of the Department of Sustainable Energy and Environmental Technologies and leads a research team on bioenergy and climate protection. Email: Achim.loewen@hawk-hhg.de

Elisa Lopez-Capel is research associate in urban soil science, School of Agriculture, Food and Rural Development, Newcastle University. Newcastle upon Tyne, UK. Elisa's expertise is in biogeochemistry, soil science and analytical method development. Her research interests include urban agriculture, carbon abatement technologies and energy biosciences. Email: Elisa.lopez-capel@ncl.ac.uk

Ondřej Mašek, School of GeoSciences, University of Edinburgh, UK. Ondřej is lecturer in engineering assessment of biochar, UK Biochar Research Centre. Key research interests include: pyrolysis, biomass conversion, biochar production and characterisation, biochar design and applications, biochar technologies and their integration into systems. Email: ondrej.masek@ed.ac.uk

Jan Mumme is an agricultural engineer, who specializes in the areas of renewable energy, carbon sequestration and biochar product development. He works as a Marie Curie fellow at the UK Biochar Research Centre at the University of Edinburgh and is appointed lecturer at Humboldt Universität zu Berlin. Email: jan.mumme@ed.ac.uk

Victoria Nelissen is research associate at the Institute of Agricultural and Fisheries Research (ILVO) in Belgium. She obtained a PhD on the effects of biochar on soil processes, soil functions and crop growth. Her key research interests are related to sustainable agriculture, soil and nutrient management, and interactions between environment and agriculture. Email: victoria.nelissen@ilvo.vlaanderen.be

Adam O'Toole, researcher, Climate and Environment Division, Norwegian Institute of Bioeconomy Research (NIBIO), Ås, Norway. Email: adam.otoole@nibio.no

Romke Postma, senior project manager, Nutrient Management Institute (NMI), Wageningen, the Netherlands. Email: postma@nmi-agro.nl

Daniel P. Rasse, researcher, Department of Soil Quality and Climate Change, Norwegian Institute of Bioeconomy Research, Norway. Email: daniel.rasse@nibio.no

Jan-Markus Rödger was employed as a research assistant at the University of Applied Science (HAWK) in Göttingen, Germany during the writing process and now works at the Technical University of Denmark, Department of Management Engineering, Division of Quantitative Sustainability Assessment. Email: januw@dtu.dk

Frederik Ronsse is professor of biomass thermochemical process technology, Department of Biosystems Engineering, Faculty of Bioscience Engineering, Ghent University, Belgium. Email: Frederik.Ronsse@UGent.be

Greet Ruysschaert is senior researcher at the Institute of Agricultural and Fisheries Research (ILVO) in Belgium. Her main research interests are related to sustainable soil management, including effects of soil amendments and soil tillage on soil quality, nutrient dynamics and crop growth. Email: greet.ruysschaert@ilvo.vlaanderen.be

Hans-Peter Schmidt, Ithaka Institute for Carbon Strategies, Arbaz, Switzerland. Email: schmidt@ithaka-institut.org

Michael Schulze is a geoecologist with further interests in the fields of geoinformatics and carbon sequestration by use of biochar. He was involved in the APECS-Project (Anaerobic Pathways to Renewable Energies and Carbon Sinks) as a diploma student at the Leibniz Institute for Agricultural Engineering. Email: michael-schulze@outlook.de

Simon Shackley is lecturer in carbon policy and programme director for the MSc in Carbon Management in the School of GeoSciences, University of Edinburgh, Scotland. He works at the interface of the natural and social sciences on carbon mitigation and abatement. Email: simon.shackley@ed.ac.uk

Franziska Srocke is a PhD student at the UK Biochar Research Centre at the University of Edinburgh and Plant Science Department at McGill University. Her research focuses on biochar-microbe interactions for bioremediation of soil and includes the analysis of the porous structure of biochar using microto mography. Email: Franziska.srocke@ed.ac.uk

Marianne Stenrød is head of department/research scientist at the Norwegian Institute of Bioecoenomy Research (NIBIO), Ås, Norway, where she works on the fate and occurrence of pesticides in the soil and water environment. Email: marianne.stenrod@nibio.no

John Stenström, Uppsala BioCenter, Department of Microbiology, Swedish University of Agricultural Sciences, Uppsala, Sweden. Email: john.stenstrom@ slu.se

Tania Van Laer, affiliated researcher, Ghent University, Belgium. Email: tania. vanlaer@ugent.be

Laura van Scholl, project manager, Nutrient Management Institute NMI. Email: Laura.vanscholl@nmi-agro.nl

Kor Zwart is a retired soil scientist from Alterra, Wageningen UR. He has been involved in many different research projects, including research on the application of biochar in agriculture. Currently he is a Guest Scientist at Alterra. Email: kor.zwart@wur.nl

Acknowledgements

This book has been long in the making, due to unpredictable hindrances that took away some contributors' time at key moments. The editors would like to thank the following for ensuring that the book became a reality. Rob van Haren initiated the Interreg IVb North Sea region project 'Biochar: climate saving soils' (2009–2014) led by the Provincie Groningen, the Netherlands, which brought together many of the book's contributors and provides the backbone for the book. We thank Frans Debets and Douwe van Noordenburg, who took over the administration of the project on behalf of the Provincie Groningen from 2012 to 2014. We are grateful to Claudia Kammann, Hans-Peter Schmidt and other colleagues not involved in the Interreg IVb project for accepting our invitation to contribute to the book, thereby bringing wider coverage and relevance to the manuscript. This spirit of collaboration was fostered by the European Science Foundations's support for a COST Action on Biochar (2012–2016). Many of the graphics in the book were designed by Jonathan Stevens of Starbit Ltd., Edinburgh, Scotland. Special thanks go to Brendan Martin (School of GeoSciences, University of Edinburgh, Scotland) who has helped in innumerable ways with the complex administrative and financial procedures entailed in running an Interreg project. Jennifer Mills and Bill Bruce from the University of Edinburgh also provided excellent support in administering the Interreg project. Caitlin Smythe did a magnificent job as copy editor of the full manuscript. Finally, Ashley Wright of Routledge, our publishers, exhibited excessive patience in waiting for the final manuscript. She and her colleague Tim Hardwick have always been extremely helpful and encouraging, and without their support the book would never have been completed.

Chapter 1

Introduction

Simon Shackley, Hans-Peter Schmidt
and Bruno Glaser

Biochar is the solid product of heating biomass in zero or very low oxygen conditions. It is a carbon-rich, porous, black material, which, in addition to storing carbon in a stable form for hundreds of years, has important potential benefits for soils and plant growth. The chemical structure of biomass changes when heated in the absence of oxygen, resulting in a loss of hydrogen, nitrogen and oxygen relative to carbon. The carbon atoms become strongly bound to one another, forming a molecular structure which makes it very hard for microorganisms to break biochar down. By contrast, most decaying biomass (that has not been converted to biochar) is broken down readily and rapidly by microorganisms, releasing the carbon to the atmosphere as carbon dioxide. Since the carbon in plant biomass comes from CO_2 in the atmosphere via photosynthesis, there is no net effect on atmospheric carbon concentrations ('carbon neutral'). By removing CO_2 as stable carbon for storage in soils, however, the concentration of atmospheric CO_2 is reduced. While biochar can contribute to mitigating climate change in this way, overall economics and constraints on the amount of spare and affordable biomass that can be used for producing biochar put limits on the operation.

To understand biochar, it is useful to distinguish the short-term carbon cycle from the long-term carbon cycle. The short-term cycle involves carbon that moves through living organisms. It can occur on timescales of a few minutes to a few hundred years, and so it includes most of the emissions of CO_2 from burning vast amounts of fossil fuels (approximately half a trillion tonnes of carbon since the industrial revolution began in around 1750). The long-term cycle occurs over thousands to millions of years, and involves carbon stored in rocks and fossils. It includes the formation of fossil fuels from vegetation (trees, other plants, algae and other living organisms) and the weathering of rocks. Over the long term, vast amounts of carbon have been sucked out of the atmosphere and locked up in rock formations, only to be rapidly released again as CO_2 when coal, oil and gas are burnt. Such carbon is laid down during the long-term cycle and its sudden release disrupts the balance of the short-term cycle.

Biochar production has the opposite effect; namely, it captures a portion of the atmospheric carbon in the short-term carbon cycle and shunts it into the long-carbon cycle. It puts into reverse the dominant trend of adding CO_2 from the long-term to the short-term carbon cycle. The continued (and growing) annual emissions of 8 billion tonnes of carbon (30 billion tonnes of CO_2) from fossil fuels into the atmosphere is creating an enormous change in the short-term carbon cycle – one from which it will not fully come into balance for thousands of years (i.e. only on the timescale of the long-term carbon cycle as the deep ocean comes into balance again with the atmosphere). Biochar stores carbon long enough to have an impact on the long-term carbon cycle, but it probably needs to be implemented continuously for hundreds of years to make a lasting impression, given the time that it takes for the atmosphere, the earth's vegetation and oceans to come into balance. If the earth is suffering a chronic risk from excessive CO_2 emissions, biochar is one of the potential medicines, but it would need to be taken continuously for the indefinite future to be effective because it removes carbon for hundreds but probaly not for many thousands of years. For this reason, biochar is not a long-term solution to climate change, but rather is part of the 'bridge' that takes us to a sustainable low-carbon energy future based upon renewable energy or other low-carbon energy supply options over the next 30 to 100 years.

Adding biochar to cultivated land – as well as to degraded land that could be brought in to use (at least as a carbon store) – can, in principle, contribute to climate change mitigation. However, it is only likely to make a large contribution where spare biomass is concentrated (e.g. aggregated over space and time), and where the costs of collection, preparation and biochar production are acceptable in relation to potential benefits. Plenty of spare biomass exists but if it is thinly spread over vast areas of land, it takes a lot of work to be collected up and used and is therefore expensive.

So what are the main benefits of biochar when added to soil? Consisting of a large proportion of carbon, biochar adds both inorganic and organic matter, in particular carbon, to soil. Many soils have very low amounts of organic carbon due to many years of culivation without sufficient biomass replacement. We know that soil organic matter improves the quality of many soils: it increases its ability to hang on to water, it reduces the soil's density, making it easier to plough and allowing better development of plant roots, it enhances its ability to provide essential nutrients to plants, and so on. Because of the high number and volume of spaces – known as pores – within biochar, it stores soil water and nutrients that are dissolved in that water. Chemical reactions occur between molecules on the biochar surface and nutrients that are vital to healthy plant growth, such as nitrates, ammonium, phosphates, potassium, magnesium, sodium, calcium and so on. Some evidence suggests that such chemical reactions act to improve the availability of nutrients to the plant by reducing loss by run-off and leaching. This means that added nutrients (such as from chemical or organic fertilisers) are utilised more efficiently by plants grown in soils containing suitable types of biochar. The ashes in biochar are alkaline and this adds a further important property, namely, countering the acidity of many soils. There is also emerging evidence that biochar in soils aggregates with other soil organic matter and with minerals with benefits for soil health and plant productivity (Scott *et al.* 2014).

The best examples of the potential for biochar to modify soils are the so-called *terra preta* soils of the Amazonia region, which were made by human societies hundreds of years ago through adding considerable amounts of charcoal, pottery, bones and other human-derived wastes to the soil. The *terra preta* soils are noticeably more fertile than surrounding soils which are highly 'weathered' (depleted of nutrients that are vital to healthy plant development and prone to erosion). *Terra preta* soils contain much higher amounts of nitrogen, phosphorus and potassium than the adjacent weathered soils and this is likely connected to the biochar-mineral complexes which hang onto the nutrients that are present and allow more productive use of them over successive generations of vegetation (which, in the case of forests, return the nutrients of leaves and, in the longer term, of rotting trunk to the soil). Put together the benefit of healthy soils with carbon mitigation – adding in the potential for sustainable energy provisioning if pyrolysis can be employed to generate syngas and useable heat – and there is the prospect of a 'double' if not a 'triple' win. Biochar has been widely debated by scientists, agriculturalists and climate policy makers and activitists from 2005 onwards (see Box 1.1).

Box 1.1 A tale of one city and two banks

Newcastle upon Tyne, UK, September 2008: two hundred people gathered from all over the world for the 2nd International Biochar Initiative (IBI) conference. Newcastle was also home and HQ to the bank Northern Rock. Exactly one year before the 2nd IBI conference, queues started forming outside Northern Bank branches across the UK as the first 'run-on-a-bank' in 150 years hit Northern Rock, signalling the start of the global financial crisis in the UK. The bank was sold at a bargain basement price a few years later, a mere rump of its former self. Just days after the IBI meeting, global financial services firm Lehman Brothers filed for bankruptcy in New York City, and other investment banks teetered on a similar dire fate – signalling the zenith of the global crisis. Central banks and senior politicians worked frantically to shore up vulnerable banks that were considered to be 'too large to fail' by injecting them with billions of dollars. With the benefit of hindsight, we can now see that the fate of Northern Rock and Lehman Brothers in 2007 and 2008 had a profound effect on the topics of discussion at the IBI conference in 2008. The presentations at the IBI conference in Newcastle were upbeat and the assorted delegates from research and development, companies, investment firms and governments (predominantly from industrialised countries) were reasonably confident that biochar would have a major role to play in tackling carbon reduction policies in the following decade. One European politician told the audience that biochar was 'too important' to leave to soil scientists alone, implying that project developers, investors and even politicians would soon come and make the issue theirs. The delegates went home confident that biochar had a promising future: not only could it store carbon in the (climate change relevant) long-term, but it would have benefits for many soil types and improve crop productivity. Biochar also reflected the shift away from exclusive focus upon 'carbon emission reduction' to thinking more in terms of 'carbon abatement', whereby carbon removal from the atmosphere and its long-term secure storage would be as equally valued as avoiding emissions from fossil fuels in the first place. This approach, enshrined in the 1992 United Nations Framework Convention on Climate Change (UNFCCC), was pioneered by the late Peter Read (1994) – who actually invented the term biochar a decade or so later – but had become enmeshed in controversy on the reversibility of carbon removal in forests and soils; an issue which biochar appeared to effectively address when the stability (recalcitrance) of a high percentage of biochar carbon was recorded (Lehmann and Joseph 2009; Spokas 2010).

From the perspective of carbon markets, biochar seemed an ideal candidate technology and practice for support under the United Nation's treaty on climate change, as explained in Box 1.2.

Box 1.2 **The role of biochar in carbon markets**

The internationally accredited carbon market was established through the Clean Development Mechanism (CDM) of the Kyoto Protocol (the 1997 Treaty which implemented elements of the UNFCCC) that was agreed in Rio in 1992). Bioenergy technology development would be incentivised, it was then assumed, by the 'reality' of depleting gas and oil supplies, hence rising fuel prices, and by new carbon emission target policies which would push up the price of using fossil fuel yet further. The rise of 'unconventional fossil fuel' supplies (shale oil and gas, tar sands, etc.) was even then underway in North America; however, together with concerns over biomass sustainability, it has since undermined the case for (and investment in) bioenergy.

About one year after the 2nd IBI conference, the Conference of the Parties (COP) to the UNFCCC) held its 15th meeting in Copenhagen (December 2009). The UNFCCC was an outcome of the Rio Earth Summit in 1992, with its most significant achievement being the signing of the Kyoto Protocol in 1997, at which industrialised countries agreed to small, but obligatory, carbon reductions (entailing a basket of six greenhouse gases including CO_2, CH_4 and N_2O) by an average over the years 2008-2012, relative to a 1990 baseline. After the USA refused to ratify and only a few countries achieved the modest carbon reductions they signed up to (an average reduction of 5.2 per cent), the Kyoto Protocol was not a success. Where reductions were achieved, it was not due to deliberate carbon reduction policies, but due to deindustrialisation (carbon leakage to Asia, Middle East, etc.) and energy policy changes resulting in an increase in use of natural gas rather than coal for electricity generation. The flaws of the Kyoto Protocol have been widely debated over the years (Verweij *et al.* 2006; Prins and Rayner 2007; Victor 2004). In the build up to COP-15 Copenhagen, however, the mood was highly optimistic ('Hopenhagen'), as many businesses, environmental organisations, researchers and other stakeholders and government delegations anticipated the barebones of an international treaty which would involve binding and more ambitious carbon reduction targets than the Kyoto Protocol had prescribed.

What happened in reality at Copenhagen was the European countries were in disarray, China asserted itself as the world's second largest economy and soon-to-be superpower, the so-called G77 countries demanded more leadership and financial assistance from industrialised countries and the USA sustained its sceptical stance towards binding commitments to carbon emission reduction. Despite some positive notes, the general tone of COPs since Copenhagen has not been positive: divisions remain pronounced and major industrialised countries such as Japan, Canada and Australia have signalled policy reversal on climate change and carbon reduction. At the UNFCCC COP-20 meeting in Lima, December 2014, the old conflicts flared-up again between 'developed' and 'developing' countries (a flawed and seriously outdated categorisation going back to the early 1990s), on who is most responsible and therefore who should do the most, how to hold countries to account for their claimed-for emission reduction policies and the ever present issue of how much richer countries are willing to give to poorer ones to respond to climate change.

The much vaunted EU Emissions Trading Scheme (EU-ETS), launched with great expectations in 2006, has also largely failed to date. The current price of a tonne of CO_2 (December 2014) under the EU-ETS is only c. €7 tCO_2^{-1}, whereas expectations were that the price should be moving towards €30 to €40 tCO_2^{-1} proceeding through the second decade of the twenty-first century. The collapse in the European carbon price – due to over allocation of emission permits by national governments in response to lobbying by industry, as well as the slowdown in the EU economies, reducing demand since 2009 – has had serious impact upon the internationally traded price of carbon reductions under the Kyoto Protocol's CDM (because the EU-ETS has traditionally been the largest buyer for emission permits generated by the CDM).

The implicit value of biochar for carbon storage was always the keystone in the biochar arch, with improved soil health and fertility and crop productivity acting as nice buttressing arguments. The high hopes of biochar entrepreneurs in the mid-2000s fell victim to the collapse of the carbon price in 2009/2010 and resulted in the flight of venture capital from low-carbon projects and innovations. Moreover, the following additional problems led to developers hitting the biochar 'pause' button in the early 2010s:

- the methodological complexities associated with verifying the long-term stability of the carbon in biochar;
- methodological complexities associated with verifying biochar's persistence within the confines of the 'project site' (as required for carbon financing under the regulated or voluntary carbon markets);
- increasing realisation of the lack of proven dominant technological designs for producing biochar;
- the high costs of producing biochar from clean biomass in competition with other uses (Shackley et al. 2011); and
- the mixed and unpredictable impacts of biochar in agronomic terms and the associated uncertainty in the incremental economic benefits for farming (Jeffery et al. 2011; Biederman and Harpole 2013; Spokas et al. 2013; Crane-Droesch et al. 2013).

The early euphoria gave way to a more nuanced and measured understanding of biochar's potential, one which recognised that there was an important resource cost to producing biochar, that feedstocks are frequently scarce and competed for, that biochar works best in particular soil and crop combinations and that commercial viability of biochar products will be key. On top of that, it became obvious that there is not one type of biochar, but many different types, each with its own properties, depending upon feedstock properties, pyrolysis conditions, soil type and crop varieties. This new understanding has resulted in a focus upon configurations of feedstock – technology – processing – soil – crop – land management which are likely to be strong candidates for successful biochar deployment. It has also led to a greater interest in getting added value out of biochar through mixing it with minerals and/or organic materials which will enhance its performance, though scientific trials are still few and far between.

Definition of biochar

> Biochar is a solid material obtained from the thermochemical conversion of biomass in oxygen-restricted conditions which is used for any purpose that does not involve its rapid mineralisation to CO_2. Biochar is commonly used for soil improvement and for the long-term storage of stable carbon.

Put simply, biochar is biomass heated in the (near) absence of oxygen. It undergoes a process called *pyrolysis* whereby carbon in the biomass is concentrated while other elements (especially oxygen, nitrogen and hydrogen) form compounds and are emitted as vapour, becoming diluted in the solid residue. Biochar is made by the same process as used in making *charcoal* and the two materials are indistinguishable in physical and chemical terms. The key difference is that charcoal is intended to be burnt as a fuel, whereas biochar is made to be added to soil (or otherwise used or deposited safely in the environment), where it remains in a stable form for hundreds or even thousands of years. Modern pyrolysis technologies allow biochar to be produced from a range of biomass feedstocks, not just the woody logs which are used to produce charcoal.

Biochar contains a lot of carbon (usually >50 per cent by mass and typically >80 per cent in a woody biochar). The carbon is in a very stable form, which means it is very resistant to being broken down by bacteria and fungi, which usually consume bio–wastes and biomass left in the environment at a rapid pace. Biochar has a very porous structure – it is full of tiny holes and cavities – giving it a very high surface area to volume or mass ratio. Amazingly, this is often several hundred metres' squared in just one gramme! Because of this huge surface area, water, and nutrients dissolved in water, are absorbed into the biochar, making it a bit like a rather rigid sponge. All sorts of chemical interactions start to take place on the surfaces of the biochar, some of which appear to improve the performance of the soil. What's more, the surface chemistry and reactions don't stand still, but change as the biochar ages and as others things in the soil change around it.

Box 1.3 Charcoal and biochar: what's the difference?

As a material, charcoal is indistinguishable from biochar. But charcoal is produced as a fuel (for cooking, most familiarly for BBQ) or as a reducing agent in preparing metals and pure silicon (which entails the conversion of the carbon to CO_2). Anything called biochar must not be degraded in the environment so that the carbon is safely and securely stored on climate change relevant time scales (hundreds to thousands of years). Charcoal (*Holzkohle* in German, *charbon de bois* in French) has traditionally been used as a soil amendment, but given the new nomenclature, we now call this biochar.

Other words exist for biochar-type materials, including 'vegetal carbon', 'carbon black', 'activated charcoal' and 'activated carbon' (in German, vegetal carbon is *Pflanzenkohle*, in French *charbon végétal* and in Spanish *carbón vegetal*). The key definitional principle, though, is straightforward – any material which does or is intended to breakdown or decompose thereby releasing carbon content as CO_2 is not biochar; anything which retains the carbon in a stable form during and after its application (on a climate change relevant timescale) is biochar.

Charcoal is traditionally produced from woody feedstocks. Biochar can be produced from a much wider range of biomass feedstocks. However, when straws, residues and other bio-wastes are converted into an energy-rich char for use as a fuel, the result is not biochar but could be called charcoal or biocoal. Vegetal carbon can be found in regulation as a food additive and colouring agent E153[1] or an additive to animal feed. Vegetal carbon is defined as a product obtained by carbonisation of organic vegetal material.

Ambition of the book

A vast number of academic journal articles have been written on the topic of biochar. There were 380 scientific papers published on biochar in the year 2013 alone, a five-fold increase in quantity from 2008. In 2009 a major academic book by Lehman and Joseph was published, titled *Biochar for Environmental Management*, and a greatly expanded and updated second edition of the book was published in 2015. This book, in both its editions, provides the most wide-ranging account of the science and technology of biochar. Contemporary science is highly specialised and highly technical, requiring considerable starting knowledge, and some experience of laboratory and field research methods and procedures, to access and fully understand the academic literature. It is generally very difficult for non-scientists to navigate their way around academic literature. On top of this, non-academics have to pay a fee to read much of the scientific literature, though research sponsors are increasingly promoting open (i.e., free) access to all.

The field of biochar has attracted a large number of non-professional proponents – gardeners, engineers, small holders, organic food and slow-food movement advocates, and so on. The non-professionals tend to be quite practically focused; they enjoy learning-by-doing and prefer individual, low-cost experimentation. In our experience, they do not usually sit down and read articles published in academic journals. These non-professional proponents are driven by curiosity, the sense of being able to make a difference ('getting on with doing low-carbon things, not just sitting around talking about it endlessly') and do not typically rely upon their interest in biochar to raise an income. However, biochar may sometimes be a sideline in a wider business enterprise, e.g. composting, organic waste management, charcoal making, and so on.

Academics working on biochar, on the other hand, are doing so as part of a professional career trajectory which may include a student project, a PhD project, technician work, a postdoctoral research project or academic faculty work. The incentives within academia are to publish new findings in the best possible journals. Being in a position to publish such work tends to impose considerable constraints on what is done and how. The research has to abide by the agreed peer standards in terms of design, methods, execution, interpretation, presentation and so on. The upside of meeting these requirements is the generally high quality of the data that gets published. Science develops mostly by incremental advances and the tried-and-tested approach of specialisation and apprenticeship (e.g. through a PhD). Pursuing a reductionist philosophy (whereby the wider problem or question is distilled down into more manageable smaller problems/questions) has survived as the (currently) most effective means of adding to the corpus of scientific knowledge, at least in specialisms such as soil science, materials science and agronomy. The adequacy of reductionist approaches to understanding complex systems such as the soil and plant development, given the multiple processes which are at play, has increasingly been questioned.

Academia has also moved increasingly over the past several decades to privilege 'pure' or 'basic' knowledge (such as underpinning theories) rather than applied knowledge. The reasons for this are numerous but include the legitimation of spending public money on science. If funding is going to applied research and development (R&D), could this be regarded as a case of the public sector paying for something that the private sector could and should pay for? If the benefits of R&D accrue predominantly to a private sector firm, spending public money is questionable unless there is a gap in what is good value from the societal perspective and what is profitable for the firm. Consider R&D on an Ebola vaccine, for example: a private firm may not spend much money on this, not because it is not a medically worthy topic (it certainly is), but because it will not profit from being able to put a high price tag on a vaccine which is the firms protected intellectual property (owing to the low incomes of the majority of customers who are based in poorer countries with poorly resourced health services). Yet, the social 'good' of an Ebola vaccine is very high indeed. This is an example of where the gap between private gain and public good is so large that governmental support for R&D is justified, though in the absence of a global government, philanthropically minded donors in the public and private sectors need to step in.

A related trend within academia in the past several decades has been the removal of applied disciplines such as agriculture and forestry and their relocation into specialist technical and agricultural institutes/rural/land-use colleges. There are hints of an academic 'hierarchy' at work here, with physics and mathematics as 'pure', underpinning and quantitative sciences occupying lofty positions, while agricultural science, seen rather as applied and less concerned with uncovering fundamental processes and mechanisms, appears lower down the hierarchy. Where agriculture and forestry departments have survived in universities, they are typically focused on the 'pure' more theoretical topics and domains.

The net result of such developments is that the biochar research community sits in one corner and the non-professional biochar proponents in another corner, with neither side talking much to one another. The non-professionals get frustrated because they see (relatively) well-paid professionals engrossed in their R&D and preoccupied by publishing in academic journals, while hardly addressing the non-professionals' questions and needs. The non-professionals have a long list of question they would like answered but academics seem unable or unwilling to take time out to answer them. The non-professionals, not unreasonably, assume that the answers to their questions must be buried somewhere in the huge volume of academic papers and become frustrated by scientists' reluctance to engage. Meanwhile, the scientists tend to view the non-professionals as amateurs: the latter don't register on the professional ladder, they can't be evaluated or 'located' with the usual academic criteria. The scientists don't regard many of the questions raised by non-professionals as the 'right' questions or as ones which can actually be answered. Hence, scientists will frequently change the question to the one they believe *should* be asked and non-professionals become further frustrated by what can be perceived as arrogance. Interestingly, there are also quite major rifts within science itself: some researchers maintain the view that scientific knowledge is superior to all other types, while others are much more open to the idea that other understandings – such as the local knowledge of farmers – are in some instances equal to that of science (e.g. Wynne 1996 for a soil science example).

There is a third group to add into this mix, though it is a lot smaller than the two already identified. This is companies, from small to multi-national, which have engaged with biochar. Large firms will usually have a history of working collaboratively with universities and, if R&D-focused, will have their own laboratories and 'research, development and demonstration' (RD&D) facilities. Smaller firms that are involved in creating biochar products such as high-value composts and soil amendments are typically much less engaged in science-based R&D.

Given the abundance of academic articles and several academic book-length studies, it is surprising how little material there is to *explain* biochar in scientifically credible terms to potential user groups and producers who are not themselves scientists. It is the ambition of this book to try and put this right and to provide an easy-to-understand account of the science and technology of biochar, including a simple-to-comprehend account of what we do and don't know and how certain or uncertain we are about what we know.

Part of the problem, we think, is the sheer range of disciplines, some knowledge and understanding of which is necessary to comprehend biochar as an interlinked system from production to application and impacts. Soil science is an integral part of the biochar story, but it is a very specialised topic that is not widely taught or known about. Likewise the topics of bioenergy and biomass feedstocks, the process of pyrolysis by which biochar is produced and the addition of organic materials to soils in order to improve their long-term health and performance – none of these topics are widely known about and are hardly the subject of everyday discourse! Our challenge has been to convey the basics of these underpinning disciplines such that the reader knows enough to be able to understand and evaluate the central ideas, hypotheses and evidence concerning biochar and its addition to soil.

This book is written by scientists from a range of disciplines and nearly all chapters have been co-authored by several scientists from different backgrounds and trainings as one way of creating an account that is comprehensible to non-specialists. In addition, each chapter has been reviewed by two – and usually more – scientists and their comments have been responded to by the chapter writing team, with the Editors ensuring that the response is adequate. We have made use of graphics and figures to try and convey the information clearly and informatively, as befits the purpose of the book. A number of these graphics were produced by a graphic designer specifically for this volume.

Where are we now?

Before we describe the book structure, it is useful to briefly detail the current status of biochar in European marketplaces. This is a 'reality check' so we can understand that the sector currently is minute and in its very early days. The import of charcoal into Germany (Europe's largest economy) in 2009 was 322,000 tonnes. Figure 1.1 compares this amount with total biochar production in Europe. It can be seen that charcoal imports into just one EU country dwarfs Europe's entire biochar production, being more than thirty times greater. Biochar production by European country is illustrated in Figure 1.2. Even the production of carbon-rich ashes, as produced by biomass gasification in Europe, is five times greater than biochar production in Europe. We have only taken the very first small step in the biochar voyage.

Book content

Chapter 2 presents an account of how to produce biochar, giving an easy-to-understand description of the process of pyrolysis by which biological material is converted into a solid, a gas and a liquid. While the process has been known about and used for centuries, its relative neglect in the fossil fuel era means that a modern, up-to-date version of the technology is still lagging. Technology development is time consuming and usually expensive. The pyrolysis process is complicated and a convincing description of exactly what happens to biomass – sufficient to create a mathematical model to predict what would happen in different conditions – is not available. The processes underway in small pieces of equipment on the laboratory

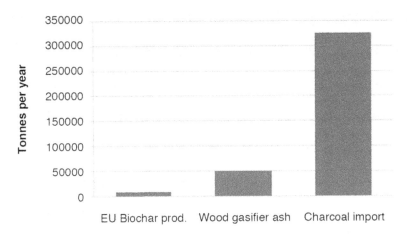

Figure 1.1 Production of biochar and wood ash from gasifiers in the EU compared with the amount of charcoal imported into Germany per year.

bench are not necessarily the same as those which take place in larger pieces of equipment such as pilot or full commercial scale. The only (currently) reliable way to work out what is happening at the larger scale is to go ahead and build the larger equipment with appropriate monitoring and collect information which can give an insight into the scale dependency of the processes.

The problem is that designing and constructing such equipment is expensive. Research organisations and institutes usually find it difficult to fund and manage pilot and larger-scale plant technologies, though there are some success stories in very well-funded areas such as particle physics, astronomy and space exploration. Private sector organisations have a stronger track record of technology development and scale-up – this being a core skill and strength of design, engineering and

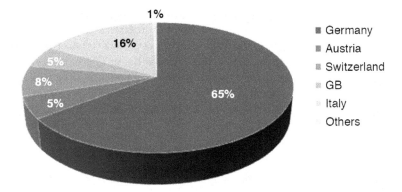

Figure 1.2 Breakdown of biochar production in Europe by country.

technology-based firms. The problem is that the private sector has, to date, shown little appetite for investing in the core technologies for producing biochar, largely because the commercial case at present is not sufficiently convincing. This creates a chicken-and-egg paradox: reliable knowledge and understanding of key issues such as scaling-up will only occur through commercial investment, but private-sector investment is deterred by the lack of understanding of pyrolysis technologies and their costs at different scales. Nonetheless, considerable knowledge of pyrolysis and its key parameters has been accumulated and Chapter 2 presents a succinct and accessible account of what is known as well as pointing to key knowledge gaps to guide future RD&D.

Chapter 3 describes what biochar is in physical and chemical terms (known as characterisation). Biochar has many different properties and these are relevant to the performance of different functions in the soil and in other contexts. This chapter therefore aims to describe *why* and *how* scientists try and describe the properties of biochar. Characterisation is also needed to make sure that biochar is not a hazard to human health and safety and to the environment, e.g. through containing toxic heavy metals and organic contaminants or due to particle characteristics (creating dust which can enter the eyes and lungs) or even excessive washing out of nutrients into streams next to fields on which biochar is added, which could result in water pollution. Such characterisation informs us of what we *don't* want biochar to be; other tests are needed to tell us what we *do* want biochar to be (i.e. its intended beneficial properties). Among these beneficial properties are storing carbon in the long term and improving soil health, and its performance in agricultural situations. Chapter 3 provides a user-friendly account of the range of properties tested for, the methods and their adaptation to biochar testing and examples of the results of biochar testing.

Chapter 4 concerns the way that biochar functions when added to agricultural soils. The characteristics of biochar as discussed in Chapter 3 are only a *potential* property of the material, which is only realised as given properties with wanted effects when biochar is added to a particular soil in a specific place and time. Chapter 4 presents the main insights and hypotheses into how biochar 'works' when added to soil – including water holding capacity, its effects upon pH, the addition and retention of nutrients, its effect on the physical properties of soil, its effect on soil biology, and so on. There remains considerable uncertainty regarding the mechanisms and processes by which biochar has its effects in the soil, and so Chapter 4 presents the main suggestions without presenting the processes with more certainty than exists and that the existing evidence can support. The ideas presented in Chapter 4 are, however, key to understanding how biochar can be applied effectively to agricultural soils and to thinking about how it can be modified to enhance its performance.

Chapter 5 presents thirty two published and unpublished European biochar field trials and draws out some general conclusions on results and underlying explanations for the effects. Chapter 5 gives special attention to the 'ring trial' which was undertaken in seven countries in North Western and Northern Europe, all using the same biochar, crops and growing seasons. All the trials reviewed in Chapter 5 refer to 'pure' biochar additions only, meaning that biochar is added as it comes, not pre-mixed with other organic matter (such as compost or manure) or other inorganic materials (such as fertilisers, though these are frequently still added in the trials as per normal practice). The results show that the effects of pure biochar addition are relatively minor in the majority of circumstances. Some practical considerations in conducting a biochar field trial are also presented in Chapter 5, drawing upon the experience of its authors and others in establishing and monitoring such trials.

In Chapter 6 the focus moves to the application of biochar mixes which combine one or more types of biochar with composts. Pure biochar applications in good agricultural soils are known to have limited effectiveness under most conditions. To enhance the effectiveness of biochar soil amendment, biochar mixtures have therefore been proposed and tested. Some interesting results have been obtained, pointing to the ability of co-composting biochar and organic matter to effectively 'age' the biochar and enhance some of its most desirable properties. The potential reasons explaining these positive results are discussed in Chapter 6 and the wider implications are drawn out.

Chapter 7 directly tackles the issue of the carbon storage properties of biochar. After all, if the carbon in biochar is not very stable, then a major rationale behind biochar as a concept and valuable practice is thrown into question. The stability of biochar can be conceptualised over a range of timescales, from short to long-term. Some of the organic matter contained within biochar will break down very rapidly (hours to days), largely due to the action of bacteria and fungi living in the environment. The type of organic matter which is readily accessible to microorganisms is chain-like, or small ring-like molecules containing atoms such as hydrogen, nitrogen and oxygen as well as carbon, the molecular bonds between the atoms being weak enough that bacteria and fungi can fairly easily break them down to acquire energy, carbon and other nutrients. The amount of such readily available organic matter is very low in most types of biochar because the high temperature of pyrolysis (>450°C) will burn away the types of organic matter which do not convert into the sheets of ring-like carbon molecules that typify biochar. The carbon and nitrogen which is accessible to microorganisms within biochar also has important impacts for soil microbiology and agronomy. In the longer term (decades to thousands of years), the strong molecular bonds linking up the carbon atoms in biochar are degraded, though the precise mechanisms are not known.

Life cycle assessment (LCA) has emerged as a key tool in environmental management and is covered in Chapter 8. It looks at the impacts of a product (or service) from 'cradle to grave', i.e. from the creation of the product to its end use and eventual disposal. In the case of biochar, this means 'from soil to soil', since (in the majority of cases) the starting biomass comes ultimately from a plant and the biochar ends up back in the soil.[2] The reason why LCA is used is that it is necessary to understand the overall effects of a new or modified product (service) in comparison with the existing offerings and/or other proposed alternatives. LCAs are also important in elucidating whether reducing impacts along one part of the supply chain can increase the impacts elsewhere. For example, if only the carbon balance of a bioenergy process is calculated, it might be concluded that this is favourable compared to fossil fuel powered energy production – and therefore a 'good thing'. What it might ignore, though, is the impact of using biomass for energy on land-use – if demand for land rises then more land may be converted to agricultural use and the price of food may also rise. What it may also ignore is other types of environmental pollution – to atmosphere, land or water – arising from the burning of biomass in a power plant. The purpose of LCA is to pick up on all these types of impacts. Chapter 8 undertakes this analysis with reference to typical situations and contexts in northern Europe.

The economics of biochar are covered in Chapter 9. Making biochar pay for itself in Europe is very difficult in current market conditions. Several possible solutions to this problem are then discussed including finding new applications, hence more lucrative markets, for biochar; promoting the multiple use of the same kg of biochar for different successive applications such that economic value can be accumulated along the product life cycle; finding ways of increasing the value of a kg of biochar by mixing it with other materials (organic and inorganic); and finding ways to reduce the cost of biochar production. The perspective which emerges from Chapter 9 is that what matters is less the present economics but more how the economics can be turned around in the future.

In Chapter 10, the focus is upon regulations and laws which control the production and use of biochar as a soil amendment. Europe shares common legislation on most environmental issues through EU Directives. These apply to waste handling, incineration and the storage and movement of waste, as well as to the safety of products and commercial practices. National laws also apply to marketing and quality assurance for fertilisers and other agricultural inputs. There are, as yet, no common European laws on soil amendments and their assessment – these issues are instead dealt with at the national level, though European action is anticipated in these domains. Chapter 10 covers the major issues for regulation of biochar and reviews how different European countries have approached the control of biochar.

Chapter 11 pulls together the most important, forward-looking applications of biochar. It identifies exciting new trends and opportunities, some with stronger underpinning evidence and other concepts that are more speculative. Yet, the ideas together represent a credible way forwards for thinking about the opportunities of biochar and, crucially, expanding out from the application to agricultural soils alone. Some of the most economically viable and useful applications of biochar may occur from non-agricultural applications, as hinted at in Chapter 11 (Schmidt 2012).

Some have referred to the change away from thinking of 'pure' biochar applications in tens of tonnes per hectare, to envisaging much smaller biochar additions (one to a few tonnes per ha) and usually in combination with other organic and/or chemical ingredients (Joseph *et al.* 2013), as a 'paradigm shift'. Some evidence suggests that synergies occur between biochar and certain other ingredients such that the 'sum is greater than the parts', though this evidence remains far from conclusive. 'Gestalt shift' is perhaps more apt than paradigm shift, meaning that a more holistic view of the place of biochar in the wider climate, carbon, food production and land-use debate is emerging.

Chapter 12 provides concluding comments and points to the possible directions for the biochar sector in Europe in the next decade. This is our attempt to 'crystal gaze' into the next decade, not because our vision will be anywhere close to 'right' but more because it will hopefully stimulate the debate that's urgently needed on biochar among technologists, growers and farmers, governments, companies, community groups and many other stakeholders.

Bibliography

Biederman, L.A. and Harpole, W.S. (2013). Biochar and its effects on plant productivity and nutrient cycling: a meta analysis. *GCB Bioenergy* 5: 202–214.

Crane-Droesch, A., Abiven, S., Jeffery, S. and Torn, M.S. (2013). Heterogeneous global crop yield response to biochar: a meta-regression analysis. *Environmental Research Letters* 8: 44–49.

Elad, Y., Cytryn, E., Meller-Harel, M., *et al.* (2011). The biochar effect: plant resistance to biotic stresses. *Phytopathologia Mediterranea* 50(3): 335–349.

Graves, D. (2013). A comparison of methods to apply biochar to temperate soils. In N. Ladygina and F. Rineau (eds), *Biochar and Soil Biota*, 202–260. London: CRC Press (Taylor and Francis).

Jeffery, S., Bezemer, T., Cornelissen, G., *et al.* (2013). The way forward in biochar research: targeting trade-offs between the potential wins. *Global Change Biology Bioenergy* 7(1): 1–13 (doi:10.1111/gcbb.12132).

Jeffery, S., Verheijen, F.G.A., Van der Velde, M., *et al.* (2011). A quantitative review of the effects of biochar application to soils on crop productivity using meta-analysis. *Agriculture, Ecosystems and Environment* 144: 175–187.

Joseph, S., Graber, E., Chia, C., *et al.* (2013). Shifting paradigms: development of high-efficiency biochar fertilizers based on nano-structures and soluble components. *Carbon Management* 4(3): 323–343.

Lehmann, J. and Joseph, S. (eds). (2009). *Biochar for Environmental Management: Science and Technology*. London: Earthscan.

Lehmann, J. and Joseph, S. (eds). (2015). *Biochar for Environmental Management: Science and Technology*, 2nd edition. London: Earthscan from Routledge.

Prins, G. and Rayner, S. (2007). Time to ditch Kyoto. *Nature* 449: 973–975.

Read, P. (1994). *Responding to Global Warming: The Technology, Economics and Politics of Sustainable Energy*. London: Zed Books.

Schmidt, H.P. (2012). 55 uses of biochar. *Ithaka Journal for Ecology, Winegrowing and Climate Farming*, www.ithaka-journal.net/55-anwendungen-von-pflanzenkohle?lang=en.

Scott, H., Ponsonby, D. and Atkinson, C. (2014). Biochar: an improver of nutrient and soil water availability – what is the evidence? *CAB Reviews* 9(19): 1-19.

Shackley, S., Hammond, J., Gaunt, J. and Ibarrola, R. (2011). The feasibility and costs of biochar deployment in the UK. *Carbon Management* 3(2): 335–356.

Spokas, K.A. (2010). Review of the stability of biochar in soils: predictability of O:C molar ratios. *Carbon Management* 1(2): 289–303.

Spokas, K., Cantrell, K., Novak, J., *et al.* (2013). Biochar: a synthesis of its agronomic impact beyond carbon sequestration. *Journal of Environmental Quality* 41(4): 973–989.

Verweij, M., Douglas, M., Ellis, R., Engel, C., Hendriks, F., Lohmann, S., Ney, S., Rayner, S. and Thompson, M. (2006). The case for clumsiness for a complex world: the case of climate change. *Public Administration* 84(4): 817–843.

Victor, D.G. (2004). *The Collapse of the Kyoto Protocol and the Struggle to Slow Global Warming.* Princeton, NJ: Princeton University Press.

Wynne, B. (1996), May the sheep safely graze? A reflexive view of the expert-lay knowledge divide. In S. Lash, B. Szerszynski and B. Wynne (eds), *Risk, Environment and Modernity: Towards a New Ecology*, 44–84. London: Sage Publishing.

Notes

1   See www.food-info.net/uk/e/e153.htm.
2   Strictly speaking the origin of the cycle is atmospheric carbon dioxide utilised by the plant during photosynthesis. The source of this CO_2 is most likely other plants and other life forms but it may be also be derived from the combustion of fossil fuels or from other non-organic sources such was weathering of certain types of rock.

Chapter 2

Biochar production and feedstock

Ondřej Mašek, Frederik Ronsse and Dane Dickinson

Introduction

This chapter reviews the key biochar production technologies, with particular focus on technologies relevant in the European context. Besides the technological aspects of biochar production, feedstock-related issues, such as homogeneity, contamination and pre-treatment requirements, are also discussed. Biochar, is discussed as a separate category of products, compared to charcoal and activated carbon, although all three of these categories can be overlapping, as shown in Figure 2.1. The challenges of controlled biochar production, such that the resulting biochar has the specified desired properties, are covered, emphasising specific problems related to measurement and control of key process parameters, and their influence on product distribution and resulting properties.

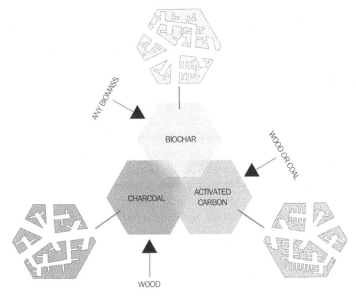

Figure 2.1 Schematic representation of the distinction between biochar, charcoal and activated carbon.

Feedstock-related issues

Many different organic materials are a potentially suitable feedstock for biochar production. This offers many advantages in terms of feedstock sourcing flexibility and subsequent biochar properties, as the different feedstock have diverse chemical and physical characteristics. However, the diversity of potential feedstock and its properties also poses challenges in preparation, handling and feeding of the material into the pyrolysis unit. The most important ones are moisture content reduction, size reduction and feeding.

Moisture content reduction

Most biomass and organic residues naturally contain a high concentration of moisture, and although it is possible to pyrolyse feedstock with high moisture content, there are two main undesirable consequences. First, energy consumption increases, as more heat is needed to dry the material within the unit. Second, water evaporated from the feedstock ends up in the gaseous and liquid co-products, diluting them, and thus reducing their heating value. Most pyrolysis systems can tolerate a moisture content in the feedstock of up to 30 per cent by weight (wt per cent), although numbers around 10wt per cent are preferred. Pyrolysis units will no longer be self-sustaining in terms of thermal energy requirements if the biomass feed moisture content exceeds a certain threshold.

There are different possible approaches to moisture content reduction. The most suitable one depends on the starting and desired material moisture content, material properties, and the possibility of integrating the drying step with the pyrolysis process, or other processes on the same site. The main technologies include: mechanical dewatering (used for biomass with very high moisture content), thermal drying and microwave drying.

Size reduction

Biomass typically comes in many different shapes and sizes, such as wood chips, logs, straw bales etc. In terms of particle size, we can distinguish between four classes: cordwood or firewood (logs, > 10cm up to 1m), chipped and shredded feedstock (>2mm up to several cm), pelletised (diameter 0.5 ~ 1cm, length several cm) and sawdust-like fine particles (<2mm) (Garcia-Perez et al. 2011). While batch pyrolysis units are often capable of handling large-sized feedstock, typically they can't process small particles. Most continuous pyrolysis units, on the other hand, require feedstock that can be easily and continuously fed into the pyrolysis unit, and as such is in the form of relatively small particles, such as chips, pellets or even sawdust-like fines.

There are numerous technologies for reducing the size and preparing feedstock, ranging from simple chipping and shredding to densification (pelleting and briquetting). The importance of feedstock preparation can't be underestimated, as the properties of feedstock – not only in terms of its chemical composition, but to a large extent also its size, shape, density and uniformity – influence the pyrolysis process and affect the properties of the resulting biochar. Feedstock uniformity is particularly important as it allows us to consistently produce biochar with pre-scribed properties. In this respect, feedstock pelleting offers numerous advantages as it improves feedstock uniformity and density (or the potential to increase pyrolysis unit throughput). On the other hand, owing to the pressure and temp-erature used during pelleting, the feedstock's original cellular structure can be changed or even lost, which can lead to reduced porosity and changes in surface chemistry in the resulting biochar.

The downside of all pre-treatment processes is the need for additional equipment, and often considerable energy consumption associated with biomass shredding or pelleting. Therefore, the choice of required pre-treatment is very important.

Feeding

Because specialised feeding equipment is required for most biomass, constructing reliable and safe feeding poses a considerable challenge to biomass projects. The situation is even more complicated when a system capable of handling very diverse feedstock is required, and in these cases further feedstock processing, such as pelleting, is often required. The challenges in biomass feeding are mainly related to how well biomass flows from storage through the feeding system and into the pyrolysis reactor. The main issue is when biomass undergoes 'arching' or 'bridging' in hoppers due to interlocking and potential coherence between particles. Owing to the relatively low density, irregular shape and often fibrous nature of biomass particles, biomass does not readily flow into hoppers, and often requires mechanical or other agitation to ensure smooth flow and uninterrupted feeding. The second and equally important challenge is to achieve continuous feeding without intro-ducing air into the pyrolysis system, which would reduce efficiency (due to combustion) and increase the risk of fire.

Feedstock contamination

The potential sources of contamination in virgin biomass feedstock can be grouped into two categories. The first group contains contaminants present in biomass before harvest, and is related to the type of biomass, the environment where it was produced and management practice (e.g. fertilisation). The most relevant elements in this category include, for example, nitrogen (N), chlorine (Cl), potassium (K), sodium (Na) and phosphorus (P). Although most of the elements in this category are considered to be beneficial when applied to soil, they pose additional challenges to the producer in terms of converting feedstock into biochar.

The second group includes contaminants introduced to biomass during and after harvest as a result of handling and transport. These operations can result in the incorporation of foreign matter, such as soil and other minerals, causing a consid-erable increase of ash content in the biomass.

Non-virgin biomass is an entirely different situation as it may contain a wide range of contaminants introduced during its lifecycle, and its discussion is beyond the scope of this chapter. However, there are some common issues related to contaminants in virgin and non-virgin biomass. Contaminants, irrespective of their origin, can be divided by their potential effects, i.e. contaminants that affect the conversion process and contaminants that affect the use of the resulting biochar. Feedstock components with low boiling or volatilisation temperature (N, S, Cl) or low melting point temperature (Na, K) can pose considerable challenges to the thermal conversion process due to their tendency to cause corrosion and/or form deposits. On the other hand, most other metals present in feedstock will be retained in biochar, and as such do not pose a challenge for the conversion process, but could limit the usability of the resulting biochar. Therefore, it is important to understand how the availability of these contaminants changes during the conversion process and how it affects the final product's safety.

Feedstock pre-treatment

We have identified a number of contaminants potentially contained in biomass feedstock. As some of these may have a negative effect on production or use of biochar, it may be desirable to reduce their content in the starting feedstock. It has been shown that washing biomass is an effective method for removing some chemicals (Cl, K and Na). Effective washing can be achieved as part of the harvesting process (Ogden *et al.* 2009) as leaving biomass in fields exposed to rain results in lower content of Cl, K and Na. Another option is to incorporate a washing step into biomass pre-treatment before conversion (Ravichandran *et al.* 2013). This requires higher investment in equipment, but it offers a more controlled and effective removal of the desired compounds.

Besides the removal of undesirable compounds, it may also be beneficial to add certain components into the feedstock to obtain biochar with the desired properties. Additives may include mineral matter such as clay or other biomass feedstock.

Biochar production

Biochar, the product from thermal conversion of biomass, can be obtained as a main product or a co-product from a number of diverse technologies ranging in scale and complexity. The choice of the most suitable process for biochar production is influenced by the scale of operation, primary target products (biochar, liquid or gaseous biofuels or electricity), required degree of control, available feedstock and a number of other factors. The following section briefly introduces the key technologies capable of producing biochar.

Table 2.1 Comparison of different technologies for biochar production

		Slow pyrolysis	Intermediate pyrolysis	Fast pyrolysis	Flash pyrolysis	Microwave pyrolysis	gasification	flash carbonisation
Temperature	range	250–750	320–500	400–750	450–650	250–1000	600–1100	400–800
	typical	350–400	350–450	450–550	500–600	400–600	850	550–650
Time	range	min.–days	1–15 min.	ms–s	ms	10–30	s–min	12–60 min
	typical	2–30 min.	4 min.	1–5 s	<1s	20–30	1–5 min	20–40 min
heating rate [C/s]	range	0.1–2	2–5	10–200	>200	1–2	1–1000	>30
pressure	range	atmos.	atmos.	atmos.	atmos.	atmos.	atmos.	0.7–3 Mpa
Yields [wt.% (d.b.)]								
Char	range	2–60	19–73	0–50	9–35	22–85	1–25	26–57
	typical	25–35	30–40	10–25	13–23	30–50	5–10	30–35
Liquid	range	0–60	18–60	10–80	30–65	10–70	0–5	–
	typical	20–50	35–45	50–70	50–60	35–45	1–3	–
Gas	range	0–60	9–32	5–60	10–59	10–65	80–99	54–73
	typical	20–50	20–30	10–30	10–25	20–40	85–95	65–70

Pyrolysis

Pyrolysis is a process of thermal decomposition of carbonaceous organic materials in the complete or nearly complete absence of oxygen. As such, pyrolysis is part of most thermochemical conversion processes, such as combustion and gasification, where it constitutes the second step following drying. The process of pyrolysis has been utilised for various outcomes, such as charcoal production, production of pyroligneous liquids (also called wood vinegar) for thousands of years. Major developments in pyrolysis of woody materials have taken place during the nineteenth and early twentieth centuries, when various industrial unit designs were developed for wood distillation (for the production of chemicals) and charcoal production. Most of those systems utilised relatively slow heating, and resulted in extended processing times (usually several hours). A new type of technology, the so-called 'fast pyrolysis', was only introduced in the second half of the twentieth century. This technology uses very fast heating rates in the order of hundreds to thousands of degrees Celsius per second to maximise the yield of liquid products. Heating rate, and related parameters, such as particle size and residence time, can be considered among the key distinctive features of different pyrolysis technologies. In the following section, we will describe three main technologies spanning the range of heating rates from slow to fast.

Slow pyrolysis

Slow pyrolysis is characterised by the relatively low heating rate (up to, but often much lower than, $100°C \cdot min^{-1}$) used to heat the feedstock from starting (ambient) temperature to its peak temperature, and consequently results in a relatively slow progress towards pyrolysis and charring. The charring process can take anywhere from tens of minutes to several days, depending mainly on the technology, and for batch processes, the amount of biomass used. The slow heating, prolonged residence time and often intensive interaction of solid, liquid and gaseous co-products during pyrolysis promotes the formation of char (also called secondary char) at the expense of pyrolysis liquids (see Table 2.1). Condensation of poly-aromatic compounds in pyrolysis vapours into larger structures results in the formation of secondary char.

Due to its high biochar yield, slow pyrolysis is a very suitable technology for biochar production. Unlike other, more recent biomass conversion technologies, such as fast pyrolysis hydrothermal carbonisation, and to some extent gasification, slow pyrolysis has been used extensively for the production of charcoal and chemicals for thousands of years (Domac et al. 2008) and could therefore be considered well understood. Although this may be true for charcoal production processes, there are many uncertainties and unknowns when it comes to production of specified/bespoke biochar. The key challenges relate to producing biochar with desirable properties and stability, as an integral part of systems for co-production of biochar, energy and/or chemicals. Historically, charcoal production was focussed on converting a relatively limited range of feedstock, mostly woody biomass, to fuel-grade charcoal, which did not require a high degree of specification, i.e. relatively simple quality control could be used. On the other hand, making biochar a useful material for agriculture, or for other specific purposes such as cleaning up waste water, requires a detailed understanding of the relationship between feedstock properties, conversion process parameters and the end use of the resulting biochar. Perhaps the only comparably demanding historic application for charred biomass was in the production of gun powder, where specific properties of char were required to ensure good product quality. However, this was typically achieved by using a limited range of carefully selected feedstock. To some extent, activated carbon industry may also be a useful analogue. Therefore, future research and development needs to focus on production of biochar with properties specifically suited for its application (carbon storage, soil amendment, etc.) in a way that is environmentally sustainable and economically viable.

Traditional charcoal kilns, such as pit and mound kilns, are not deemed suitable for biochar production in areas like Western Europe because of their high demands in manual labour, which add significantly to the production cost. However, more modernised types of kilns may be suitable as a cost-effective alternative: small-scale steel ring kilns, such as the New Hampshire kiln (Figure 2.2) are being produced in the UK by companies including Pressvess (Kingswinford, UK). However, the fact that pyrolysis gases (containing toxic gases) are released into the environment, and therefore endanger the operator and the environment, is a disadvantage of this type of kiln. Carbon Gold offers a steel kiln – dubbed 'SuperChar 100' – with the ability to produce 150kg of biochar from a batch of 500kg of biomass, following a carbonisation process of eight hours. These steel ring kilns are autothermal (i.e. heat needed for reaction is generated by partially combusting the biomass and char load, and hence, lower biochar yields). They operate in batch and do not offer the possibility to valorise (beneficially use) the resulting pyrolysis vapours. Their small size and low weight make them transportable.

Traditional steel kilns offer the ability to produce small quantities of biochar at low cost. For large production, industrial scale, automated and continuously operated kilns are required. Such machinery is manufactured by Carbon Terra (Germany); its Schottdorf kiln (Figure 2.3) is able to continuously produce two tonnes/day of biochar (starting from six tonnes/day of biomass with a maximum moisture content of 33wt per cent), while producing a combustible gas at a thermal capacity of 300kW. Although this process shares a lot in common with updraft moving bed gasification, in the Schottdorf kiln, the temperature is regulated to stay below 700°C (by adjusting the air infiltration); hence, at this low temperature, the process is not considered to be true gasification. The RomChar used in the field trials reported in Chapter Five was produced in the Schottdorf kiln. Another example of a continuously operated kiln is the Herreshoff multiple-hearth furnace, of which several units have been commercially built by Black is Green (Australia).

Other examples of established technology include different types of retorts for industrial-scale production of charcoal. For example, the O.E.T. Calusco tunnel retort relies on the use of open train wagons to transport batches of wood into a

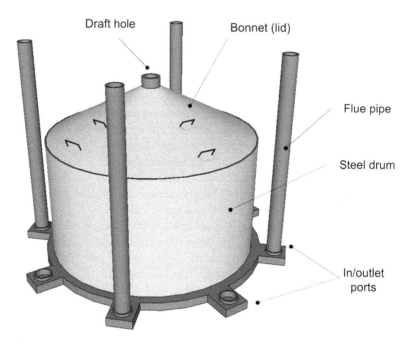

Figure 2.2 The New Hampshire kiln.

retort, which is indirectly heated using the combusted pyrolysis off-gases (Grønli 2005). After pyrolysis, the wagons convey the hot charcoal into a cooling chamber. Other retorts still in use today include the Reichert retort, developed by Evonik (formerly known as Degussa, Germany), which is based on batch-wise pyrolysis in a directly heated 100m³ retort using combusted pyrolysis off-gases (Grønli 2005). Completing carbonisation in a Reichert retort takes approximately 12 to 18 hours (Dahmen *et al.* 2010). Charcoal yield, on a feedstock dry basis, is 34wt per cent and annual charcoal production in Reichert retorts can reach 30,000 tonnes (Nachenius *et al.* 2013). Both the Calusco and Reichert retorts are batch or semi-continuous units. Continuously operated industrial-scale retorts exist as well, such as the Lambiotte retort, which was capable of an annual production of 13,500 tonnes of charcoal per year.

Two examples of modern continuous slow pyrolysis systems are the rotary drum and screw pyrolyser. Rotary drum pyrolysis units move feedstock through an externally heated, horizontal (or slightly inclined) cylindrical reactor by rotating the reactor or by paddles moving inside a stationary cylindrical shell. Both of these technologies can achieve a wide range of pyrolysis temperatures, good product quality control, long residence times (in the order of tens of minutes) and residence time distribution (Klose and Wiest 1999). In addition, rotary drum pyrolysers can operate on a wide range of scales; the largest unit at present treats 100,000 tonnes of feedstock per year.

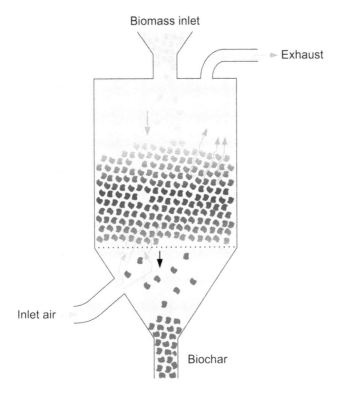

Biomass inlet

Exhaust

Inlet air

Biochar

Figure 2.3 **The Schottdorf kiln as used by Carbon Terra (Germany) for commercial production of biochar.**

Screw pyrolysers use a rotating auger to move material through a tubular reactor that can be heated either externally or internally by introducing a heat carrier material such as sand or metal balls. These units also offer good control of the production conditions over a wide range and are well suited for small to medium-scale applications. However, unlike rotary drum kilns, screw pyrolysers are less amenable to large scale applications beyond a few tonnes per hour (in a single unit) as beyond this scale the size of the screw becomes too large to withstand the high temperatures without considerable deformation. A well-known example of a screw-based biochar production system is the PYREG 500 reactor from PYREG GmbH (Germany) – operating on a similar principle as outlined in Figure 2.4, and capable of producing one tonne biochar per day.

In both cases (i.e. the rotary drum and screw pyrolysis units), the heat needed to drive the conversion process is most often provided by gaseous and liquid by-product combustion during the pyrolysis process. The hot flue gas then heats the pyrolysis reactor, either directly or indirectly, and depending on the complexity of the process, it is possible to direct the hot gas into different sections of the reactor to achieve the desired temperature profile. Any excess heat can be used in external

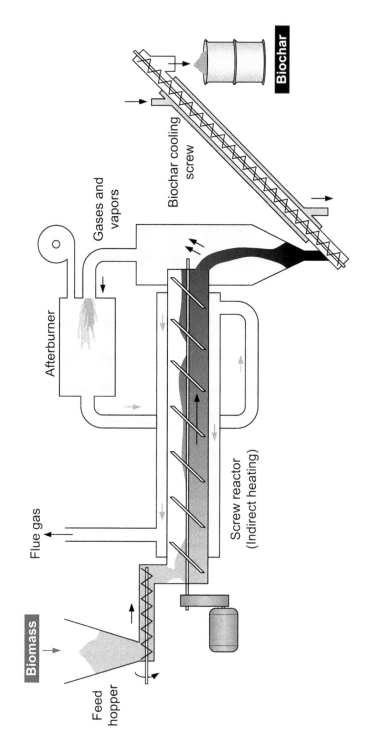

Figure 2.4 Screw reactor by Pyreg (Germany).

applications (e.g. heating, feedstock drying or integration with other processes on the same site). In some cases, part of the gaseous and liquid stream can be used for combined heat and power (CHP) generation. For example, the cleaned pyrolysis gases are burned in a combustion engine driving a generator for power or electricity production and the heat contained in the flue gases is used for heating applications. In certain cases, the heat released by the combustion of by-products is not sufficient to drive the pyrolysis process (particularly when using feedstock with high moisture content) and additional fuel needs to be provided.

Fast pyrolysis

Fast pyrolysis is a process which is characterised by reaction conditions, selected to obtain a maximum yield of bio-oil (Czernik and Bridgwater 2004; Balat *et al.* 2009; Bridgwater 2012). Typically fast pyrolysis conditions include moderate temperatures (400–600°C) and rapid heating rates (> 100°C min⁻¹) combined with short residence times of the biomass particles (0.5–2s) (Demirbas 2004). In order to support these high heating rates, the biomass feedstock needs to be ground down to a small particle size, typically less than a few millimetres. Under these process conditions, bio-oil yields of up to 70–75wt per cent, expressed on a dry biomass feedstock basis, could be obtained while the char and gas yields are limited to around 12 and 13wt per cent respectively (Bridgwater 2012; Nachenius *et al.* 2013).

A key feature of fast pyrolysis is keeping the vapour residence time below a few seconds by ensuring rapid quenching or cooling. As such, one can avoid unwanted secondary vapour phase decomposition reactions, which give rise to additional char and non-condensable gases at the expense of the yield in bio-oil. In this respect, fast pyrolysis differs from slow pyrolysis, in that the latter process aims to achieve a maximum yield of char (up to 35wt per cent) by employing long vapour residence times and slower heating rates (Bridgwater and Peacocke 2000; Lu *et al.* 2009; Brown 2009).

To date, several reactor designs for fast pyrolysis have been developed and scaled up to pilot or even commercial scale (Venderbosch and Prins 2010). The most popular reactor design is the fluidised bed (Figure 2.5a) as it is properly understood in terms of design, process control and scale-up. In a fluidised bed, a preheated solid material (or heat carrier, often sand) is suspended in a stream of hot, inert gas. The resulting fluidisation ensures optimal mixing and high rates of heat and mass transfer. Biomass particles are added to the fluidised bed, and contact with the heat carrier drives the fast pyrolysis reactions. Downstream of the reactor, cyclones separate the entrained char particles from the pyrolysis gases, while electrostatic precipitators or spray condensers condense the pyrolysis vapours into bio-oil. This type of fluidised bed is commonly known as the bubbling fluidised bed. A second type, the circulating fluidised bed (Figure 2.5b) does not separate the char particles, but instead uses higher fluidisation gas velocities to circulate the biomass/heat carrier mixture upwards through a narrow riser. At the top of the riser, the vapours are separated from the char/heat carrier mixture, the latter is sent to a secondary

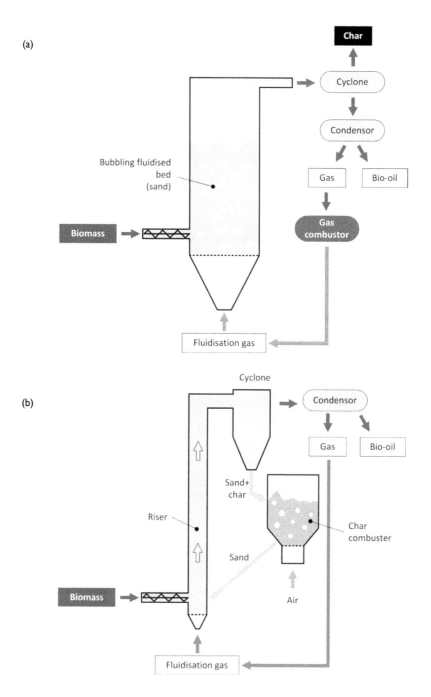

Figure 2.5 Pyrolysis reactor types: (a) bubbling fluidised bed and (b) circulating fluidised bed.

reactor in which the char is combusted and the heat carrier is reheated prior to recirculation to the bottom of the riser.

More recent fast pyrolysis reactor designs include the rotating cone reactor, developed by BTG and the University of Twente (both in the Netherlands) (see Figure 2.6a). This reactor type is based on the mechanical mixing of biomass with an inert heat carrier using centrifugal forces. No carrier gas is required, which in turn results in less dilution of the pyrolysis vapours, and hence bio-oil condensation is supposedly more efficient. Another recent technique to reduce dependency on the use of carrier gas is to employ a screw auger (Figure 2.6b). Either a single screw or two counter-rotating screws ensures that the biomass/heat carrier mixture moves properly through the reactor. Auger or screw reactors can handle more difficult feedstocks (i.e. irregular shape, heterogeneous particle size, low density) but their drawback is that the longer vapour residence times (5 to 30s) – compared

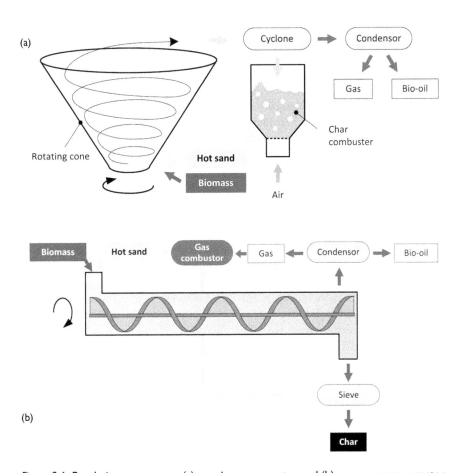

Figure 2.6 Pyrolysis reactor types: (a) rotating cone reactor and (b) auger or screw reactor.

with fluidised bed reactors – usually result in lower bio-oil yields (typically 10 per cent lower). Next to the four discussed reactor types, there are also less common reactor configurations, such as the ablative pyrolysis and vacuum pyrolysis reactors.

When pyrolysis vapours condense, a dark, viscous liquid called bio-oil forms. Bio-oil is highly oxygenated and approximates the elemental composition of the biomass feedstock. Bio-oil typically has a heating value ranging from 17 to 20MJ kg^{-1} and is highly corrosive due to the presence of organic acids and water. Despite the name 'bio-oil', it is immiscible with petroleum and other hydrocarbons. Bio-oil also tends to change in composition when stored over prolonged periods of time, a process known as 'ageing', which is detrimental to its quality (Bridgwater 2012). Despite the drawbacks of bio-oil in terms of its physic-ochemical properties, it can serve in different applications, including combustion in boilers, turbines and internal combustion engines, upgrading into drop-in biofuels or to be used for the extraction and isolation of green chemicals. Adding value by producing both biochar and bio-oil (co-valorisation) by means of fast pyrolysis appears to be an economically viable route, provided producers are able to overcome some technical challenges in commercial-scale valorisation and bio-oil upgrading (Anex et al. 2010). It is important to note, however, that the primary decomposition of biomass by means of fast pyrolysis is inherently an endothermic process (Venderbosch and Prins 2010) and thus during the required process heat is usually obtained, on a commercial scale, by complete or partial combustion of the non-condensable gases or char. The latter case is not a suitable option if both biochar and bio-oil production is the aim.

Intermediate pyrolysis

As the name suggests, intermediate pyrolysis fits in between slow and fast pyrolysis (i.e. it features relatively high heating rates and short vapour residence times, but longer solids residence times). As a result, the product yield distribution of gas, liquid and solid is also somewhere between those typical for fast and slow pyrolysis (see Figure 2.7). Intermediate pyrolysis combines some of the advantages of fast pyrolysis (i.e. high yield of liquid products) with those of slow pyrolysis (i.e. the ability to use larger particles compared with fast pyrolysis and a lower content of char in liquid products). Intermediate pyrolysis is the most recent development and several different technologies have been adopted (Hornung and Seifert 2006; Henrich et al. 2007).

Gasification

Gasification is a thermochemical conversion process in which solid carbonaceous materials are converted into gaseous compounds (H_2, CO, CH_4, CO_2, etc.) by reacting the solid fuel with steam, CO_2, O_2 or air at relatively high temperatures (McKendry 2002) in an atmosphere with a low concentration of oxygen. The main product of gasification is a flammable gas (product gas), that can be used as a

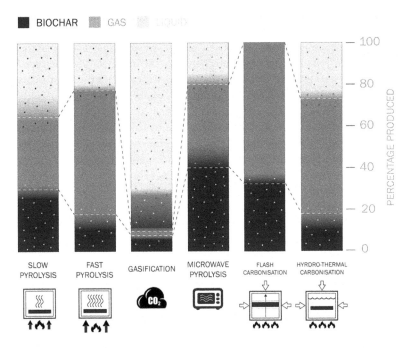

Figure 2.7 Distribution of co-products for different biomass thermochemical conversion technologies.

feedstock for production of liquid fuels and/or chemicals, or as fuel for power generation (using combustion engines, gas turbines or fuel cells). With carbon conversion levels in most commercial biomass gasification systems in the order of 94–99 per cent, there is only a relatively small yield of co-products (char/ash and tar). As a result of this high conversion efficiency, the ash/char contains only small amounts of carbon, and is therefore less suitable for use as biochar compared with biochar from pyrolysis, at least from a carbon sequestration point of view (Leiva *et al.* 2007). One possible exception may be fly-ash from fluidised bed gasification, as its carbon content can be up to 70 per cent (Pels *et al.* 2005); however, close attention would need to be paid to the contaminant content of this material, as both organic and inorganic contaminants can be present in high concentrations. In addition, for certain feedstock (e.g. rice husk and rice straw) the high processing temperatures in gasification can lead to the formation of crystalline silica, which poses health risks.

Hydrothermal carbonisation

Hydrothermal carbonisation (HTC) is a thermal treatment process that was discovered in the early twentieth century by Bergius during research on coal formation mechanisms. The concept has only recently been rediscovered as a method to achieve a very high conversion rate of biomass carbon into solid carbonaceous residue. The HTC process consists of treating organic materials in an aqueous environment at temperatures between 150 and 350°C under autogenous pressure (pressure generated within the closed vessel by evaporation of water) to form a condensed carbonaceous solid product, sometimes referred to as hydrochar or bio-coal.

Published studies have shown that it is possible to achieve carbon retention in the solid as high as 80 per cent (Sevilla and Fuertes 2009). The HTC process is exothermic, and approximately one-third of the energy contained in the feedstock is released (Titirici *et al.* 2007) during the conversion. This energy can be utilised internally within the process to heat the reactor, or to dry the solid product, as well as externally for heating and power generation. These characteristics, together with the fact that wet starting material is not only acceptable, but required, make the HTC process potentially very attractive for production of solid carbonaceous products from various organic residues with high moisture content (e.g. sludge, food waste, manure). The resulting solids are distinct from products of pyrolysis, both physically and chemically. Most applications of HTC solids have focussed on use as fuel substitutes for fossil fuel or higher value biomass, but their use in soil as biochar has also been under investigation. So far there are only very few examples of HTC systems developed for the production of biochar (e.g. CarbonSolutions, www.cs-carbonsolutions.de; AVA-CO2, www.ava-co2.com), as most attention has been focused on conversion of various wet waste streams into fuels (Libra *et al.* 2010). Several studies have investigated the potential to use HTC solids as biochar for carbon sequestration in soil and for soil amendment. The results showed a clear difference between biochar produced by pyrolysis and HTC solids, especially in respect of its stability. This is due to the fact that HTC solids are mostly composed of aliphatic hydrocarbons, while biochar consists mostly of condensed aromatic hydrocarbons, which are more recalcitrant (Fuertes *et al.* 2010; Schimmelpfennig and Glaser 2011). Besides stability, there are also differences in environmental performance.

Other technologies

The pyrolysis modes described above represent the more mature, state-of-the-art methods of producing biochar from biomass pyrolysis. Aside from these and gasification, there are several other pyrolysis technologies that have been proposed for biochar production.

Microwave pyrolysis

Microwave pyrolysis can be operated either as fast, slow or intermediate pyrolysis. However, it uses a different means of transferring heat into the pyrolysing solid. Whereas in standard pyrolysis units heat is mostly transferred by conduction or convection from reactor walls and hot gases, in microwave pyrolysis the energy that drives the pyrolysis process is delivered in the form of microwave radiation, to which the feedstock is exposed on a specially designed reactor. This results in a considerably different heating pattern than in any of the other pyrolysis systems. In conventional pyrolysis, biomass particles are heated from outside and therefore the pyrolysis process progresses from the surface of the particle inwards. In contrast, in the case of microwave pyrolysis, the microwave radiation heats the internal volume of biomass particles (volumetric heating), thus pyrolysing them from the inside (but not necessarily the centre). Microwave pyrolysis has been studied for the production of liquid biofuels (Domínguez et al. 2006; Tian et al. 2011) and also for biochar production (e.g. CarbonScape, http://carbonscape.com). The advantage of microwave pyrolysis is its high efficiency; however, this is at least partially offset by the need to use a high-quality (and therefore costly) energy source, i.e. electricity, rather than heat from combustion of pyrolysis co-products, as in the case of pyrolysis units using convection and conduction.

Flash carbonisation

Flash carbonisation is a process developed at the Hawaii Natural Energy Institute which is unique in combining a pressurised system with the batch-style, partial combustion methods of traditional kilns and charcoal making (Antal et al. 2003). Flash carbonisation uses pressurised air (<10atm) in the biomass bed to start a flash fire at one end of the reactor. The heat released by the fire as it moves in the form of a front through the volume of the reactor is used to rapidly pyrolyse the remaining biomass. The gases produced in the process are further combusted to maintain the reactor temperature at around 800°C until the pyrolysis process is completed (< 30min). The advantage of the system, compared with traditional production methods, is in using an endogenous heat source and the use of pressure to decrease reaction time, while achieving high char yields, as a result of enhanced conversion of volatiles into char.

Vacuum pyrolysis

Vacuum pyrolysis, which uses low vacuum conditions (2–20kPa absolute pressure), is at the other end of the pressure scale. The vacuum assists in the rapid removal of volatiles from the pyrolysing feedstock, which reduces secondary reactions and therefore allows production of a bio-oil of comparable quality to that produced by fast pyrolysis, but without the need for very small sized feedstock particles. The product distribution is similar to that of slow pyrolysis, making it suitable for biochar production. Vacuum pyrolysis is presently progressing from academic interest into commercial scale-up, and looks to be a promising new development in pyrolysis technology.

Combined processes

Besides the different individual thermochemical conversion technologies, there also exists the potential to combine several technologies together. An example of such a combined process is staged gasification. Staged gasification is a process in which pyrolysis and gasification are physically separated (usually both processes would take place within the same reactor). Separation of the pyrolysis step from gasification offers several advantages, such as improved feedstock flexibility, as the pyrolysis step serves as a pre-treatment, making the material reaching the gasification chamber more uniform. In terms of biochar production, the two-stage process offers another advantage: flexibility in terms of prioritising production of biochar or heat/electricity, depending on the market situation. When it is desirable to produce biochar, part of the char destined for the gasifier can be diverted and used for biochar instead; when more heat/power is required, on the other hand, all char would go to the gasifier. Such units would be particularly suitable where diverse feedstock is likely to be used (e.g. straw, woody biomass, organic residues).

Other combined processes are also possible (e.g. HTC–pyrolysis and HTC–gasification), but these are still in a research phase.

Use of liquid and gaseous pyrolysis co-products

The liquid and gaseous co-products of pyrolysis present both challenges and opportunities for the successful development and operation of biochar production technologies. Not only does the type of pyrolysis technology selected for biomass processing have a significant impact on the product distribution (Figure 2.6), but it also impacts the quality and potential value of the product streams. In general terms, when a pyrolysis process is optimised for biochar production (e.g. slow pyrolysis), the liquid and gas co-products suffer in quality and pose special extraction and/or handling problems. The converse is also true (e.g. fast pyrolysis), which makes it especially important to select a pyrolysis technology that is most appropriate for the overarching application.

Pyrolysis gases are predominately composed of carbon monoxide and carbon dioxide, with smaller amounts of hydrogen, methane and other gaseous hydrocarbons. Although the gas composition varies with feedstock and pyrolysis mode, generally the overall energy content is low (LHV of 2–12MJ kg^{-1}) (Crombie and Mašek 2014), meaning there is limited scope for product valorisation. If the production process has employed an inert carrier gas such as nitrogen, then it will also present in the producer gas, diluting the product stream. For these reasons, and to ensure safe final emissions, up-scaled pyrolysis processes almost always combust the gaseous product stream immediately with a thermal oxidiser afterburner or other combustion system, which then works to provide part of the necessary process heat for the whole pyrolysis system. After complete combustion the oxidised gases can then be safely exhausted. In some systems, part of the final exhaust stream (CO_2, H_2O and inert carrier gas) may be reused in the pyrolyser as the oxygen-free carrier gas.

The liquids obtainable from biomass pyrolysis contain a complex mixture of organic compounds that are highly dependent on feedstock and pyrolysis mode. Water is also present, often in large quantities (15 to 80 per cent by weight). Efficient collection and handling of the liquids can be challenging as highly viscous tars frequently foul condensing systems, and organic acids can corrode vulnerable materials. Sometimes, pyrolysis liquids contain readily separable organic and aqueous phases or can be similarly fractionated in a heterogeneous condensing system, but often they are simply condensed into a single interspersed mixture.

Taken as a whole, the liquids have a low energy content and further exhibit a distinct aging phenomenon whereby their viscosity steadily increases over time and exponentially with storage temperature. This degradation process is a significant hurdle because even when a viable use is found for the liquids, they demand cool storage conditions and must be used within a few weeks or months of production.

Despite the mentioned problems, pyrolysis liquids have been the subject of much research since the late 1970s as potential alternatives to liquid fossil fuels and petrochemicals. Indeed, prior to the petrochemical revolution, pyrolysis liquids produced in the 'destructive distillation of wood' were the backbone of the then fledgling chemical industry. Modern focus now directs pyrolysis technology away from chemical manufacture towards energy resourcing. And while the bio-oils produced in fast pyrolysis can be substituted for heavy fuel oils in some applications where fuel quality is less important, generally they require significant refinement before they can be placed into existing product lines. Unfortunately, the traditional petrochemical distillation modes of refinement are unsuitable for pyrolysis liquids owing to their thermal instability, meaning that more exotic and energy intensive refinement processes are necessary – refinements which are generally too expensive for bio-oils to be competitive with their fossil fuel counterparts.

Although research is ongoing for improving the value of pyrolysis liquids, in lieu of commercial demand the default option is energy recovery through complete combustion. While the energy content of the liquids is not high due to the presence of water, if the pyrolysis facility is sufficiently large, on site combustion can potentially be used for power generation. Where facility size is insufficient for supplying grid electricity, heat recovered in combustion can still be used for feedstock drying and/or process energy. Direct combustion schemes have an added benefit of reducing the complexity of the pyrolysis technology by eliminating any need for specialised liquids condensation systems; the hot pyrolysis vapours are simply combusted altogether with the non-condensable gases in a single stage.

Mode of operation

Many of the thermal conversion units can be operated in either a batch or continuous mode. Traditional kilns as well as older retort designs operate in batch mode, i.e. the reactor is operated in a sequence. A typical sequence consists of feedstock loading into a cold kiln, heating, carbonisation, cooling down and unloading of cold char. Batch operation is simpler to implement and requires lower capital expenditure, but can have a number of drawbacks when compared with continuous systems. Product heterogeneity may exist between different batches, or even within a batch, especially if the kiln or retort is poorly controlled (i.e. in the case of traditional charcoal kilns). Heat is not used optimally due to the sequential nature of the heating and cooling stages. Also, the composition of the mixture of gases and vapours changes throughout the carbonisation process (Kandpal and Maheshwari 1993), which renders the processing or valorisation of these vapours more difficult (i.e. to achieve thermal or recovery through condensation). Some of these limitations can be addressed by operating several batch units in a sequence, forming a semi-continuous process, where each reactor is at a different stage of the carbonisation process (i.e. loading, drying, heating, carbonisation, cooling and unloading). By sharing a common afterburner, heat utilisation is optimised in between the different retorts/kilns.

Control of biochar production and biochar quality control

Control and monitoring of the quality of biochar is an important aspect of its commercial production, from the perspective of its performance as a standalone product or ingredient of a product. However, frequent analysis of a number of biochar characteristics is impractical and can be expensive. Therefore, it is important to establish an understanding of the relationship between production conditions and how they influence the properties of biochar, as achieving good control of the production process within pre-set limits would reduce the frequency and extent of biochar analysis required.

For any given feedstock, two of the most important biochar production parameters, and often the most readily available control measures, are the peak temperature and the residence time of material at that temperature (Zhao *et al.* 2013). A good pyrolysis unit should allow these two parameters to be reliably monitored and controlled (Cantrell and Martin 2011), to ensure that the processing conditions can be maintained within prescribed limits, as deviating from these limits are likely to affect the characteristics of the resulting biochar. However, monitoring and maintaining control of these two parameters (especially temperature) is often not a straightforward task, and suitable solutions need to be built into the unit design. There are several reasons why temperature control is difficult: the dynamics of the process; difficulty installing thermocouples to desirable measurement locations due to moving mechanical parts within the reactor; the opaque and aggressive atmosphere inside the pyrolysis chamber; and the use of challenging optical temperature measurement methods.

Even in a well-controlled production process, it is necessary to monitor product characteristics at regular intervals. To get representative data, it is important to use suitable sampling and analytic methods. The importance of representative sampling of feedstock and biochar should not be underestimated, and suitable standard methods should be followed (Bucheli *et al.* 2014).

Environmental impacts

In terms of biochar production, there are two key areas of environmental impact: gaseous emissions to air, and liquid effluents; not considering the application of biochar itself. Gaseous emissions from biochar production can contain numerous pollutants and even toxic gases and vapours, especially from traditional production units where pyrolysis gases are often released directly without combustion/ thermal oxidation (Stassen 2002). These emissions not only contribute to GHG emissions (e.g. N_2O, CO_2 and CH_4), in the form of black soot particles that contribute to climate change, they can also have a negative impact on the local environment and the health of the kiln operators (in particular from inhalation of particulates). Appropriately managing gaseous emissions is key, and even simple kilns can adopt effective measures to reduce emissions. The Adam retort is one of the best examples, of which there is at least one in Europe and recent measurements on the even simpler Kon Tiki kiln show even low levels of atmospheric emissions than for the Adam retort (personal communication, Hans-Peter Schmidt, February 2015).

Liquid effluents fall into two categories: condensed pyrolysis liquids and tar, and contaminated water if direct cooling of biochar is used (or if water is used for cleaning/scrubbing of pyrolysis gases). Both of these effluents present environmental risks if released into soil or water streams without suitable treatment. Appropriate treatment of liquid effluents depends on the type and scale of operation, and can range from simple collection and external disposal (e.g. incineration) to sophisticated systems using thermal cracking, scrubbing and wastewater treatment. It is also very important to minimise contact between produced biochar and pyrolysis vapours and liquids, as this can result in the production of contaminated char (Buss and Mašek 2014).

Related to the liquid and gaseous emissions is the issue of odour, which can be of particular concern in densely populated areas, and special attention to proper management of odours is a necessity for such installations.

Conclusions

This chapter provided a basic introduction of the different technologies that can be used for production of biochar from a variety of feedstock. Different aspects of conversion of biomass to biochar were discussed (e.g. feedstock-related issues, advantages and limitations of different technologies), as well as environmental issues. It is clear that the choice of a biochar production process is not a straightforward one, and it is very much context dependent. Therefore, it is not possible to make general recommendations or statements regarding best technologies for biochar production. It is, however, clear that to maximise the efficiency of use of biomass feedstock, as many of the co-products of pyrolysis as possible should be utilised.

Bibliography

Anex, R.P., Aden, A., Kazi, F.K., Fortman, J., Swanson, R.M., Wright, M.M., Satrio, J.A., Brown, R.C., Daugaard, D.E., Platon, A., Kothandaraman, G., Hsu, D.D., Dutta, A. (2010). Techno-economic comparison of biomass-to-transportation fuels via pyrolysis, gasification, and biochemical pathways. *Fuel* 89: S29–S35.

Antal, M. J., Mochidzuki, K. and Paredes, L.S. (2003). Flash Carbonization of Biomass. *Industrial & Engineering Chemistry Research* 42(16): 3690–3699 (doi:10.1021/ie0301839).

Balat, M., Balat, M., Kirtay, E. and Balat, H. (2009). Main routes for the thermo-conversion of biomass into fuels and chemicals. Part 1: Pyrolysis systems. *Energy Conversion and Management* 50: 3147–3157.

Bridgwater, A.V. (2012). Review of fast pyrolysis of biomass and product upgrading. *Biomass and Bioenergy* 38: 68–94.

Bridgwater, A.V. and Peacocke, G.V.C. (2000). Fast pyrolysis processes for biomass. *Renewable Sustainable Energy Reviews* 4: 1–73.

Brown, R. (2009) Biochar production technology. In J. Lehmann and S. Joseph (eds), *Biochar for Environmental Management: Science and Technology*, 127–146. London: Earthscan.

Bucheli, T.D., Bachmann, H.J., Blum, F., Bürge, D., Giger, R., Hilber, I., Keita, J., Leifeld, J. and Schmidt, H-P. (2014). On the heterogeneity of biochar and consequences for its representative sampling, *Journal of Analytical and Applied Pyrolysis* 107: 25–30 (doi:10.1016/j.jaap.2014.01.020).

Buss, W. and Mašek, O. (2014). Mobile organic compounds in biochar – a potential source of contamination: phytotoxic effects on cress seed (Lepidium sativum) germination, *Journal of Environmental Management* 137: 111–119.

Cantrell, K.B. and Martin, J.H. (2012). Stochastic state-space temperature regulation of biochar production. Part II: Application to manure processing via pyrolysis. *Journal of the Science of Food and Agriculture* 92(3): 490–495 (doi:10.1002/jsfa.4617).

Crombie, K. and Mašek, O. (2014). Investigating the potential for a self-sustaining slow pyrolysis system under varying operating conditions. *Bioresource Technology* 162: 148–156 (doi:10.1016/j.biortech.2014.03.134).

Czernik, S. and Bridgwater, A.V. (2004). Overview of applications of biomass fast pyrolysis oil. *Energy and Fuels* 18(2): 590–598 (doi:10.1021/ef034067u).

Dahmen, N., Henrich, E., Kruse, A. and Raffelt, K. (2010). Biomass liquefaction and gasification. In A. Vertes, N. Qureshi, H. Blaschek and H. Yukawa (eds), *Biomass to Biofuels: Strategies for Global Industries*, 91–122. Chichester: John Wiley and Sons.

Demirbas, A. (2004). Effects of temperature and particle size on bio-char yield from pyrolysis of agricultural residues. *Journal of Analytical and Applied Pyrolysis* 72: 234–248.

Domac, J., Trossero, M. and Siemons, R. (2008). *Industrial Charcoal Production: Water Management*. FAO TCP 3101. Rome: Food and Agriculture Organization of the United Nations.

Domínguez, A., Menéndez, J.A., Inguanzo, M. and Pís, J.J. (2006). Production of bio-fuels by high temperature pyrolysis of sewage sludge using conventional and microwave heating. *Bioresource Technology* 97(10): 1185–1193 (doi:10.1016/j.biortech.2005.05.011).

Fuertes, A.B., Arbestain, M.C., Sevilla, M., Fiol, S., Smernik, R.J., Aitkenhead, W.P., Arce, F. and Macias, F. (2010). Chemical and structural properties of carbonaceous products obtained by pyrolysis and hydrothermal carbonisation of corn stover. *Australian Journal of Soil Research* 48: 618–626.

Garcia-Perez, M., Lewis, T. and Kruger, C.E. (2011). *Methods for Producing Biochar and Advanced Biofuels in Washington State. Part 1: Literature Review of Pyrolysis Reactors.* First project report. Pullman, WA: Department of Biological Systems Engineering and the Center for Sustaining Agriculture and Natural Resources, Washington State University.

Grønli, M. (2005). *Industrial Production of Charcoal.* Energy Research paper N-7465. Trondheim: Sintef.

Henrich, E., Dahmen, N., Raffelt, K., Stahl, R. and Weirich, F. (2007). The Karlsruhe 'Bioliq' process for biomass gasification. In *2nd European Summer School on Renewable Motor Fuels.* Warsaw. Available at http://citeseerx.ist.psu.edu/viewdoc/download?doi=10.1.1.453.4591&rep=rep1&type=pdf ˙

Hornung, A. and Seifert, H. (2006). Rotary kiln pyrolysis of polymers containing hetero-atoms. In J. Scheirs and W. Kaminsky (eds), *Feedstock Recycling and Pyrolysis of Waste Plastics: Converting Waste Plastics into Diesel and Other Fuels,* 549–567. Chichester: Wiley & Sons.

Kandpal, J. and Maheshwari, R. (1993). A decentralized approach for biocoal production in a mud kiln. *Bioresource Technology* 43: 99–102.

Klose, W. and Wiest, W. (1999). Experiments and mathematical modeling of maize pyrolysis in a rotary kiln. *Fuel* 78(1) 65–72 (doi:10.1016/S0016-2361(98)00124-0).

Leiva, C., Gómez-Barea, A., Vilches, L.F., Ollero, P., Vale, J. and Fernández-Pereira, C. (2007). Use of biomass gasification fly ash in lightweight plasterboard. *Energy and Fuels* 21(1): 361–367 (doi:10.1021/ef060260n).

Libra, J.A., Ro, K. S., Kammann, C., Funke, A., Berge, N.D., Neubauer, Y., Titrici M., Fühner C, Bens O., Kern J. and Emmerich, K.-H. (2010). Hydrothermal carbonization of biomass residuals: a comparative review of the chemistry, processes and applications of wet and dry pyrolysis. *Biofuels* 2(1): 71–106 (www.future-science.com/doi/abs/10.4155/bfs.10.81).

Lu, Q., Li, W.-Z. and Zhu, X.-F. (2009). Overview of fuel properties of biomass fast pyrolysis oils. *Energy Conversion and Management* 50: 1376–1383.

McKendry, P. (2002). Energy production from biomass (part 3): gasification technologies. *Bioresource Technology* 83(1): 55–63.

Nachenius, R.W., Ronsse, F., Venderbosch, R.H. and Prins, W. (2013) Biomass pyrolysis. In D.Y. Murzin (ed.), *Advances in Chemical Engineering,* vol. 42, 75–139. Burlington, VA: Academic Press.

Ogden, C.A., Ileleji, K.E., Johnson, K.D. and Wang, Q. (2009). In-field direct combustion fuel property changes of switchgrass harvested from summer to fall. *Fuel Processing Technology* 91(3): 266–271 (doi:10.1016/j.fuproc.2009.10.007).

Pels, J.R., De Nie, D.S. and Kiel, J.H.A. (2005). *Utilization of Ashes from Biomass Combustion and Gasification.* October. Petten: ECN. Available at www.ecn.nl/docs/library/report/2005/rx05182.pdf.

Ravichandran, P., Gibb, D. and Corscadden, K. (2013). Controlled batch leaching conditions for optimal upgrading of agricultural biomass. *Journal of Sustainable Bioenergy Systems* 3: 186–193 (doi:10.4236/jsbs.2013.33026).

Schimmelpfennig, S. and Glaser, B. (2011). One step forward toward characterization: some important material properties to distinguish biochars. *Journal of Environmental Quality* 32 (doi:10.2134/jeq2011.0146).

Sevilla, M. and Fuertes, A.B. (2009). The production of carbon materials by hydrothermal carbonization of cellulose. *Carbon* 47(9): 2281–2289 (doi:10.1016/j.carbon.2009.04.026).

Stassen, H.E. (2002). Developments in charcoal production technology. *Unasylva* 53(211): 34–35.

Tian, Y., Zuo, W., Ren, Z. and Chen, D. (2011). Estimation of a novel method to produce bio-oil from sewage sludge by microwave pyrolysis with the consideration of efficiency and safety. *Bioresource Technology* 102(2): 2053–2061 (doi:10.1016/j.biortech.2010.09.082).

Titirici, M. M., Thomas, A., Yu, S.-H., Müller, J.-O. and Antonietti, M. (2007). A direct synthesis of mesoporous carbons with bicontinuous pore morphology from crude plant material by hydrothermal carbonization. *Chemistry of Materials* 19(17): 4205–4212 (doi:10.1021/cm0707408).

Van Loo, S. and Koppejan, J. (eds). (2010). *The Handbook of Biomass Combustion and Co-firing*, 2nd edition. London: Earthscan.

Venderbosch, R.H. and Prins, W. (2010). Fast pyrolysis technology development. *Biofuels, Bioproducts and Biorefining* 4: 178–208.

Zhao, L., Cao, X., Mašek, O. and Zimmerman, A. (2013). Heterogeneity of biochar as a function of feedstock sources and production temperatures. *Journal of Hazardous Materials* 256: 1–9 (doi:10.1016/j.hazmat.2013.04.015).

Chapter 3

Biochar properties

Elisa Lopez-Capel, Kor Zwart, Simon Shackley,
Romke Postma, John Stenstrom, Daniel P. Rasse,
Alice Budai and Bruno Glaser

Summary

The search for meaningful and desirable biochar properties is still under way. Nevertheless, there are certain chemical and physical properties that are widely considered relevant to the behaviour and function of biochar in soil. In this chapter, some of the more accessible properties are described, giving the reader the necessary tools and understanding to grasp the interaction of biochar in the soil environment covered in Chapter 4.

In this chapter, 'what biochar is' is described in general terms, and then we go on to take a closer look at the desirable material properties of biochar. The most important biochar properties are its polycondensed aromatic carbon structure with functional groups and its porous physical structure, which can be translated into stability (C sequestration), and nutrient and water holding capacity, respectively.

This chapter illustrates this theoretical approach by empirically examining a range of specific biochars and property variation across biochars, including a range of biochars produced under the European context in order to achieve European Biochar Certification.

Characterising and mapping desirable properties of biochar

Why do we need to characterise biochar properties?

A bespoke process of regulating and monitoring of biochar production, use and disposal is currently missing in the European Union Member States, as well as in the USA, Australia and other countries which might come to use considerable amounts of biochar in agriculture (see Chapter 11). In many cases, use of biochar is neither regulated nor explicitly forbidden, though it is worth noting that Japan enacted legislation to permit use of charred biomass in soil in 1983, while, since 2013, Switzerland has been the first country in the world to have an explicit regulatory framework for biochar under the Federal Ministry of Agriculture (Schmidt 2013). Therefore, it is of utmost importance to agree on what biochar is and to come up with a minimum set of desired properties and acceptable limits to avoid unintended harmful effects.

The reason why we urgently need clear and easy-to-use quality guidelines, or even standards, for biochar is that without such clear guidance, there is nothing to stop a firm claiming that it is producing biochar without meeting quality and safety requirements, as no accepted mandatory standards for biochar exist (only voluntary ones).

Currently, there are three main voluntary schemes available for assessment of biochar products. The International Biochar Initiative (IBI), the European Biochar Certificate (EBC) and the Biochar Quality Mandate (BQM). Information provided by these schemes has been compiled into the Biochar Testing Protocol (BTP) to provide information on the analytical methods for measuring the properties of biochar materials and biochar products.

Mapping properties through biochar protocol and (voluntary) biochar standards

What could science do for practitioners?

Currently there are two biochar labels associated with voluntary standards: the IBI Guidelines and EBC, referring to basic material properties such as elemental composition (content of carbon, nitrogen, hydrogen, oxygen and ash), lack of contaminants such as heavy metals and polycyclic aromatic hydrocarbons and absence of ecotoxicological impacts upon plants and soil organisms.

In the European context, key requirements under EBC include: maximum permissible distance for transporting feedstocks; clean production conditions; material definition (as having a minimum stable organic carbon content of 50 per cent); basic utility material properties (such as for use in soil); safety and possible toxicants (metals, organic contaminants, seed germination); optional advanced property testing (potentially beneficial soil enhancement properties); protocol for consistency testing; and details of accredited and modified analytical tests (grading biochar as basic or premium).

Key biochar functions (see Figure 3.1) can be evaluated through a set of biochar property testing requirements (as shown in the list of material properties in Table 3.1). This information allows the user to describe and define the properties of the biochar product, as summarised in the Biochar Testing Protocol (BTP 2013), and illustrated with a European biochar case study.

How is biochar certified?

The European Biochar Certificate has two biochar categories (basic and premium), differing by the level of contamination (positive list). The IBI Biochar guidelines differentiate between three different biochar categories based on the level of organic carbon. A third scheme, the BQM, differentiates between two biochar grades: 'high' and 'standard', according to quality threshold values specified in the form of maximum permissible limits (MPLs) of toxicants present in biochar; this is similar to the EBC but the values are calculated differently.

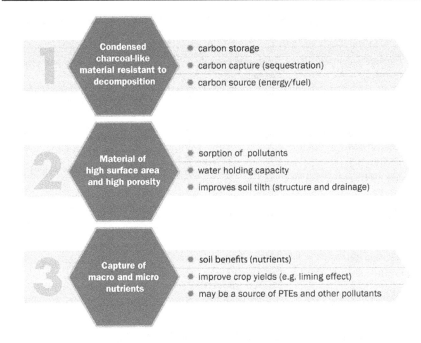

Figure 3.1 Summary of what biochar is (boxes) and what biochar does (bullet points).

Biochar properties and functions are very much dependent on the production process and the type of feedstock. Guidelines require that feedstocks are produced from biomass grown in a sustainable manner from: wood from forests or short rotation forestry; a mixture of wood and waste (e.g. manure); and biomass from waste streams, such as anaerobic digestion, and sewage sludge (under restricted use). The feedstock contaminant level will influence the contaminant level of biochar that is produced; therefore, feedstocks are required to be free of paint, solvents and other non-organic contaminants, and non-organic wastes (such as plastic, rubber, electronic scrap) have to be removed. Examples of biomass feedstocks and their influence on biochar production processes are presented and discussed in detail in the European biochar (Interreg project) case study section.

It should be mentioned that, although biochar may be expected to improve soil quality, this is not directly assessed under basic properties by regulatory authorities. Biochar labels reflect material properties but are not related to desired effects of biochar on organisms or ecosystems. The mechanisms by which biochar affects soil properties for agricultural and ecological benefit are complex and remain a subject of research. This research has to be performed separately and is more difficult to assess. However, the IBI biochar guidelines do include analysis of optional advanced property testing for the use of biochar as a soil enhancement. Biochars fulfilling such properties are certified by IBI under category C. For details, see Table 3.1 (and appendix G in BQM 2014).

Table 3.1 Summary of the basic and optional Biochar Testing Protocol property testing requirements

Basic material properties (for use in soil)	Safety and possible toxicants	Optional advanced property testing
pH	Potentially toxic elements (PTEs)	Nutrients: Zn, Ca, Fe, Ni, Se, Mo, Mg, Mn, Fe
Moisture at time of delivery	Heavy metals: arsenic,	
Organic carbon (Corg)	cadmium, chromium,	Neutralising capacity (liming if pH above 7)
Inorganic carbon	cobalt, copper, lead,	
Total carbon (C)	mercury, manganese,	Liming (carbonate value)
Hydrogen (H)	molybdenum, nickel,	
Oxygen (O)	selenium, zinc, boron,	Electrical conductivityCation exchange capacity (K, Ca, Mg, Na)
H:Corg ratio	chloride, sodium	
O:Corg ratio		
Total nitrogen (N)	Organic compounds:	Porosity Specific surface area/total surface area
Total sulphur (S)	PAHs	
C/N ratio	PCBs	
Total ash	PCDDs (dioxins)/	Labile carbon content/volatile matter
Total phosphorus (P)	PCDFs (furans)	
Total potassium (K)		Long-term stable carbon
Water holding capacity	Seed germination (plant toxicity)	Available P
Bulk density		
Particle size distribution		
Heavy metals		
PAHs		
PCBs		

Source: Biochar Testing Protocol (2013).

Biochar characterisation in the European context

The Interreg IVB North Sea Region biochar case study

There is a fair degree of overlap between IBI and EBC schemes (properties, certification, labelling), since both rely upon existing practices and regulations in major jurisdictions, including Europe, North America and Australia. This chapter examines a range of specific biochar properties and their variations and evaluates whether a set of biochar types from the Interreg IVB North Sea Region biochar case study meet the requirements of the European Biochar Certification. From Table 3.2 it is evident that properties and thresholds for the existing voluntary systems are similar. Biochar properties arising from the case study and recommendations on their potential use are made in the following sections.

Six biochar types were produced from locally available and sustainable feedstocks commonly found in northern Europe. The samples were produced by slow pyrolysis using a lab-scale pyrolysis unit (with a high heating temperature of HHT of 550°C and a residence time of 20 minutes (for more details see Brownsort and Dickinson 2012 and www.biochar-interreg4b.eu/). These were compared with the certification by the EBC of one of the biochar types, named RomChar, produced by European Charcoal AG from mixed wood at 550°C (for details see www.swiss-biochar.com/eng/biochar.php).

RomChar biochar was used as a soil amendment in the IVB Interreg Biochar project's field trials (see Chapter 5). Basic and optional physical and chemical properties were analysed, and their characteristics compared against EBC and IBI biochar label criteria (Table 3.2).

Biochar carbon properties

For characterisation and certification of biochar, adequate properties and thresholds are necessary to unambiguously identify materials as biochar, and this information will be sufficient to designate material as biochar even if a detailed description of the production process is not available (Schimmelpfennig and Glaser 2012).

Since the predominant interest in and rationale for biochar has been to reduce or avoid carbon emissions, we first discuss the carbon storage properties of biochar in relation to biomass feedstocks and production processes, and continue with carbon composition and stability.

Biochar carbon storage

The carbonisation process and carbon conservation

Most woody biomass contains approximately 50 per cent carbon by dry weight while straw contains approximately 45 per cent carbon. The rest of biomass is made up of the other 'building blocks' of life – hydrogen (H), oxygen (O), nitrogen (N). In addition, there are smaller quantities of essential elements including sulphur (S), phosphorus (P), potassium (K) and trace amounts of elements such as iron (Fe), zinc (Zn), magnesium (Mg), manganese (Mn), copper (Cu), chromium (Cr) and molybdenum (Mo) among others.

During the carbonisation process, which takes place during pyrolysis, oxygen, hydrogen and nitrogen are driven off from the biomass through heating and reorganisation of the chemical structure of the plant material from long molecular chains of C, O, N and H to ring-like structures of six carbon atoms (Schimmelpfennig and Glaser 2012). The loss in O and H is relatively higher than the loss in C (Amonette and Joseph 2009). As a result, the carbon content increases disproportionately compared to the other constituents and some biochar (and charcoal) contains more than 80 per cent carbon by weight with a range of values of about 50–95 per cent. Hence carbon is concentrated in biochar (charcoal) relative to the feedstock. This can be assessed by calculating the carbon conservation, a measure of the quantity of carbon in the feedstock that is retained in the biochar.

Table 3.2 Comparison of current biochar labels: European Biochar Certificate (EBC) and International Biochar Initiative Biochar Guidelines (IBI) against properties of feedstock and biochars

Category	Property	EBC	IBI	OSRP biochar	MW biochar	AD biochar	GW biochar	SWP biochar	WP biochar	Romchar biochar
Feedstock description				Oilseed rape pellets	Mix wood	Anaerobic digestate	Green waste	Softwood pellets	Willow pellets	Romchar mix wood
Feedstock label		Positive list	No contaminants	Positive list	Positive list	Positive list	Positive list	biomass	biomass	Positive list
Feedstock	C			41.5	44.6	39.1	40.9	45.8	45.1	44.6
Material properties	stability	basic/ premium	A/B/C grade	A	A	B	B	A	A	A
	TOC	> 50%	>60/30/10% (A/B/C)	67.95	82.44	44.09	17.79	88.08	83.8	67.61
	H/C	<0.6	<0.7	0.03	0.03	0.03	0.03	0.03	0.03	0.02
	O/C	<0.4		0.13	0.06	0.19	0.14	0.05	0.06	0.13
	Black Carbon	10–40% of TOC		n.a.	n.a.	n.a.	n.a.	n.a.	n.a.	n.a.
	pH		report	9.05	7.69	11.21	9.67	6.79	8.46	8.76
	P mg/kg	Report (total)	Report (total P)	15.4	6.3	560.4	64.3	2.6	131.2	34
	K mg/kg	Report (total)	Report total K	12991	1060	62443	1783	120	806	3108
	porosity m2/g	BET surface area (report)	BET surface area (report)	1.23	4.82	3.55	2.10	Below detection	0.20	295

Table 3.2 Continued

Category	Property	EBC	IBI	OSRP biochar	MW biochar	AD biochar	GW biochar	SWP biochar	WP biochar	Romchar biochar
Contaminants	Heavy metals	National law	National law	very below	very below	below	below	very below	very below	very below
	PAH	<12 mg kg^{-1}	<20 mg kg^{-1}	3.24	3.8	3.71	2.51	2	0.68	<0.48
	PCB	<0.2 mg kg^{-1}	<0.5 mg kg^{-1}	n.a.	n.a.	n.a.	n.a.	n.a.	n.a.	<0.01
	PCDD/PCDF	<20 ng kg^{-1}	<9 ng kg^{-1}	n.a.	n.a.	n.a.	n.a.	n.a.	n.a.	3.81
Ecotoxicology	germination		Pass	n.a.	n.a.	n.a.	n.a.	n.a.	n.a.	n.a.

Note: TOC (total organic carbon), H/C (hydrogen-to-carbon atomic ratio), O/C (oxygen-to-carbon atomic ratio), PAH (polycyclic aromatic hydrocarbon), PCB (polychlorinated biphenyl), PCDD/PCDF (dioxin), Total Potassium (K); Total Phosphate (P); porosity surface area by BET method; n.a.= not analysed.

Source: Biochar Testing Protocol (2013).

The net stable carbon conservation in biochar per oven dry tonne (odt) feedstock is calculated as follows:

Net stable carbon stored (as tC per odt feedstock
    = char yield (t per odt feedstock)
    × carbon content of char × Carbon Stability Factor

Where the Carbon Stability Factor (CSF, as described in Chapter 7) is the proportion of biochar carbon which is stable on a 100-year time scale is considered to be appropriate for climate mitigation).

Calculating the carbon conservation from feedstock to char/biochar is important since one of the ways that biochar can be defined is to prescribe a minimum level of carbon conservation to it. A minimum value which has been discussed for a material to qualify as biochar is (the arbitrary) 30 per cent carbon conservation. This way of defining biochar allows the different carbon contents of the feedstocks to be taken into account.

Because some feedstocks have lower amounts of carbon, the biochar produced from such materials is inevitably lower in carbon and may not meet an arbitrary absolute threshold of carbon content (e.g. that a biochar should have a minimum of 50 per cent carbon).

Since the yield of biochar is typically one-third by weight, assuming a stable carbon content of 80 per cent, the overall conservation of carbon from a feedstock containing 50 per cent carbon is:

$$\text{Carbon conservation (woody feedstocks)} = \frac{(0.33 \times 0.8)}{(0.5)} = 0.528$$

or approximately 50 per cent. The remaining carbon in the feedstock is released to the atmosphere during pyrolysis. This one-off release of C as CO_2 is lower than after simply burning the biomass (whereby nearly 100 per cent of the carbon is lost), lower than when biomass is incorporated into soil (mineralisation, time-scale depending on soil conditions), but higher in the short-to-medium term than when biomass is used for construction (e.g. timber).

The non-carbon component of biochar contains inorganic matter such as minerals, metals, metal oxides and salts. This mixture is commonly known as ash. The ash content of biochar can range from 0.7wt per cent in pine wood to 45wt per cent in rice husk (Antal and Grønli 2003; Asian Development Bank 2011). Since most of the mineral material in the biomass will persist during pyrolysis, the ash content concentrates when biomass is converted into biochar. So, if the mineral content of feedstock increases, the yield of solid product from pyrolysis also increases because most metals and minerals are not converted into gases but remain as solids.

Carbon storage in the Interreg IVB North Sea Region case study

The biochar feedstock materials examined, including RomChar mix wood, softwood pellets, and willow pellets, contained approximately 45 per cent carbon by dry weight, and yielded an 82 per cent carbon content biochar, giving a conversion of carbon from woody feedstock to biochar carbon of approximately 50 per cent. On the other hand, feedstocks containing lower carbon contents (40 per cent), such as oilseed rape pellets, anaerobic digestate and green waste, yielded biochars of lower carbon content (18–44 per cent), and therefore lower carbon conservation.

Typical yields of char for waste materials such as sewage sludge and anaerobic digestion digestate (ADD) using slow pyrolysis are 40 per cent and 45 per cent, respectively (Karve *et al.* 2011). The carbon content in the sewage sludge char is approximately 30 per cent and in the ADD it is approximately 27 per cent, so the carbon conservation can be calculated as:

$$\text{Carbon conservation of sewage sludge} = \frac{(0.4 \times 0.3)}{(0.43)} = 0.28$$

$$\text{Carbon conservation of anaerobic digestion digestate} = \frac{(0.45 \times 0.27)}{(0.33)} = 0.37$$

The carbon conservation for sewage sludge char is, accordingly, 28 per cent while that for ADD char is 37 per cent. The conservation percentage may be lower for waste feedstocks compared to virgin woody feedstocks, because of the structural shielding that the existing plant structure exerts upon the hydrocarbon molecules and the relatively higher labile C content (Singh *et al.* 2012). Animal feeds such as grass, soymeal and other organic matter are processed in the digestive system of the animal, undergoing partial physical and chemical decomposition and loss of carbon as CO_2. As an example, the carbon content of cow manure is roughly 35 per cent (dry weight or DW), while the carbon content of the char is lower at around 21 per cent DW (Singh *et al.* 2012).

Table 3.3 shows carbon content, stable carbon content and carbon conservation percentages for a range of feedstocks and corresponding biochar products. Notice that column two presents the initial carbon content of the biomass feedstock; column three, the (oven dry) char yield (expressed as a percentage of oven dry weight of feedstock); column four, the carbon content of the char expressed as a percentage of dry weight; column five, an estimate of how much carbon in the char is stable on climate-relevant timescales (≥ 100 years) (calculated using an accelerated ageing method described in Chapter 7); column six shows the percentage of the carbon conservation. Note that carbon conversion percentages can be expressed as CO_2 by conversion ($CO_2 = C \times 44/12$).

One sample that stands out in Table 3.2 and Table 3.3 is the Edinburgh green waste sample, with very low char carbon content and a high solid char yield. This

Table 3.3 Carbon content (CC), stable carbon content and carbon conservation percentages of a range of feedstocks and corresponding biochar products

Feedstock	CC of biomass feedstock (%)	Char yield (%)	CC of char (%)	Stable carbon content (%)	Carbon conservation (%)
Oilseed rape pellets	41.5	30.4	69	100	51
'RomChar' woodchip	44.6	25.2	87.1	87.2	43
Digestate from AD	39.1	44.0	51.6	98.4	57
Edinburgh green waste	40.9	60.6	18.1	97.7	26
Commercial softwood pellets	45.8	23.9	88.9	86.8	40
Willow Pellets	45.1	24.9	85.7	92	44
Sewage sludge (Singh *et al.* 2012)	43.3	40.4	29.9	88.72	25
Cow manure (Karve *et al.* 2011)	35.24		21.61		

is due to the large quantity of non-organic contamination in the green waste sample – mostly soil – which has a low organic content but contains large quantities of minerals. These remain in the solid fraction, hence the high percentage of char yield. The higher the ash and mineral content in the feedstock, the higher the ash content in the biochar and the lower the carbon content.

Carbon composition and stability

Total carbon, labile carbon and black carbon content

The resistance of biochar to mineralisation in soil (to produce carbon dioxide) depends largely on its molecular structure. All biomass contains labile molecules, (which tend to be easily decomposed by microorganisms) and more recalcitrant molecules. An example of a highly labile molecule is glucose (Figure 3.2), found in, plants such as sugarcane and sugar beet. If glucose is added to soil, it is very readily and hastily broken down by microbes as a source of energy and carbon, stimulating a proliferation of microbial activity. This activity sometimes facilitates the break down of other less easily decomposed organic matter in the soil.

Recent research has shown that plant residues that have not undergone pyrolysis contain little or no highly stable molecules (Schmidt *et al.* 2011). Thus, a farmer applying fresh biomass or compost to his fields will see little of that carbon remaining after a number of months or years (Singh *et al.* 2012). Only one percent or less of the carbon added will remain as 'recalcitrant' soil organic carbon (sometimes known as humus), persisting for decades or even hundreds of years.

By contrast, biochar is composed predominantly of stable structures and only small quantities of labile molecules. Biochar becomes stable during the pyrolysis process as the carbon backbone of biological materials is rearranged. Aliphatic carbon chains (chain-like structures whose chemical bonds are readily attacked by microbial enzymes) are converted to aromatic rings (usually six carbon such as benzene, or occasionally other atoms such as nitrogen, linked in a ring structure by strong chemical bonds, resistant to microbial break down). Further restructuring leads to the linkage of these aromatic groups into large complexes (aromatic sheets) through a process known as condensation. In practice, well pyrolysed biochars are composed of sheets of condensed aromatic rings of various sizes, together with ash, and traces of smaller molecules (Figure 3.2). Both aromaticity and degree of condensation have been linked to the stability of biochars (Nguyen *et al.* 2010).

During pyrolysis the mass of biomass decreases as components such as water and organic compounds are removed. The most rapid decrease in weight takes place between 200 and 400°C, followed by a slower rate of decrease between 400 and 700°C. Consequently, the amount of biochar (solid material, as mass fraction percentage) produced decreases with increasing temperature, whereas the amounts of liquids and gases increase.

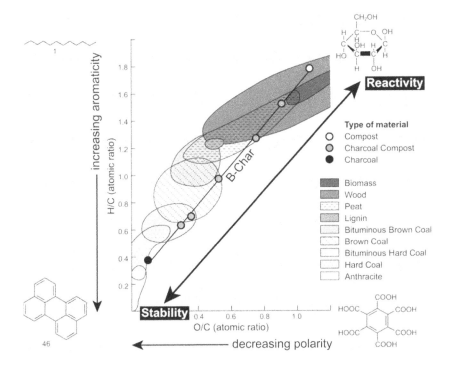

Figure 3.2 Basic interpretation of a Van Krevelen diagram. There is a trade-off between reactivity and stability of biochar which means the more stable a biochar is (being in the lower left of the Van Krevelen diagram), the lower its reactivity.

Source: Bruno Glaser.

These relationships can be observed through analysis of char yield, stable carbon content of the char, the highest treatment temperature (HTT) and the duration of heating (see Figure 3.3). At a lower HTT (<350°C), the char yield is higher but the carbonisation process is less extensive and there is more unstable carbon in the char. This unstable carbon can be used as a source of energy and carbon by microorganisms as it is converted into CO_2 fairly rapidly and will therefore not contribute to carbon sequestration. As the HTT and duration of heating increases, the char yield goes down as more C, H, N and O are released as vapours under the more aggressive conditions. The carbon that remains is more recalcitrant, meaning that it has become more stable and resistant to decomposition by microorganisms. The optimal temperature in terms of char yield and recalcitrance is approximately 450°C for woody feedstocks as shown in Figure 3.3 (Antal and Grønli 2003).

Because carbonisation is associated with a relatively greater loss of O, H, N and S compared to carbon, it is possible to use elemental ratios to characterise the extent of carbonisation. Elemental composition can be obtained by combustion of the oven dry biochar followed by specific detection of carbon, nitrogen, hydrogen and oxygen, known as elemental analysis. Ratios of hydrogen to carbon (H/C) and oxygen to carbon (O/C) are, consequently, key properties for characterising biochars. Ratios are used because the absolute concentrations of C, H, and O are strongly influenced by the ash fraction, composed of elements that are not lost (volatilised) during the pyrolysis process. The Van Krevelen diagram plots the H/C ratio against the O/C ratio. While the H/C ratio indicates the degree of aromatisation (extra molecular stability), the O/C ratio indicates the degree of polarity (reactivity or electrical charge) (Figure 3.2). Therefore, while aromaticity increases with temperature (Figure 3.3), polarity decreases with temperature during pyrolysis irrespective of the biomass feedstocks' compositions as illustrated in Figure 3.4.

The Van Krevelen diagram was developed to evaluate the properties of different types of coal and has also proven a useful way of mapping the relative carbonisation of different types of biochar, allowing comparison with other types of material (lignite, coal, wood, etc.). Glaser *et al.* (2012) proposed that biochar should have an H/C of ≤0.7 and an O/C of ≤0.4.

A range of European biochar and feedstock samples is plotted on a Van Krevelen diagram in Figure 3.5. Four out of the six European biochar materials produced could potentially be certified EBC premium labelled. Biochar produced from anaerobic digestion and Edinburgh green wastes are highly stable with high aromatisation (H/C ration) and low polarity (O/C ratio). However, they have organic carbon content values below the 50 per cent required by EBC to qualify as biochar. Nonetheless, they can be classified under the other two voluntary schemes (as grade B by IBI and by the BQM, both of which use a minimum stable organic carbon content of 10 per cent by weight).

The O/C ratio is also a reasonable proxy for the longevity of biochar in soil. The lower the ratio, the longer the biochar is likely to remain in the soil (see Table 3.4). Half-life is a measure of how long it takes for half of the biochar to break down into simpler chemicals that are utilised by microorganisms or released from the soil as gases following reaction with oxygen, conversion to methane or some other reaction.

Other chemical and physical techniques are available for studying biochar aromaticity. Some of these provide a more in-depth analysis of biochar chemical composition, such as nuclear magnetic resonance (NMR) spectroscopy. Our understanding of the link between elemental ratios and aromaticity largely comes

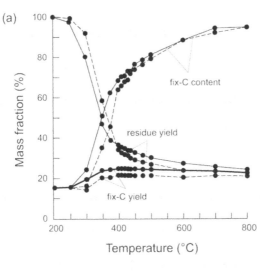

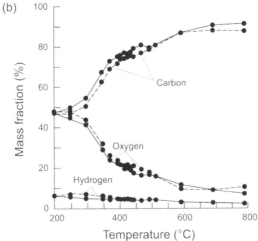

Figure 3.3 Optimal temperature in terms of char yield and recalcitrance is approximately 450°C for woody feedstocks.

Source: Antal and Grønli (2003).

Table 3.4 Estimated half-life in relation to the O/C ratio of biomass, two types of biochar and graphite

Material	O:C ratio	Estimated half-life time (years)
Biomass	>0.6	<100
Biochar type A (IBI category)	0.4–0.6	100–1000
Biochar type B (IBI category)	<0.2	>1000
Graphite	0	>1000

Note: Graphite is a crystalline form of carbon which consists nearly entirely of carbon and is formed by heating carbonaceous materials to a very high temperature (several thousands of degrees centigrade). It is the most stable form of carbon under standard conditions.

Source: Spokas (2010).

from NMR investigations (Calucci *et al.* 2013; Krull *et al.* 2009). In practice, NMR is a time consuming and costly method, most often used at the research level to calibrate cheaper and more accessible operational methods. In addition to elemental ratios, development of infrared spectroscopy for rapid estimates of biochar quantity and characteristics is under way (Zheng *et al.* 2014).

While elemental ratios are widely used as the standard approach to categorising material as biochar, the technique cannot be used when biochar has been applied to soils, as the indigenous soil organic matter will have a large influence upon these ratios. More sensitive methods are therefore needed for measuring properties of soil-incorporated biochars. Black carbon analysis could be a valuable tool for quantification of biochar in soil. Methods based on the detection of molecular

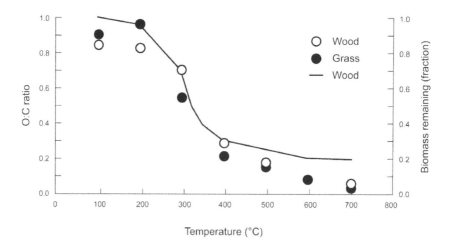

Figure 3.4 Loss of biomass (Reed and McLaughlin North American Biochar 2009) and changes in O/C ratio during pyrolysis.

Source: based upon data for wood and grass presented in Keiluweit *et al.* (2010).

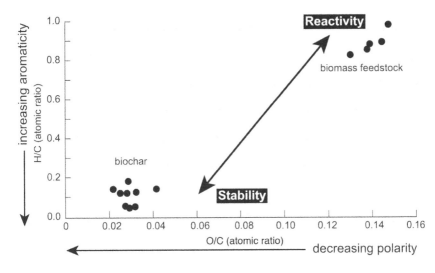

Figure 3.5 Basic interpretation of a van Krevelen diagram of European feedstocks and corresponding biochar types (case study).

markers, such as BPCA (named for the benzene polycarboxylic acids that are liberated when aromatic sheets are broken down under specific laboratory conditions), have been developed (Glaser *et al.* 1998). The concentration and composition of BPCAs are known to correlate with the quantity of aromatic structures (i.e. benzene rings and their degree of condensation; Manning and Lopez-Capel 2009). The methods applied in biochar research have thus far shown that elemental composition, chemistry and stability are interlinked even though not in a fully understood way.

Desirable properties

Biochar properties relevant for soil and agriculture

Desired biochar properties are associated with their beneficial effect on agricultural and environmental systems. As discussed in Chapter 4, they potentially include: increase in crop yields (through increased availability of nutrients and water); improved soil quality (soil structure, organic matter content, drainage); soil remediation (sorption of pollutants, reduction of nitrate and phosphate leaching); and carbon capture and storage (long term carbon stability) (see Figure 3.6).

Since the late 2000s, the carbon storage properties of biochar have become less valued due to collapse in the carbon price and general political intransigence on achieving a global deal on limiting carbon emissions (Chapter 9). As a result, attention has moved increasingly to the other potential applications of biochar, especially its use in agriculture and horticulture. In this chapter, we will briefly discuss biochar as it relates to the above properties. In Chapter 4 we will discuss how these properties relate to agronomic functions through influencing the soil into which the biochar is incorporated.

Nutrient value of biochar

One of the agronomic properties arises from the nutrient content of the biochar, which is found in the ash. Woody feedstocks have roughly 5–10 per cent ash, straw feedstocks have slightly more at 10 to 15 per cent and waste feedstocks more again (e.g. mixed green waste has 19 per cent ash and digestate from anaerobic digestion has 21 per cent ash), as shown in Table 3.2.

Ash is concentrated in the char. Pyrolysis of three tonnes of feedstock typically produces one tonne of char. The main plant nutrients of interest are nitrogen (N), phosphorus (P), potassium (K), calcium (Ca), iron (Fe) and very small (trace) amounts of other elements. There is an important difference between the total amount of these nutrients present in the ash and that which is actually available to the plant (i.e. total versus available).

Nitrogen is one of the few minerals lost as a gas during heating or volatilisation (about 50 per cent of N is locked up in the biochar matrix), while minerals such as calcium, potassium and phosphate, and metals such as cadmium and nickel, concentrate in the biochar during the process of pyrolysis (Singh *et al.* 2012). An overview of C, N, P and K contents of various biochars reported in literature is provided in Table 3.5.

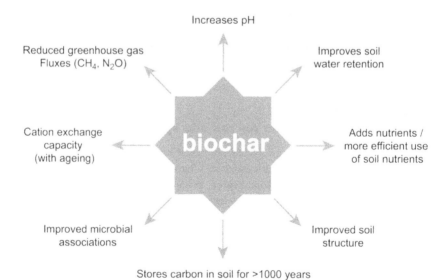

Figure 3.6 Key properties of biochar in soil.

The C content is relatively high in biochar from wood chips and green wastes, but the N, P and K contents are much higher in poultry litter biochar. These differences result in a much higher C/N ratio in biochars from wood chips and green wastes in comparison to that from poultry or pig litter or cattle manures. The same is true of these feedstocks prior to pyrolysis.

Where the feedstock composition is highly variable, the biochar produced will also be variable. Green waste, for example, is composed of whatever materials are delivered to a municipal waste facility by local residents or small companies that manage gardens and estates, and this will change on a daily basis. As an example of this variability, in Table 3.2 we provided information on the properties of a green waste char with a carbon content of only 18 per cent, yet the green waste char in Table 3.5 is much higher at 68 per cent. The green waste and anaerobic digestion biochar tested could be considered soil enhancers owing to their high nutrient content (potassium and phosphate, see Table 3.2).

Surface area and porosity of biochar

Biochar is a bit like a hard sponge. It contains a large number of tiny tunnels or pores, some of which connect to the outside, others of which are closed off when the biochar is just newly produced. The specific surface area (SSA) is a measure of the size of all the external and internal spaces that are accessible (i.e. open to the exterior). The surface area of biochar varies from less than a few cm² per g to hundreds of m² per g. The highest surface areas tend to be virgin wood-based biochars, owing to the retention of structure of the biomass. Woody biochar surface areas can be 400 to 600m²/g (by means of comparison, chemically activated carbon

Table 3.5 Total nutrient contents of biochars from different feedstocks

Biochar feedstock	C, g C/kg	N, g N/kg	C/N ratio	P, g P/kg	K, g K/kg	Reference
Wood chips	708	10.9	65	6.8	0.9	Lehmann et al. (2003)
Green wastes	680	1.7	400	0.2	1.0	Chan et al. (2007)
Poultry litter	380	19	19	25.2	22.1	Chan et al. (2007)
Cow manure		6.46	19.1	3.51	13.6	CoalTech
Maize cobs	645[1]	5.8	111	0.34	1.7	ADB (2011)
					3.8	Karve et al. (2011)
Maize straw				0.46	2.36	ADB (2011)
Pig manure		21.1		28.5		Marchetti et al. (2012)
Wood chips		10.8		1.3		Marchetti et al. (2012)

Note: Notice that total nutrient contents are not large. The units in columns 5 and 6, g P/kg, are the same as kg per tonne. To convert to fertiliser units, multiply P by 2.29 (P₂O₅) and K by 2.5 (K₂O).

used as filters in industrial settings typically have surface areas of 1000m²/g but can be as high as 2500m²/g. Less structured feedstocks can still exhibit reasonable surface areas; one cow manure biochar has a surface area of 260m²/g. A range of values relating to the surface area, pore volume and pore size of biochars are shown in Table 3.6.

RomChar, produced commercially (in retorts, see Chapter 2) and used in the field trials reported in Chapter 5, showed a relatively high surface area. The surface area of the biochars produced in the lab unit (as described in Chapter 2) were all very low, even for woody materials anticipated to have a higher surface area. It is unknown why these biochars have such low surface areas, but this would have a major impact upon their characteristics.

Biochar pore structure consists of multiple interconnected networks of micropores, mesopores and macropores, as illustrated in Figure 3.7. There are large (macro) pores which are >50 micrometres in diameter (1/1000th of a millimetre) or open canals/pores which are much longer. These larger pores are often derived from xylem and phloem vessels, the main conduits for water and nutrients. There are also smaller macropores, in the range of 5 to 20 micrometres, which were once plant cells. Smaller (meso) pores also occur in biochar (0.1–0.5 micrometre). Microorganisms such as bacteria are typically 0.5 to 5 micrometre in length. They can, therefore, move around the macropores and some mesopores. Plant roots and fungal hyphae can also penetrate the macropores and access water (and ions dissolved in the water).

However, there are also much smaller pores in biochar, micropores, which are at the nanometre (nm) scale (1000nm is equal to 1 micrometre). Micropores are typically 10s of nm in width and are sometimes smaller than larger molecules.

Table 3.6 Surface area, pore volume and average pore size of European biochar samples analysed using the Brunauer, Emmett, and Teller (BET) equation

Char sample	Surface area m²/g	Pore volume cm³/g	Pore size Å
Oil seed rape pellet char (lab)	1.23 ± 0.15	0.003 ± 0.000	427 ± 32
Mixed wood biochar (lab)	4.82 ± 5.20	0.002 ± 0.003	1112 ± 320
AD digestate char (lab)	3.55 ± 0.15	0.008 ± 0.001	173 ± 9
Edinburgh green waste char (lab)	2.10 ± 1.55	0.006 ± 0.004	371 ± 110
Softwood pellet char (lab)	Below d. level	Below d. level	Below d level
Willow pellet char (lab)	0.20 ± 0.01	0.002	–
RomChar (field trial) (retorts)	295 ± 6.74	0.163 ± 0.027	42 ± 5
Commercial mixed woodchip (retorts)	0.99 ± 0.03	0.004 ± 0.004	557 ± 144
Waste wood pellets (retorts)	3.51 ± 1.40	0.009 ± 0.004	208 ± 52

Note: Below d level = below detection level. The unit Å is an angstrom and is 10^{-10} m or 0.1 nanometre.

Biochars produced at high temperatures have a higher surface area, but with a larger proportion of micropores and surface areas, ranging from 2 to 800m²/g (Illingworth *et al.* 2013). Figure 3.8 illustrates how RomChar biochar pore size distribution is dominated by micropores (71 per cent of the pore volume), and Figure 3.9 how surface area changes with increasing pyrolysis temperature.

There is good evidence that pores in biochar become blocked by ash, tar and oils. During the process of pyrolysis, hot vapours rise through the biomass and, as it cools, some of the larger molecules in the vapour may condense and settle onto the biochar surface. Alternatively, condensation may occur above the biomass on the metal surfaces of the reactor from which oil and tars drip down onto the hot biochar. Some of this tarry material gets lodged into the pore space, reducing the connections between adjacent pore spaces. When this biochar is added to soil, however, microbes attack the tars and oils as a food and energy source, eventually exhausting the supply and opening up the pores. The nutrients in the ashes in blocked pores can then be washed out of the biochar, becoming available to support growth of plants and microbes.

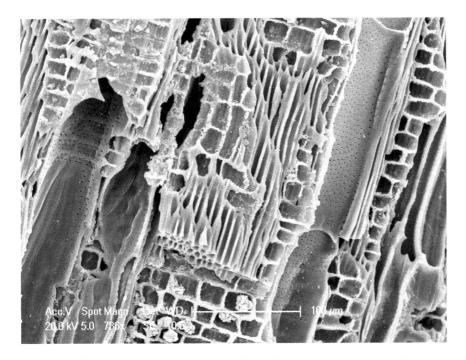

Figure 3.7 SEM of white charcoal produced from Cratoxylon spp. showing the inter-connecting network of macro, meso and micropores.

Source: image courtesy of Simon Shackley.

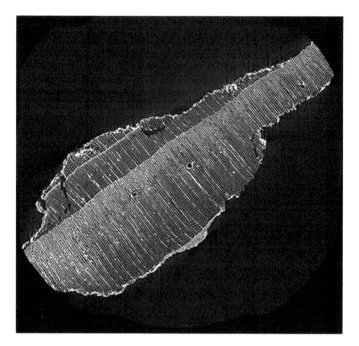

Figure 3.8 3D reconstruction of Romchar biochar sample by XRCT scan analysis. A threshold of grey level distribution is chosen to demonstrate that biochar pore size distribution is dominated by micropores.

Source: image courtesy of XRCT Durham University facility, Elisa Lopez-Capel.

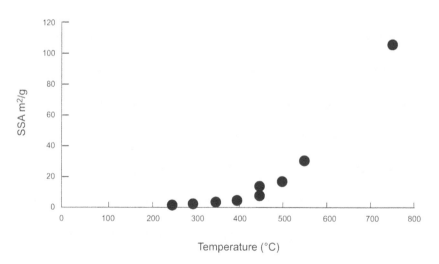

Figure 3.9 Effect of pyrolysis temperature on the specific surface area (SSA) of biochar produced from corn cob and wood.

Source: data from Zheng *et al.* (2010).

Water holding capacity, adsorption and reduced leaching of nutrients

Biochar can hold a lot of water – easily its own weight and sometimes up to six times its own weight when water is slowly added. Because biochar is incorporated into the soil, measuring how much water can be added to biochar without it leaking back out is not a reliable way of calculating its capacity to hold water in the soil. This changes its physical form, the pathways through the soil by which water drains and moves sideways, and the surface tension that holds water within the biochar. We return to this issue in Chapter 4.

Biochar tends to absorb solutions in which it is placed. When water enters micropores, the small dimensions mean that biochar tends to strongly hold the water molecules, making it difficult for plant roots to extract it. Just because water can be held within biochar does not mean that it is necessarily available to a plant.

The biochar surfaces preferentially attract certain kinds of molecules and these are adsorbed into the biochar. *Ad*sorption refers to the binding of molecules onto surfaces, while *ab*sorption means that a fluid is dispersed throughout a material (as in a paper towel being used to clean up a spilt liquid). In reality, both adsorption and absorption occur when a liquid is poured onto biochar, since there is selective binding of molecules onto surfaces but also movement of liquid to fill pore spaces within the biochar particle. An abbreviation to cover both processes is sorption with the verb form being 'to sorb' and we will use these terms in this book.

Biochar materials have a high capacity for sorbing particular types of molecules (particularly hydrophobic molecules like fats and oils), and they are said to have a high sorption capacity with respect to these substances. Examples include pollutants, such as two to five carbon ring molecules (known as PAHs), chlorine-containing molecules (PCBs), heavy metals such as copper (Cu), zinc (Zn), cadmium (Cd), selenium (Se) and lead (Pb), pesticides and herbicides. Because of biochar's capacity to sorb these pollutants, its use in land remediation has been widely advocated. It has also been suggested that biochar might be able to reduce the loss of nitrogen from agricultural fields where applied as synthetic fertiliser by holding on to ammonium (NH_4^+) and/or nitrate ions (NO_3^-), thereby reducing conversion of ammonium to gases (nitrogen and nitrous oxide) or to nitrate (Zheng *et al.* 2013) and reducing leaching of nitrates from topsoil. Further information on nutrients and sorption of pollutants is presented in Chapter 11.

Biochar frequently repels water (it is hydrophobic); however, biochar hydrophobicity decreases at higher production temperatures owing to aliphatic functional groups volatising and being lost (lower O/C and O/H ratios; Gray *et al.* 2014), providing greater water holding capacity. Biochars exhibit large ranges in porosity (providing a volume for water uptake; see section above) and surface chemistry (Baldock and Smernik 2002), depending on feedstock selection and production temperature. It might be supposed that the hydrophobic nature of biochar would have a negative effect on its capacity to adsorb water, yet, as noted above, biochar has a high water holding capacity. This apparent contradiction may be explained in a similar way as for dry sand, which is also very water repellent. A short while after wetting dry sand, however, it will start to take up water. This might be because water vapour advances into the sand or dry soil, settles onto internal surfaces and changes the material from water repellent to water absorptive (DeBano 1981). Similar preconditioning with water vapour has been observed to increase water uptake of biochar (Gray *et al.* 2014).

Biochar alkalinity

pH measurement and liming capacity

pH is a measure of the acidity (or its opposite, alkalinity) of a water-based solution. Pure water has a pH of 7.0 and is termed neutral. A pH below 7.0 is termed acidic and a pH above 7.0, alkaline. Lemon juice and vinegar is acidic with a pH of 2. Coffee is also acidic with a pH of 5 while sea water is slightly alkaline with a pH of 8 and household cleaning fluids such as bleach and oven cleaner are strongly alkaline with a pH of 13–14. Acids have a high concentration of hydrogen ions (H^+).

The pH of a solid can only be measured by dissolving it in water. It is not possible to measure pH in solid biochar as the equipment used requires a liquid to obtain a reading. As a very resistant carbon matrix, however, biochar cannot simply be dissolved. Biochar does not have a liquid phase and, if strongly heated, turns into a gas. It is not, however, the carbon matrix which determines the pH of biochar but rather the ashes within the char. When biochar is placed into water, the ashes will wash out. They are typically alkaline, with a pH of between 8 and 11. It is therefore not surprising that biochars with a high ash content, such as anaerobic digestion and green waste biochar, show pH values of 11 and 10, while biochar from soft wood pellets exhibit a pH value of 6.8 (see table 3.2)

pH is an important factor influencing soil chemistry and the availability of plant nutrients. Acid soils tend to have a high concentration of aluminium (Al^{3+}) and manganese (Mn^{2+}). Al^{3+} ions disrupt cell division and reduce plant root elongation. High acidity also encourages iron to oxidise, which binds to phosphorus, reducing the availability of this vital nutrient to plants.

Alkaline soils have a high concentration of sodium (Na^+), calcium (Ca^{2+}), potassium (K^+) and magnesium (Mg^{2+}) – so-called 'base cations' – which are vital for healthy plant growth. The ideal pH for a soil growing arable crops is 6.5–7.0. Outside of this range many crops' potential productivity may not be realised.

Agricultural soils tend to become acidic over time because the base cations – which counter the H^+, causing acidity – are removed with the crop and straw. Farmers counter this by applying lime to soils. Lime reduces H^+ while increasing Ca^{2+} and binding Al^{3+} with OH^- to form a mineral called gibbsite, reducing toxicity to plant roots.

The liming and neutralising capacity of biochar can be estimated (Crombie *et al.* 2015) if its mineral composition is known by using the formula of Sluijsmans (1966):

$$1*\%CaO + 1.4*\%MgO + 0.6*K2O + 0.9*Na2O - 0.4*P2O5 - 0.7*SO3 - 0.8*Cl - n$$

* N = kg lime (CaO)

Where n is between 0.2 and 2. Therefore, the biochar pH and also its mineral composition is important regarding its effect on soil pH.

Biochar surfaces and their importance when added to soil

The surface chemistry of biochar is important in understanding its properties. The surface chemistry occurs due to the very large external and internal surface areas of biochar. Biochar surfaces tend to be negatively charged which means that they attract positively charged cations such as Na^+, K^+, Ca^{2+}, Mg^{2+} and Al^{3+} as well as charged molecules such as NH_4^+. The chemical bond between the surface and ions is relatively weak and easily reversible, meaning that plant roots and fungal cells (known as hyphae) can easily take up the cations and use them to build new plant and fungal matter. Since many of these cations are essential for healthy plant growth, the soil's ability to hold them in reserve is an important indicator of soil 'health'. This ability of a material such as biochar (or soil) to sorb cations is measured and expressed as cation exchange capacity (CEC). In addition to CEC, there is emerging evidence that biochar in soil also exhibits higher than anticipated anion exchange capacity (AEC), which could account for the retention of negatively charged ions such as nitrate (see Chapter 6). This enhanced AEC is most noticeable for biochar that has been co-composted with organic matter rather than for 'pure' biochar added to soils.

Clayey soils have the highest CEC of all soil types due to the large negatively charged surfaces on and within clay particles. However, the CEC varies substantially between different clays. Some have a low value of 3–10cmolkg^{-1}, while a clay such as montmorillonite has a CEC of between 70 and 100cmolkg^{-1}. Vermiculite, a particular type of clay used as a plant growth media, has an even higher CEC of 100 to 150cmolkg^{-1}. Humus, that part of organic matter remaining after normal decomposition of organic material in soil or compost piles, has a still higher CEC of over 1000cmolkg^{-1}, a consequence of the large, negatively charged surface area (Ashman and Puri 2002).

The CEC of biochar is not as high as that of humus. Typically it is in the range of several cmolkg^{-1} to a few tens of cmolkg^{-1} for freshly produced biochar with occasionally higher values (up to a few hundred cmolkg^{-1} for aged samples that have been exposed to air and have undergone slow oxidation (Hammond *et al.* 2013).

So-called 'functional groups' may be attached to the biochar surfaces – these are molecules which are bound to the carbon matrix of the biochar but which are, to some extent, reactive and able to combine with other molecules close-by or able to promote certain reactions. The functional groups also react with cations. Hydrogen and oxygen are especially prevalent in functional groups of organic molecules.

An example of an organic acid functional group is COOH, known as carboxylic acid. As biomass is heated during pyrolysis, a process known as decarboxylation takes place which involves elimination of H_2O and CO_2. With the loss of functional groups, the overall carbon concentration of biochar increases while oxygen and hydrogen concentrations decrease. The surface chemistry changes, becoming less reactive. Changes in H/C and O/C ratios give important clues about production temperature and biochar reactivity (Figure 3.2).

When biochar is added to soil, a slow process of oxidation occurs. Oxidation at the surface can be assisted by microbes and/or occurs through exposure of biochar to the very small concentration of oxygen that is present in the soil. Slow oxidation results in a return of the carboxyl groups and an increase in surface reactivity and CEC (and potentially AEC).

Stabilisation of soil organic matter and resistance to microbial degradation

The addition of biochar to soil has been reported to increase the long term stability of humic substances (Lehmann and Joseph 2009). Humic acids are a very large group of chemicals (many thousands) which differ in their resistance to decomposition by microbes. There is evidence that biochar can protect humic acids from decomposition. This may be due to a physical process of 'shielding' whereby the organic matter is physically protected from microbes, which cannot access the biochar pores, either because they are too small or because clay (mineral) particles form a physical barrier. Soil from sites where there is a long-term history of charcoal making and addition of charred organic materials to soil have been shown to have a higher than anticipated soil organic matter (SOM) content, such as the terra preta soils from Amazonia and charcoal blast soils from USA – this is due to the protective characteristic of biochar (Cheng *et al.* 2008).

Biochar is resistant to microbial degradation, but the relationship between biochar C/N ratio and mineralisation/immobilisation is different to that of other organic materials (Cayuela *et al.* 2013). If organic material with a high available C/N ratio is added to soil, microbes will flourish, breaking organic matter down quickly to take advantage of the energy and nutrient content. While this is good for microbes, it deprives plants of nitrogen in the accessible mineral form – at least until microbes die and decompose, releasing nitrogen as ammonium (NH_4^+) and nitrate (NO_3^-). This locking-up of nitrogen is known as 'immobilisation' and is explained in more detail in Chapter 7.

A carbon to nitrogen (C/N) ratio in organic matter of 20 is optimal to support microbial growth. At lower C/N ratios there is surplus nitrogen, which is released in mineral form into the environment (nitrogen mineralisation). At higher C/N ratios, N immobilisation occurs as the availability of nitrogen becomes a limiting factor in microbial (and plant) growth. Many biochars have very high C/N ratios (e.g. up to 500) and, on this basis, rapid N immobilisation may be expected. Simply measuring the total C and N of biochar, however, is misleading since much of the C is not available to microbes as an energy or carbon source. Furthermore, much of the N is also locked up in the ring structures of biochar and thus cannot be accessed by microorganisms. The 'available' C/N ratio, representing only the C and N in smaller organic molecules loosely attached to the surfaces of biochar or minerals within the biochar and available to microbes, gives a more accurate measure of the suitability of the material to support microbial life forms. Since the fraction of labile C and N is low in most biochar (Table 3.3), biochar is of little value to soil microorganisms.

Potentially Negative Properties of Biochar

It is important to consider possible negative influences on the environment from biochar addition to soil. These might arise from the presence of toxic substances such as heavy metals, polycyclic aromatic hydrocarbons (PAH) or dioxins. While the origin of heavy metals is restricted to the feedstock, PAHs and dioxins can be produced during biochar production as by-products of incomplete combustion processes. Therefore, their presence needs to be measured and controlled.

Heavy metals

While oxygen, hydrogen and (to a lesser extent) carbon and nitrogen are lost during pyrolysis, other elements with a higher boiling point become concentrated in the material. This includes the so-called 'heavy metals' – cadmium (Cd), lead (Pb), chromium (Cr), arsenic (As), etc. – all of which are toxic to humans and many other life forms. There are many uncertainties around the potential mobilisation of heavy metals in biochar by plants and other life forms and the potential for subsequent entry of these into the animal or human food chain. As discussed in Chapter 10, government regulators tend to take a risk-averse approach to soil additions and amendments whereby they look at the total metals added not just that which is bioavailable to, for example, plants. The prescribed limits on heavy metal content in biochar require use of feedstocks that are low in heavy metals (Freddo and Reid 2012), although a case could be made to regulators on the legitimacy of applying a biochar containing higher amounts of heavy metals of low bioavailability. The biochar samples from virgin or clean biomass analysed in this chapter have very low metal content, while the waste material biochars analysed (produced anaerobic digestion and green waste) have low metal content (see Table 3.2). It is possible that the higher metal concentrations detected in urban green waste could be due to metal contamination in urban soils.

Polycyclic aromatic hydrocarbons

Polycyclic aromatic hydrocarbons (PAHs) are a large group of chemicals composed of two to six benzene-type rings. PAHs are commonly found in the environment but above certain concentrations they can exhibit carcinogenic, mutagenic and teratogenic properties. They are formed during the incomplete combustion, gasification or pyrolysis of organic material. They originate mostly from incomplete combustion in industry, car, truck and other vehicle engines and energy production. In nature, they are formed during forest fires and volcanic eruptions.

The US Environmental Protection Agency (USEPA) has identified 16 PAHs of greatest concern, and threshold levels for the total content of these have been provided by biochar standards (expressed in milligrams per kilogram – equivalent to parts per million) as discussed in Chapter 10. The biochar samples analysed in this chapter all have low total PAH concentrations which are in the range of PAH concentrations found in UK soils (average UK rural and urban soil values, Creaser *et al.* 2007).

Some biochars can contain higher amounts of PAHs (Freddo and Reid 2012; Fabbri *et al.* 2013; Kloss *et al.* 2012) with concentrations influenced considerably by pyrolysis conditions such as temperature and time, production process and feedstock. The highest PAH concentrations have been observed in biochar produced at lower temperatures and shorter pyrolysis times (Hale *et al.* 2012).

PAH concentrations in biochar produced from corn (maize) stover were up to four times greater than for coconut shells or sawdust (Hale *et al.* 2012) and 3–4 times greater in grass biochar than for wood (Keiluweit *et al.* 2012). Bioavailable PAH levels are not as high as the total PAH content of biochar; yet, as with metals, regulators tend to require a measurement of total content due to adoption of a precautionary approach. Furthermore, there is disagreement among experts on both the correct method for measuring PAH content in biochar and on what the appropriate 'safe' levels of PAHs in biochar should be. In order to avoid any risk, PAH concentration in biochar should always be carefully tested before application to the soil.

Dioxins, furans and polychlorinated biphenyls

Dioxins, furans and polychlorinated biphenyls (PCBs) are highly toxic. They can be produced during pyrolysis if the feedstock contains more than trace amounts of chlorine, for instance, due to the presence of salts like sodium chloride or potassium chloride. Provided chlorine levels are very low in the feedstock, production of dioxins, furans and PCBs during pyroylsis should not be an issue. Since plastics usually contain chlorine, careful screening of feedstocks used for producing biochar should be undertaken to ensure their absence.

Conclusions

Voluntary schemes, labels and associated certificates are in place to: assess biochar products; provide information on the analytical methods for measuring the properties of biochar materials and biochar products; and to provide guidelines in accordance with the latest findings and developments, ensuring verifiable monitoring and quality assurance of biochar and biochar-based products. Schemes and guidelines are available and regularly updated, however, they are only voluntary, and biochar producers do not have to comply and may wish to ignore them. Beyond the technological development, one of the key objectives of biochar projects is to provide a strong policy support in the revision of the Fertiliser Regulation (Reg. EC No. 2003/2003) and possible inclusion of biochar as organic fertiliser and soil additive (http://refertil.info/biochar-policy).

As an example, the European biochars presented in this chapter complied with regulatory requirements of the IBI Guidelines and the European Biochar Certificate – assuring their safe use. However, biochars from waste materials, such as anaerobic digestion and green waste, are classified as a lower grade (IBI grade B) with respect to stable biochar carbon content. Nevertheless, they could be regarded as soil enhancers owing to their high nutrient content and liming properties.

There is strong dependence between biochar properties, biomass feedstock and production methods, which influence the positive and negative properties of biochar. Depending on feedstock selection and production temperature, biochars exhibit large ranges in carbon stability, nutrient content, pH, CEC, porosity and surface chemistry. It should also be noticed that these properties may differ between freshly produced biochars in the laboratory and biochar in soil owing to soil weathering processes over time and associated changes in physical and chemical properties, further discussed in Chapter 4. The present chapter describes biochar properties (in isolation or in the soil), and provides evidence of the following:

1   Biochar carbon storage, stability and resistance to mineralisation in soil is linked to both aromaticity and degree of condensation. Biochar is composed predominantly of stable structures. However, there is a trade off between reactivity and stability of biochar which means the more stable a biochar is, the lower its reactivity, and thus, for instance, its nutrient holding capacity, and vice versa.

2   The nutrient value of biochar is found in the ash and its mineral composition has an effect on soil pH. Biochars of high carbon content have low ash contents and exhibit a pH value of 6.8, while biochars from waste origin have higher ash and nutrient contents and show pH values of 11 and 10. The nutrient content present in the ash, however, may not necessarily be available to plants and organisms owing to ash, oil and tars that sometimes block the pores in biochar.

3   Surface area and porosity is higher in biochars produced at high temperatures with a larger proportion of micropores , as opposed to macro and mesopores, with the latter being more accessible to plants and microbes.

4   While biochar is known to be hydrophobic, able to absorb solutions and reduce leaching of nutrients, biochar hydrophobicity decreases at higher

production temperatures owing to aliphatic functional groups volatising and being lost, providing greater water holding capacity.

5   Biochar has very large external and internal surface areas. These tend to be negatively charged and attract positively charged cations (as CEC), and stabilise functional groups when added to soil. The high sorption capacity of biochar may protect humic acids from decomposition and increase soil organic matter content.

6   The prescribed limits on heavy metal content in biochar require use of feedstocks that are low in heavy metals. Biochar produced from pyrolysis of some organic wastes generally retains heavy metals. However, it is suggested that high levels of biochar addition would be needed to potentially contaminate soil, surface water and crops.

7   PAHs can be produced during biochar production as by-products of incomplete combustion processes. The highest PAH concentrations have been observed in biochar produced at lower temperatures. There is also disagreement among experts on both the correct method for measuring PAH content in biochar and on what the appropriate 'safe' levels of PAHs in biochar should be. In order to avoid any risk, PAH concentration in biochar should always be carefully tested before application to the soil and their presence measured and controlled.

8   Provided chlorine levels are very low in the feedstock, the production of dioxins, furans and PCBs during pyrolysis should not be an issue. Nevertheless, careful screening of feedstocks used for producing biochar should be undertaken to ensure the absence of dioxins, furans and PCBs, and low metal and PAHs concentrations.

In conclusion, although there are a high number of studies highlighting the benefits of using biochar and providing characterisation methodology, the relationship between biochar properties and its potential to enhance agricultural soils is still undergoing research. Further efforts are needed to establish an appropriate formulation of desired biochar properties. Knowledge gaps still remain, such as: appropriate measurement protocols; knowledge of ecotoxicology; uncertainties around the potential mobilisation of contaminants; and total versus available concentrations of heavy metals and PAHs. There needs to be greater collaboration among researchers working on production and characterisation on one hand; and measurement of both environmental and agronomical benefits linked to the addition of biochar to soils, on the other hand.

Bibliography

Amonette, J.E. and Joseph, S. (2009). Characteristics of Biochar: Microchemical Properties. In J. Lehmann and S. Joseph (eds), *Biochar for Environmental Management*, ch. 3. London: Earthscan Publishers.

Antal, M.J. and Grønli, M. (2003). The art, science, and technology of charcoal production. *Industrial and Engineering Chemistry Research* 42(8): 1619–1640.

Ashman, M. and Puri, G. (2002) *Essential Soil Science: A Clear and Concise Introduction to Soil Science*. Chichester: Wiley-Blackwell.

Asian Development Bank (2011). *Biochar Sample Analysis: Knowledge Product*. ADB TA-7833. Manila: Asian Development Bank.

Baldock, J.A. and Smernik, R.J. (2002). Chemical composition and bioavailability of thermally altered *Pinus resinosa* (Red pine) wood. *Org Geochem* 33(9): 1093–1109.

Basso, A.A., Miguez, F.E., Laird, D.A., Horton, R. and Westgate, M. (2013). Assessing potential of biochar for increasing waterholding capacity of sandy soils. *GCB Bioenergy* 5: 132–143.

Biochar Testing Protocol (2013) Biochar: climate saving soils. Interreg IVB North Sea Region project, version 1.0, October. Available at www.biochar-interreg4b.eu (accessed 27 September 2014).

Bird, M.I., Ascough, P.L., Young, I.M., Wood, C.V. and Scott, A.C. (2008). X-ray microto-mographic imaging of charcoal. *Journal of Archaeological Science* 35: 2698–2706.

Boateng, A.A., Manuel Garcia-Perez, M., Mašek, O., Brown, R. and Bernardo del Campo, B. (2015) Biochar Production Technology. In J. Lehmann and S. Joseph (eds), *Biochar for Environmental Management*, 2nd edn, ch. 4. London: Earthscan from Routledge.

BQM (2014). *Biochar Quality Mandate*, v.1. Available at http://biocharbraf.files. wordpress.com/2013/06/bqm-v-1-0-version-for-public-consultation1.pdf (accessed 16 October 2014).

Brownsort, P. and Dickinson, D. (2012). *Mass and Energy Balances for Continuous Slow Pyrolysis of Six Feeds and Production of Biochar for Characterisation*. Biochar-interreg4b project report. Available at www.biochar-interreg4b.eu/images/file/WP32%20-%20Mass%20and%20 Energy%20Balances.pdf (accessed 20 March 2014).

Calucci, L., Rasse, D.P. and Forte, C. (2013). Solid-state nuclear magnetic resonance char-acterization of chars obtained from hydrothermal carbonization of corncob and miscanthus. *Energy Fuels* 27: 303–309.

Cayuela, M.L., Sanchez-Monedero, M.A., Roig, A., Hanley, K., Enders, A. and Lehmann, J. (2013). Biochar and denitrification in soils: when, how much and why does biochar reduce N_2O emissions? *Scientific Reports* 3: 1732 (www.nature.com/srep/2013/130425/ srep01732/pdf/srep01732.pdf, accessed 26 March 2014).

Chan, K.Y. and Xu, Z. (2009) Biochar: nutrient properties and their enhancement. In J. Lehmann and S. Joseph (eds), *Biochar for Environmental Management*, 67–84. London: Earthscan.

Chan, K.Y., Van Zwieten, L., Meszaros, I., Downie, A. and Joseph, S. (2007) Assessing the agronomic values of contrasting char materials on Australian hardsetting soil. In *Proceedings of the Conference of the International Agrichar Initiative*, 30 April–2 May 2007. Terrigal, NSW, Australia: International Agrichar Initiative. Available at www.biochar-international.org/images/IAI_2007_Conference_Booklet.pdf.

Cheng, C.H., Lehmann, J. and Thies, J.E. (2008) Stability of black carbon in soils across a climatic gradient. *Journal of Geophysical Research: Biogeosciences* 113(2): 1–10.

Creaser, C.S., Wood, M.D., Alcock, R., Copplestone, D., Crook, P.J. and Barraclough, D. (2007) *UK Soil and Herbage Pollutant Survey: Environmental Concentrations of Polycyclic Aromatic Hydrocarbons in UK Soil and Herbage*. Report no. 9. Bristol: Environment Agency.

Crombie, K., Masek, O., Cross, A. and Sohi, S. (2015). Biochar – synergies and trade-offs between soil enhancing properties and C sequestration potential. *GCB Bioenergy* 7: 1161–1175 (http://onlinelibrary.wiley.com/doi/10.1111/gcbb.12213/pdf, accessed 27 September 2014).

DeBano, L.F. (1981). *Water Repellent Soils: A State of the Art*. General technical report PSW-46. Washington, DC: USDA.

EBC (2013) *European Biochar Certificate*, version 4.6. Arbaz: Delinat Institute. Available at www.european-biochar.org/en/ct/1-Guidelines-for-the-European-Biochar-Certificate (accessed 16 October 2013).

Fabbri, D., Rombolà, A.G., Torri, C. and Spokas, K.A. (2013) Determination of polycyclic aromatic hydrocarbons in biochar and biochar amended soil. *Journal of Analytical and Applied Pyrolysis* 103: 60–67.

Freddo, A., Cai, C. and Reid, B.J. (2012) Environmental contextualisation of potential toxic elements and polycyclic aromatic hydrocarbons in biochar. *Environmental Pollution* 171: 18–24.

Glaser B., Lehmann J. and Zech W. (2002) Ameliorating physical and chemical properties of highly weathered soils in the tropics with charcoal – a review. *Biology and Fertility of Soils* 35: 2219–230.

Gray, M., Johnson, M. G. and Dragila, M.I. (2014) Water uptake in biochars: the roles of porosity and hydrophobicity. *Biomass and Bioenergy* 61: 196–205.

Hale, S.E., Lehmann, J., Rutherford, D., Zimmerman, A.R., Bachmann, R.T., Shitumbanuma, V., O'Toole, A., Sundqvist, K.L., Arp, H.P.H. and Cornelissen, G. (2012) Quantifying the total and bioavailable polycyclic aromatic hydrocarbons and dioxins in biochars. *Environmental Science and Technology* 46(5): 2830–2838.

Hammond, J., Shackley, S., Prendegast-Miller, M., Cook, J., Buckingham, S. and Pappa, V. (2013). Biochar field testing in the UK: outcomes and implications for use. *Carbon Management* 4(2): 159–170.

Hilber, I., Blum, F., Leifeld, J., Schmidt, H-P. and Bucheli, T.D. (2012) Quantitative determination of PAHs in biochar: a prerequisite to ensure its quality and safe application. *Journal of Agricultural and Food Chemistry* 60(12): 3042–3050.

IBI (2012) *International Biochar Initiative Guidelines for Biochar that is Used in Soils*, version 1.1. Published May 2012, updated 11 April 2013. Available at www.biochar-international.org/characterizationstandard (accessed 16 October 2013).

Illingworth, J., Williams, P.T. and Rand, B. (2013). Characterisation of biochar porosity from pyrolysis of biomass flax fibre. *Journal of the Energy Institute* 86(2): 3–70.

Jeffery, S., Verheijen, F.G.A., van der Velde, M., Bastos, A.C. (2011). A quantitative review of the effects of biochar application to soils on crop productivity using meta-analysis. *Agriculture, Ecosystems and Environment* 144: 175–187.

Karve, P., Shackley, S., Prabhune, R., Anderson, P., Sohi, S., Cross, A., Haszeldine, S., Haefele, S., Knowles, T., Field, J. and Tanger, P. (2011). *Biochar for Carbon Reduction, Sustainable Agriculture and Soil Management (BIOCHARM)*. Final Report for APN Project, ARCP2009-12NSY-Karve. Kobe, Japan: Asia-Pacific Network on Global Change.

Keiluweit, M., Nico, P.S., Johnson, M.G. and Kleber M. (2010). Dynamic molecular structure of plant biomass-derived black carbon (biochar). *Environmental Science and Technology* 44: 1247–1253.

Keiluweit M., Kleber M., Sparrow M.A., Simoneit B.R.T. and Prahl F.G. (2012) Solvent-extractable polycyclic aromatic hydrocarbons in biochar: influence of pyrolysis temperature and feedstock. *Environmental Science and Technology* 46(17): 9333–9341.

Kloss S., Zehetner, F., Dellantonio, D., Hamid, R., Ottner, F., Liedtke, V., Schwanninger, M., Gerzabek, M.H. and Soja G. (2012). Characterization of slow pyrolysis biochars: effects of feedstocks and pyrolysis temperature on biochar properties. *Journal of Environmental Quality* 41(4): 990–1000.

Krull, E.S., Baldock, J.A., Skjemstad J.O. and Smernik, R.J. (2009). Characteristics of biochar: organic-chemical properties. In: J. Lehmann and S. Joseph (eds), *Biochar for Environmental Management*, ch. 4. London: Earthscan.

Laird, D., Fleming, P., Davis, D.D., Horton, R., Wang, B. and Karlen, D.L. (2010) Impact of biochar amendments on the quality of a typical Midwestern agricultural soil. *Geoderma* 158: 443–449.

Lehmann, J. and Joseph, S. (eds). (2009). *Biochar for Environmental Management: Science and Technology*. London: Earthscan.

Lehmann J., Da Silva J.P., Steiner C., Nehls T., Zech W. and Glaser B. (2003) Nutrient availability and leaching in an archaeological Anthrosol and a Ferralsol of the Central Amazon basin: Fertilizer, manure and charcoal amendments. *Plant and Soil* 249: 343–357.

Liu, J., Schulz, H., Brandl, S., Miethke, H., Huwe, B. and Glaser, B. (2012): Short-term effect of biochar and compost on soil fertility and water status of a Dystric Cambisol in NE Germany under field conditions. *J. Plant Nutr. Soil Sci.* 175: 698–707.

Manning, D.A.C. and Lopez-Capel, E. (2009) Test procedures for determining the quantity of biochar within soils. In: J. Lehmann and S. Joseph (eds), *Biochar for Environmental Management*, ch. 17. London: Earthscan.

Marchetti, R., Castelli, F., Orsi, A., Sghedoni, L. and Bochicchio, D. (2012) Biochar from swine manure solids: influence on carbon sequestration and Olsen phosphorus and mineral nitrogen dynamics in soil with and without digestate incorporation. *Italian Journal of Agronomy* 7(2): 189–195.

Nguyen, B.T., Lehmann, J., Hockaday, W.C., Joseph, S. and Masiello, C.A. (2010). Temperature sensitivity of black carbon decomposition and oxidation. *Environmental Science and Technology* 44: 3324–3331.

Novak, J.M., Lima, I., Xing, B., Gaskin, J.W., Steiner, C., Kas, K.C., Ahmedna, M., Rehrah, D., Watts, D.W., Busscher, W.J. and Schomberg, H. (2009). Characterization of designer biochar produced at different temperatures and their effects on a loamy sand. *Annals of Environmental Science* 3: 195–206.

Peng, X., Ye, L.L., Wang, C.H., Zhou, H. and Sun B. (2011). Temperature- and duration-dependent rice straw-derived biochar: Characteristics and its effects on soil properties of an Ultisol in southern China. *Soil and Tillage Research* 112(2): 159–166.

Qian, K., Kumar, A., Patil, K., Bellmer, D., Wang, D., Yuan, W. and Huhnke, R. (2013) Effects of Biomass Feedstocks and Gasification Conditions on the Physiochemical Properties of Char. *Energies* 6: 3972–3986 (doi:10.3390/en6083972).

Schimmelpfennig, S. and Glaser, B. (2012). One step forward toward characterization: some important material properties to distinguish biochars. *Journal of Environmental Quality* 41(4): 1001–1013.

Schmidt, H.P. (2013). Switzerland: the first European country to officially approve biochar. *Journal for Ecology, Wine-Growing and Climate Farming* 4 May. Available at www.ithaka-journal.net/schweiz-bewilligt-pflanzenkohle-zur-bodenverbesserung?lang=en (accessed 25 October 2013).

Schmidt, M.W.I., Torn, M.S., Abiven, S., Dittmar, T., Guggenberger, G., Janssens, I.A. Kleber, M., Kogel-Knaber, I., Lehmann, J., Manning, D.A.C., Nannipieri, P., Rasse, D.P., Weinder, S. and Trumbore, S.E. (2011). Persistence of soil organic matter as an ecosystem property. *Nature* 478(7367): 49–56.

Singh, B.P., Cowie, A.L. and Smernik, R.J. (2012). Biochar carbon stability in a clayey soil as a function of feedstock and pyrolysis temperature. *Environmental Science and Technology* 46 (21): 11770–11778.

Sluijsmans, C.M.J. (1966) Effect of fertilizer on the lime requirement of the soil. *Agri Digest* 8: 10–16.

Spokas, K.A. (2010) Review of the stability of biochar in soils: predictability of O:C molar ratios. *Carbon Management* 1(2): 289–303.

Sohi, S.P., Lopez-Capel, E., Bol, R. and Krull, E. (2010). A review of biochar and its use and function in soil. *Advances in Agronomy* 105: 47–82.

Sohi, S., Lopez-Capel, E., Krull E. and Bol R. (2009) *Biochar, Climate Change and Soil: A Review to Guide Future Research*. Land and Water Science Report 05/09. Canberra: CSIRO.

Sun, H., Masiello, C.A. and Zygourakis, K. (2011). Characterizing the pore structure of biochars: A new approach based on multiscale pore structure models and reactivity measurements. AIChE Annual Meeting. Available at www.researchgate.net/publication/267313076_Characterizing_the_Pore_Structure_of_Biochars_A_New_Approach_Based_On_Multiscale_Pore_Structure_Models_and_Reactivity_Measurements.

Verheijen F., Jeffery S., Bastos A.C., Van der Velde M. and Diafas I. (2009) *Biochar Application to Soils: A Critical Scientific Review of Effects on Soil Properties, Processes and Functions*. EUR 24099 EN. Luxembourg: Office for the Official Publications of the European Communities.

Wiedner, K., Naisse, C., Rumpel, C., Pozzi, A., Wieczorek, P. and Glaser, B. (2013). Chemical modification of biomass residues during hydrothermal carbonization: what makes the difference, temperature or feedstock? *Organic Geochemistry* 54: S91–100.

Zheng W., Sharma, B.K. and Rajagopalan N. (2010). *Using Biochar as a Soil Amendment for Sustainable Agriculture*. Sustainable Agriculture Grant's research report series. Springfield, IL: Illinois Department of Agriculture. Available at www.ideals.illinois.edu/bitstream/handle/2142/25503/Using%20Biochar%20as%20a%20Soil%20Amendment%20for%20Sustainable%20Agriculture-final%20report%20from%20Zheng.pdf?sequence=2 (accessed 2 October 2014).

Zheng, H., Wang, Z., Deng, X., Herbert, S. and Xing, B. (2013). Impacts of adding biochar on nitrogen retention and bioavailability in agricultural soil. *Geoderma* 206: 32–39.

Zheng, Q.F., Wang, Y.H., Sun, Y.G., Wang, Z.M. and Zhao, J. (2014). Study on structural properties of biochar under different materials and carbonized by FTIR. *Spectroscopy and Spectral Analysis* 34(4): 962–966.

Chapter 4

The role of biochar in agricultural soils

Andrew Cross, Kor Zwart, Simon Shackley
and Greet Ruysschaert

Introduction

The primary function of agricultural soils is for food, feed, fibre and fuel production. Management measures in agricultural soils should therefore focus on improving and/or sustaining this function, in both the short and long term in a sustainable manner. Both agricultural crop production and crop quality are closely linked to soil quality (i.e. soil fertility). This is largely determined by factors including nutrient and water supply, but the soil's suitability as a growth medium for roots and as a habitat for soil organisms is also important. These factors are all dependant on a multitude of bio-physical, chemical and environmental factors. Nutrient supply is dependent on fertilisation (e.g. synthetic supplies of essential plant nutrients) and decomposition (via mineralisation by soil microbes) of soil organic matter and organic amendments (e.g. animal manure, compost and crop residues) and the buffering capacity of nutrients (Box 4.2). Water supply is determined by weather conditions, irrigation and soil physical properties including soil texture and soil structure (see Box 4.1). Soil organic matter is crucial to both nutrient and water supply in many soils, while a good soil structure can enhance both root growth and the ability of the plants to utilise available nutrients and water in the soil. Soils containing low amounts of soil organic matter require higher inputs of chemicals and irrigation, and the long-term sustainability of high-input farming is now under scrutiny. It is widely hypothesised that the high-input farming typical of much of agriculture (largely reliant on plentiful water supplies and chemical fertilisers whose manufacture uses a lot of energy) will not be sustainable in the long term and that new approaches to farming are now required.

Box 4.1 What is soil?

Soils are the skin of the earth, consisting of solids (a mixture of minerals and organic matter) with spaces (called pores) in between the solid particles. These spaces hold water and also gases. Many different chemicals (and some minerals) can be dissolved in the pore-water. Soils are biologically active, containing hundreds of thousands of types of microorganism (bacteria, viruses and fungi), as well as thousands of inverte-brates such as insects and worms. The solid particles, water, gases and microorganisms are shown schematically in Figure 4.1.

The solid mineral particles are distinguished by their particle size. Stones and gravel are anything larger than about 2mm. Sand particles are the next largest particles (1/20 to 2mm). Next are silt particles which go down to 1/500mm. The smallest particles in soil are clay. Soil scientists measure the quantities of sand, silt and clay particles and use the different proportions of each in defining different soil types in standard soil classifications (for more information on soil texture types see Chapter 5).

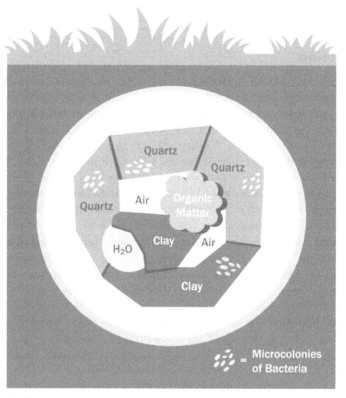

Figure 4.1 Schematic representation of soil and its components (quartz = sand particles).

Soils serve as the growing medium for much of the world's vegetation, including the majority of food crops, grasses and bushes that are grazed by animals that produce meat, dairy and other products for human use. Soils also serve as an important medium for water storage and purification. Soils interact with the atmosphere through exchange of gases, some of which are released into the atmosphere as greenhouse gases (such as methane, carbon dioxide, carbon monoxide and nitrous oxide). Soils are also a major repository of the earth's biodiversity and contain many unique communities of microbes, insects, worms and other invertebrates, as well as some mammals.

> **Box 4.2** Buffering capacity for nutrients in soil – the concept of cation and anion exchange capacities (CEC and AEC, respectively)
>
> A soil's capacity to buffer nutrients (i.e. its ability to both retain and release nutrients) is based on the property that both soil particles and nutrients have a weak electrical charge. Nutrients may have either a weak positive or weak negative charge. Examples of positively charged nutrients (cations) include ammonium, potassium and calcium, while negatively charged nutrients (anions) include nitrate and phosphate. Similar to a magnet attracting iron filings, positively and negatively charged particles attract each other, while particles with similar charges repel each other. Positively charged nutrients are weakly and reversibly bound to negatively charged soil particles (e.g. abundant in clay particles), while negatively charged nutrients are repelled from negatively charged clay particles.
>
> Sand particles are neutral and do not buffer nutrients, while soil organic matter is mostly negatively charged, although some positively charged sites may be present. The degree to which a soil can reversibly bind positively charged nutrients is called its cation exchange capacity (CEC), and the degree to which it can bind negatively charged nutrients is called its anion exchange capacity (AEC). AEC of soil organic matter is much lower than CEC, and although there are relatively few anions that are restrictive to plant growth, phosphate is one of these. Both CEC and AEC are closely linked to soil pH; CEC generally decreases when pH drops, and increases when pH rises, while the opposite is true for AEC.

In this chapter, we will discuss the potential for biochar to support crop production. Farmers would only consider using biochar if its benefits for crop production (and/or soil function leading to greater productivity) can be irrefutably demonstrated. Another potential selling point of biochar is its highly stable organic carbon content (see Chapter 3), and it could help farmers increase the organic matter content of their soils. It is therefore not surprising that many studies on biochar are related to the functions that soil organic matter deliver in soils, e.g. nutrient supply, decomposition of organic matter (mineralisation), nutrient buffering capacity (CEC and AEC; see Box 4.2), moisture management, microbial activity, soil structure and texture. In this chapter, we will discuss biochar properties (introduced in Chapter 3) and their effects on soil functions, and where appropriate compare and contrast these with the properties (and effects on soil functions) of soil organic matter. This approach highlights the following questions:

1   What is the potential role for biochar in agricultural soils, how is it comparable to soil organic matter (and how would soil organic matter benefit from biochar addition)?
2   Under what conditions (soils, crop and climate) is biochar application likely to have a positive effect?
3   Is it possible to improve the properties of biochar – biologically, chemically, physically (or in combination)?

First, we will introduce the key functions and processes of soil with respect to agricultural productivity. We will then look into these functions in more detail, and compare and contrast how these functions relate to both biochar and soil organic matter properties.

Key functions and processes in agricultural soils and the role of soil organic matter

The key functions of soil in agriculture are related to processes that create an environment that is favourable for plant growth. These are determined by soil's physical, biological and chemical characteristics, which are in turn affected by:

- the underlying rock beneath the soil, which defines many of the chemical and structural qualities of soil;
- environmental factors such as the weather and topography of the land, which determine the turnover and weathering of organic and mineral constituents, and inputs or removals via leaching;
- the amount and rate of chemical inputs, such as synthetic fertilisers, pesticides, herbicides; and
- the amount and rate of organic inputs such as crop residues and animal manures forming soil organic matter.

Soil organic matter (humus and soil organisms) plays a vitally important role in the biological, physical and chemical processes in every soil, and will be the basis for the comparison with biochar later in this chapter. Soil consists of four major components: mineral particles, air, water and organic matter. The mineral particles are partitioned into sand, silt and clay particles, based on their respective particle size. This ratio of the respective particles determines the specific soil 'texture' (see Chapter 5). Organic matter and mineral particles bind together, thereby producing soil aggregation and thus, structure. Good soil structure generally leads to a good balance between air and water in the soil matrix.

Important biological functions of soil organic matter are the supply of carbon, nutrients and energy to all soil organisms. This supply may be direct or indirect. Direct supply takes place via the degradation of soil organic matter by bacteria and fungi and other organisms like earthworms and termites, which have the capacity to degrade complex organic molecules of soil organic matter. Indirect supply takes place by microorganisms or earthworms which are consumed by organisms higher in the food web – such as mammals and invertebrates – most of which cannot directly consume soil organic matter but first need the complex molecules to be broken down into simpler ones by fungi and bacteria.

By mineralising soil organic matter, soil organisms do not only supply nutrients to other organisms but also deliver nutrients, which are readily available for plants which, like animals and insects, cannot access nutrients until organic matter is broken down by microorganisms. Soil organisms are also important for soil structure as, for example, through the action of earthworms, they can make pores and channels through which excess water can drain and roots can easily grow deeper. Moreover, worms as well as other soil organisms produce a good soil structure by in effect 'gluing' soil particles together. A diverse microbial soil life is known to make plants more resilient against soil-borne plant diseases (Jaiswal *et al.* 2014), and in most cases measures which stimulate soil organisms and their diversity are beneficial.

Soil structure and moisture retention are important physical properties which help to determine plant water supply and root growth functions, while nutrient buffering capacity (CEC; Box 4.2) and regulation of the degree of acidity or alkalinity (pH) are important chemical functions of soil organic matter in many soils. Soil organic matter helps to buffer the pH of the soil. However, other, more effective methods for regulating pH (e.g. liming) are presently widely used in farming.

Soil organic matter is also far lighter than mineral particles, and so the overall soil density (bulk density, defined as the mass of soil per volume unit) decreases with increasing contents of soil organic matter, resulting in a decrease in energy needed to plough the soil, while making it easier for roots to penetrate the soil profile and for water to move readily down the soil profile, enabling soils to drain well.

Many of these different functions have an effect on each other (Figure 4.2), for example, soil structure affects soil aeration and thereby microbial activity. A lower soil bulk density results in a higher soil pore space, and thereby a higher capacity to buffer and transport air and water. The latter in turn, affects the activity of soil microorganisms. Degradation of soil organic matter may increase owing to increased activity of microorganisms, which can then decrease soil pH, requiring more inputs of fresh organic matter to retain existing properties.

Soil functions do not always carry the same importance for every soil type. Sand particles, for instance, do not contribute to the nutrient buffering capacity of soils, since they have no electric charge (see Box 4.2). Clay particles, on the other hand, are negatively charged and therefore have a strong effect on CEC. For that reason, an increase in soil organic matter content in clay soils will only have a minor impact on the nutrient buffering capacity of the soil, whereas in sandy soils this effect will be large, though it will have other benefits such as reducing bulk density.

The regulating processes in which soil organic matter is important, and their interrelationships are summarised in Figure 4.2.

Soil organic matter, biochar and their effects on soil functions and processes relevant to crop growth

Plant nutrient delivery and availability

The availability of nutrients for plants is dependent on a number of factors, including the capacity of the soil to exchange nutrients with the soil solution, soil acidity (pH) and mineralisation (decomposition) of soil organic carbon.

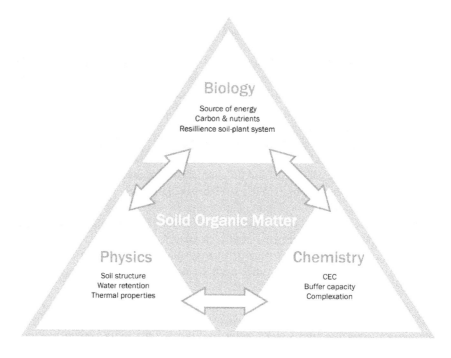

Figure 4.2 The role of soil organic matter in biological, physical and chemical functions in soils.

Cation exchange capacity

Plants can only take up nutrients as mineral salts. In solution, such salts are present in ionic form. Mineral salt ions are electrically charged, either positively or negatively. Positively charged ions include sodium, potassium, calcium, magnesium and ammonium (Na^+, K^+, Ca^{++}, Mg^{++}, NH_4^+ and all other metal ions). Negatively charged ions include chloride, sulphate, phosphate and nitrate (Cl^-, SO_4^{--}, PO and NO_3^-). Positively charged ions (cations) are attracted weakly by negatively charged groups in the soil and vice versa. A number of micronutrients (e.g. iron, zinc, copper) become more soluble – and can be taken up by plant roots more easily – when they are present as chelates, which are combinations of these metal ions and specific organic molecules. Negatively charged carboxyl and hydroxyl groups on the organic molecules play an important role in chelation.

An important soil property is the cation exchange capacity (CEC; Box 4.2). Soils with a high CEC have a large buffering capacity for positively charged nutrients (cations) as they retain cations and exchange them with the soil solution so that they become available for plants. The CEC is dependent on soil organic matter content and the clay content of the soil as they both contain negatively charged groups that attract these cations (Figure 4.3). Soil organic matter and pH

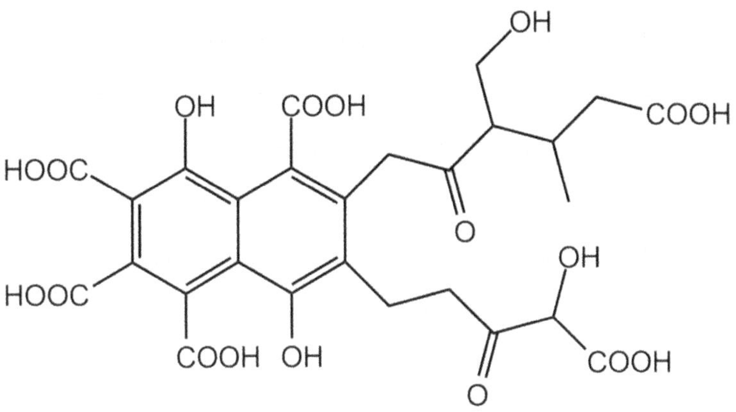

Figure 4.3 Molecular structure of fulvic acid (a component of soil organic matter) showing the high fraction of negatively charged carboxyl (COOH) and hydroxyl (OH) groups that can attract positively charged ions (cations).

Source: Michał Sobkowski, via Wikimedia Commons.

are the main determinants of CEC in sandy soils with low clay and silt contents (Chapter 5). In soils with higher clay contents, CEC is determined by negatively charged hydroxyl groups on the outside of clay particles.

The CEC per kg soil or organic matter (specific CEC) is shown in Figure 4.4. Fresh biochar generally has a low CEC – much lower than humic or fulvic substances which make up a large proportion of soil organic matter. The effect of biochar applications on soil CEC is therefore relatively small, particularly in heavy soils with a high clay but low sand content. This is not surprising as most biochar contains no or only little negatively charged (carboxyl or hydroxyl) groups. However, the CEC of biochar may increase over time following application owing to oxidation processes in the soil creating new carboxyl groups, alongside the adsorption of negatively charged organic molecules to hydrophobic biochar surfaces (Glaser *et al.* 2001). This may result in biochar applications gradually increasing CEC in soil over time. It is important to remember that although biochar applications will have a minimal effect on heavy soils with a high CEC, application of a biochar with a high CEC (e.g. 60cmol per kg) to a soil with an intrinsically low CEC (e.g. sandy soils) could have a small beneficial effect, although relatively high application rates would be needed. Increasing biochar CEC prior to application, which could potentially be reached by pre-treatment, ageing or co-composting, could make biochar somewhat more attractive in this regard. Any increases in pH associated with biochar application could also potentially increase CEC, as the number of available cations in solution decreases and negative charges on colloids increases, thereby increasing CEC.

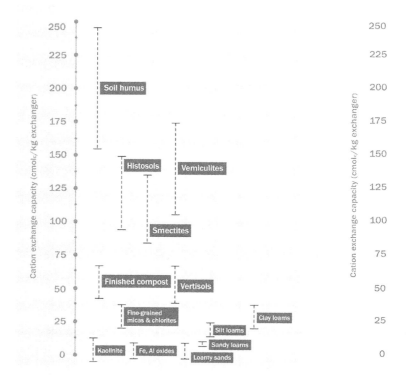

Figure 4.4 Specific cation exchange capacity (CEC) (in cmol per kg) for different textured soils and organic components, including biochar. Fulvic and humic acids are components of soil organic matter, smectite and vermiculite types of clay minerals.

Source: inspired by TerraGIS, University of New South Wales (www.terragis.bees.unsw.edu.au/terra GIS_soil/sp_cation_exchange_capaity.html).

Soil acidity

The availability of nutrients is also a function of the soil acidity or soil pH (Figure 4.5). Addition of lime to soils increases soil pH. Liming is necessary in an agricultural system as soil becomes gradually more acid through addition of mineral fertilisers. The buffering capacity of the soil is its ability to resist changes in pH. Soils with a large CEC have a large buffering capacity (i.e. ability to counteract a pH change). CEC largely determines the pH buffering capacity of soils. In order to neutralise acidity, more lime would therefore be required for highly buffered soils (e.g. silty clay loams).

One of the main differences between soil organic matter and biochar is the liming effect of biochar. This is due to the presence of mineral components in biochar ash that have a liming effect (e.g. potassium and calcium ions), and application of biochar usually results in an increase in the pH of the receiving soil. The magnitude of the increase will be determined by application rate, the ash content of the biochar (biochar with a higher ash content will generally increase pH more) and the buffering capacity of the soil. Given the positive effects of liming in acidic soils related to soil structure, biological activity and availability of plant nutrients, biochar applications would have positive effects on soil fertility when applied to a soil where low pH was limiting crop growth and development (Figure 4.5). It is worth noting that the liming effect of biochar is a short-term effect as several studies have indicated that the liming effect of biochar diminishes over time. That is similar to other liming agents as they have to be applied regularly to maintain the same pH level.

Mineralisation of organic matter or biochar

During mineralisation, which is microbial decomposition of organic compounds, carbon and nitrogen are released and used by microbes as a source of energy and nutrients, respectively. Nitrogen, in its mineral form (nitrate or ammonium) is also essential for crop growth. During mineralisation of soil organic matter, mineral nitrogen can be released, which makes nitrogen available for plants. Whether this happens or not, however, depends on the carbon-to-nitrogen ratio (Box 4.3).

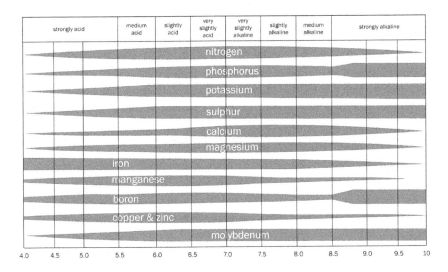

Figure 4.5 The effect of soil pH on the availability of essential plant elements. Greater nutrient availability is indicated by thickened lines, whereas narrow lines indicate a decrease in availability.

Box 4.3 The importance of carbon-to-nitrogen ratio for nutrient availability

The ratio of carbon to nitrogen in soil (C:N) varies according to soil type/region. However, it is usually in the range of 7–20. The carbon to nitrogen ratio (C:N) of soils is a key indicator of soil health and plant nutrient availability. Decomposition of soil organic matter or organic amendments with a C:N of below 20 to 24 will temporarily release mineral nitrogen (mineralisation). This nitrogen is then available for plants. If the C:N of soil organic matter or the organic amendments is above 20 to 24, microorganisms consume mineral nitrogen in the soil for their own cell production. This nitrogen is then incorporated in microbial tissue. This process is called microbial immobilisation and the result is that less mineral nitrogen is available for the plants, at least in the short-term. As a consequence, plants may suffer from nitrogen shortage and stunted crop growth. The C:N of organic amendments to soil, therefore, needs to be carefully adjusted.

Composted manure has a C:N of typically 20, sawdust has a C:N of 400, while microorganisms have a C:N between 5 and 10. Addition of sawdust could have a highly negative effect on crop yield, as the soil microorganisms that break down this organic matter would quickly run out of nitrogen and start consuming the nitrogen in the soil that is required for crop growth. Any additions of organic inputs should therefore be carefully considered with respect to the C:N of the receiving soil. Extra mineral N is sometimes added to soils receiving organic amendments with a high C:N.

The carbon-to-nitrogen ratio (C:N) of biochar will also determine if mineral nitrogen is released for plant growth or if it's immobilised during its decomposition, although the situation is more complicated. During production of biochar, the majority of feedstock nitrogen is volatilised, resulting in biochar with a low nitrogen content and consequently high C:N. The C:N of biochar can range greatly, and can be as high as 1700 in some cases (Figure 4.6). One would expect very high nitrogen immobilisation if a biochar with a C:N of 400 was added to the soil. However, given the large, highly recalcitrant C fraction, this carbon is effectively 'unavailable' as a food source for soil microbes. As microbes are not able to decompose this biochar-carbon, nitrogen immobilisation will not necessarily happen. However, there is a generally small fraction of biochar that is considered to be labile and readily available for microbial decomposition. If this labile fraction also has a high C:N, some nitrogen immobilisation will occur. The importance of this process is dependent on the magnitude of the labile carbon fraction and thus on the biochar type. Lower temperature biochars usually have a higher labile carbon fraction. So, if a low temperature biochar with a labile fraction of ~5 per cent was added to a soil, this 5 per cent would be readily available, and depending on the rate of application (and the C:N of the receiving soil), it could lead to immobilisation of mineral soil nitrogen. A more realistic and useful way of expressing the C:N of biochar is to consider the C:N of the labile fraction only. However, it is important to consider that immobilisation would only be likely under high, and often unrealistic application rates (e.g. 10–100t/ha).

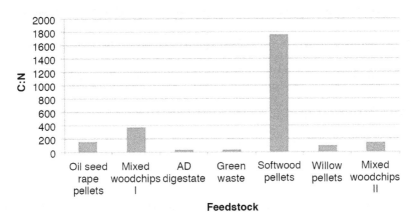

Figure 4.6 The C:N ratio of a range of biochars from the Interreg IVB North Sea Region project. All biochar was produced at 550°C using the same equipment (see Chapter 3 for more details).

Direct nutrient addition

Biochar contains an ash fraction. The magnitude and composition of this ash fraction is dependent on the feedstock and pyrolysis temperature. The ash content of biochar produced from grasses, agricultural residues like straw, husks and shells, and from animal wastes (manure) and digestates (the solid waste from animal waste decomposition in controlled conditions), is higher than in biochar produced from woody feedstocks. High temperature biochars will usually have a larger ash fraction. This ash fraction contains some nutrients that are directly available for plants. Although biochar does have some nutrient properties similar to fertiliser, it does not contain the key macro-nutrient nitrogen, and not much phosphate, making it more comparable to a soil conditioner. Still, where there are micro-nutrient deficiencies in soil not otherwise addressed, biochar addition can benefit plant development and growth. Just like other fertilisers, this effect is short term and disappears after nutrients have been used by plants or microbes, or have been lost through leaching processes. Where large quantities of biochar are applied per hectare (>10t/ha), farmers need to measure the nutrients supplied with the ash in calculating acceptable fertiliser amendments and to avoid over application.

Effects on soil structure and soil physical properties

Box 4.4 What is soil structure?

Soil structure is one of the most important soil properties relevant to plant growth, and is defined by how individual soil mineral particles and soil organic matter bind together (known as aggregation). This determines the nature and arrangement of the soil pores between them, which affects e.g. water and air flow and root penetration. Soil structure is primarily determined by the underlying soil characteristics, and then by management practices. Different agriculture practices can both destroy and enhance soil structure. Ploughing and soil tillage generally leads to a decline in soil structure through the breakdown of aggregates from mechanical mixing. Practices that increase soil organic matter content, such as incorporating crop residues, usually enhance a favourable soil structure. Soil structure is often assessed through a number of physical properties and processes such as bulk density and aggregate stability.

Bulk density

The bulk density of soil is an important agronomic property, and is expressed as the mass of oven-dry soil per unit volume of soil. Bulk density determines (among other things) the ease with which plant roots can penetrate the soil profile, and as a consequence of rooting depth, the energy needed for soil tillage (i.e. soils with a higher bulk density will require more energy for tillage) and water drainage. Bulk density is directly related to soil porosity, and it decreases with higher soil organic matter contents, as organic matter is lighter when compared to mineral soil constituents. As a result, soil porosity increases with greater soil organic matter content. Table 4.1 shows the specific gravity (how much heavier or lighter a material is than water, defined as the ratio of the mass of solid matter of a given soil sample compared to an equal volume of water) of a number of organic and other products which are used as soil amendments, including biochar. It has been shown in many soils that biochar additions have a tendency to decrease bulk density, which can largely be attributed to its low specific gravity (e.g. specific gravity for biochar is 0.8; Table 4.1) relative to the receiving soil, as the specific gravity for sandy soils is approximately 2.65, and between 2.35–2.7 for silty and clayey soils.

Table 4.1 Soil amendment products and their specific gravity

Product	Specific gravity (unitless)	Reference
Biochar	0.81	http://geoserver.ing.puc.cl/info/conferences/PanAm2011/panam2011/pdfs/GEO11Paper684.pdf
Biochar	0.08 (bagasse-char) 0.4 (hardwood char)	www.biomedsearch.com/article/Influence-biochar-use-sugarcane-growth/241179271.html
Biochar	0.42–0.64 (Gmelina arborea char)	Okoroigwe *et al.* (2012)
Bark compost	0.221	Shaaban (2012)
Peat moss	0.156	Shaaban (2012)
Wood chips	264–368	www1.agric.gov.ab.ca/%24department/deptdocs.nsf/all/agdex8875
Straw	58–357	www1.agric.gov.ab.ca/%24department/deptdocs.nsf/all/agdex8875

Aggregate formation

Soil organic matter plays an important role in the formation of stable soil aggregates. This happens through the bonding and adhesion properties of organic materials, including bacterial excretion products, root exudates, worm secretions, casts and fungal hyphae. Fungal hyphae and mucous bacterial cell walls have the capacity to effectively 'glue' mineral soil particles together, and in the process create larger particles and the so-called aggregates. These aggregates enhance soil structure and stability, enabling greater water and air provision to plants. Enhanced aggregate formation is usually linked to biodegradable soil organic matter, and stable soil organic matter has poor aggregate forming properties. This suggests at first glance that biochar is unsuitable for promoting aggregate formation; however, there is evidence to show that this is not necessarily the case (Liu *et al.* 2012), and evidence from *terra preta* soils indicates indirect effects on aggregation, given the high levels of additional soil organic matter associated with these sites.

Although the mechanisms regarding biochar effects on soil aggregation are poorly understood and not well demonstrated, there is the potential for biochar application to improve soil aggregation and stability, which might additionally protect soils from wind and water erosion. This could be the case if biochars have a high proportion of biologically available labile carbon. Another mechanism suggested is that fungal hyphae are able to grow into biochar pores, which might lead to enhanced stable aggregate formation.

Effects on hydraulic properties

Box 4.5 Definitions

Water holding capacity:	the amount of water that a particular soil can hold. That is the amount of water in the soil after two days of free drainage of a saturated soil. This is also called field capacity.
Permanent wilting point:	the moisture content of the soil below which the roots cannot extract water from the soil anymore.
Plant available water:	the amount of water available for plants (i.e. the difference in water content between field capacity and permanent wilting point).
Hydraulic conductivity:	the ease with which water can move through soil pore spaces, determined by the permeability of the soil matrix and degree of saturation.

Water is essential for plant growth, and alongside soil nutrients, the largest limiting factor to crop growth. The amount of water that can be extracted from soils by plants is one of the most important soil properties. Soils play a key role in both regulating and supplying water to plants for growth. Soil pores (the spaces between soil particles) provide the pathway for water to travel in the soil profile, and a soil's ability to retain water is closely linked to its soil texture (Figure 4.1), with water molecules binding more tightly to fine clay particles and less so to course sandy particles. Addition of any organic amendment, which is likely to change the physical structure of a soil, will have knock-on effects on the hydraulic properties of the soil.

Effects on water holding capacity and plant available water

Soil organic matter is well known to have a positive effect on the hydraulic properties of soils. It improves the water holding capacity of soil and ultimately the amount of water available to plants in almost all soils, although this effect is larger in sandy soils than in soils with a high loam or clay content.

Box 4.6 Plant available water and water retention curves

The amount of water that can be extracted is not the same as the total amount of water that a soil can hold (total water holding capacity). The relevance of this becomes clearer when considering the concept of the water retention curve (Figure 4.7), and how water behaves in soil. The pF is a measure that indicates how much force has to be exerted on soil water to remove or extract it from the soil, e.g. by plant roots. When the soil water content is very low (e.g. towards the left of the *x*-axis in Figure 4.7), the pF required to extract water is very large (upper end of the *y*-axis), and as a result, plants are not able to extract this water. When the soil water content increases (moving from left to right on the *x*-axis) and passes wilting point (i.e. the minimal point of moisture required by plants not to wilt) the force required to extract the water decreases (moving down the *y*-axis) to the point where plant roots are able to start extracting water from the soil. The more the soil water content increases, the less force is required to extract this water from the soil. Field capacity is then defined as the moment when the soil is effectively saturated, and any excess water drains freely (if it can) from the soil profile by gravity (and is therefore also not available for plant roots). The water in a soil that is available to plants is therefore the amount held between wilting point (pF 4.2) and field capacity (pF 1.8).

The positive effect of soil organic matter is largely explained by its swelling and shrinking properties, and positive influence on soil porosity. Soil organic matter and many other organic soil amendments can take up large volumes of water, in some cases up to 20 times their own mass. The effects of soil organic matter content on plant available water (in sandy soils) is shown in Figure 4.8.

It has been shown that biochar can take up to 11 times its own mass of water (Kinney *et al.* 2012), which is lower than the amounts reported for soil organic matter. This is primarily due to the rigid structure of biochar, meaning that its swelling capacities are far less than those of soil organic matter. It is important to remember though that these measurements are also not necessarily representative of field conditions (but can give helpful insights when screening a range of

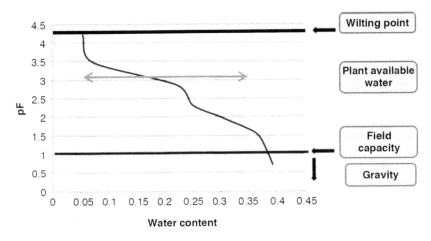

Figure 4.7 Water holding capacity (WHC) and plant available water (PAW).
Note: pF = force needed to extract water.

samples), and pF curves give a better indication of in-situ processes. However, application of biochar has been shown to increase the water holding capacity of soils, although not in every soil type, and the response is not always proportional to the amount of biochar added. The effect of biochar application on soil water holding capacity is shown for a range of different biochar and soil types in Figure 4.9. It demonstrates a positive effect of biochar addition on water holding capacity,

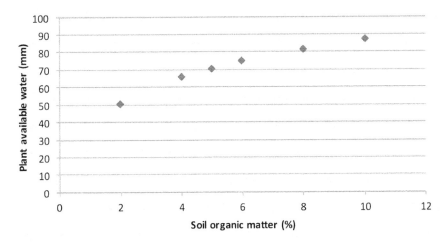

Figure 4.8 Relation between soil organic matter content and plant available water in sandy soils.
Source: after Boekel (1962).

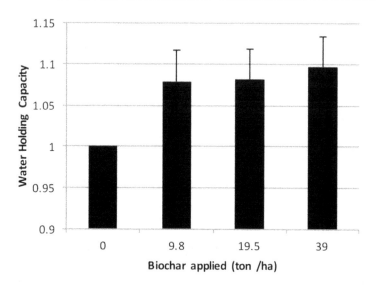

Figure 4.9 Effect of biochar application on soil water holding capacity relative to unamended soil (control = 100 per cent).

Source: after Streubel *et al.* (2011).

with increases in the region of 10 per cent. The effects are not proportional to the amount of biochar applied, which implies that significant improvements to water holding capacity with relatively low biochar application are possible.

An improvement in soil moisture availability following biochar application has been shown in several studies. These effects are partly indirect (e.g. due to improvements in soil structure), but also due to direct effects such as the capacity of biochar to take up water and release it in drier conditions. The effects of biochar on water retention are likely to be most prominent in the low pF ranges (and therefore higher water contents – see Box 4.6). This can be explained mainly by the extent of large pores present in biochar. Biochar with a large proportion of micro-pores (smaller than 50 nm) is likely to have no effect on plant water availability, as plants cannot overcome the high capillary forces needed to withdraw water from such pores. Although most results from field trials have indicated relatively small effects of biochar application on water retention, an experiment investigating the effect of biochar mixed with pure sand showed a strong increase in plant available water amounts above application rates of 10 per cent (Figure 4.10). It should be noted however that these application rates are impractical under field conditions (being too high) but might be more realistic for small scale cultivation of high value crops.

While the effects of biochar on soil are qualitatively similar to those of soil organic matter, there is much uncertainty regarding the quantitative differences between the two. Soil organic matter and other organic matter additions might

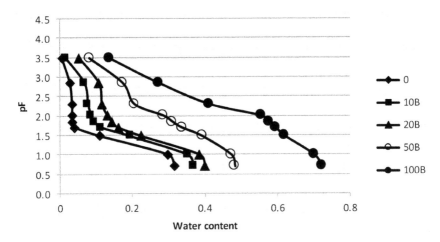

Figure 4.10 The effect on water retention (pF) of biochar mixtures with pure sand (0, 10B, 20B, 50B and 100B = 0, 10, 10, 50 and 100 per cent biochar by mass, respectively).

Source: K. Zwart, unpublished results.

have much larger effects (Mukherjee and Lal 2013); however, this could be a function of much higher application rates when compared to biochar. At this moment in time, there is insufficient evidence to draw conclusions as to the merits of organic matter additions compared to biochar, and experiments comparing the direct effects of both are needed.

Effects on hydraulic conductivity

As a result of its highly porous structure and effects on soil porosity, biochar would be expected to have positive effects on water infiltration in soils. The effect of biochar on the hydraulic conductivity (the capacity for soil to transport water) was examined in northern European countries. Results from field experiments on sandy soils in the Netherlands (which already had a high soil organic carbon of 7 per cent) showed no effect, similar to results from a Belgian field experiment on sandy loam with a low carbon content (0.7 per cent) (Nelissen *et al.* 2015). Several other publications describe a positive effect of biochar on soil hydraulic properties; however, when compared to the effects of soil organic matter additions, these positive effects are relatively small, and should be considered alongside other potential benefits of biochar addition to soils.

Thermal properties

Soil organic matter has a large capacity to absorb heat (owing to its dark brown or sometimes even black colour), and it is widely assumed that soils with a high soil organic matter content will warm up sooner in spring, compared with soils that have lower amounts of soil organic matter. Since biochar is also very dark and has a high thermal conductivity, it is expected that it will have a similar effect on soil temperature. For this reason, biochar has historically been used in Japan to accelerate melting of snow and prolong the crop growing season. Verheijen *et al.* (2013) showed an increased reflection of light and therefore a greater heat absorption with increasing application rates of biochar to soils in laboratory experiments, particularly after surface application without mixing into the soil profile. Similar results were observed in Italian field experiments (Genesio *et al.* 2012). Field experiments have also shown increases in soil surface and sub-surface temperatures following biochar application (Ventura *et al.* 2013; Bruun *et al.* 2011), although this is not always observed.

Effects on soil fauna

The diversity of soil organisms plays a large role in many important soil functions – particularly the breakdown and recycling of soil organic matter. This has further effects on important functions mentioned earlier in the chapter (e.g. soil aggregation). The amount and quality of soil organic matter has an effect on the abundance (although not necessarily on the diversity) of soil bacterial communities (Hirsch *et al.* 2009). Biochar has also been reported to have both direct and indirect effects on soil microorganisms (e.g. Steinbeiss *et al.* 2009). The direct effects are related to provision of an energy source (and nutrients) and refuge for soil organisms, while the indirect effects are related to changes in effects on hydraulic properties (e.g. availability of water) following biochar addition. Some of these effects (e.g. provision of energy and nutrients) are only short term and transient, while others (e.g. effects related to physical changes in soil following biochar addition) could be more long term.

Energy source for microorganisms/ soil respiration

Biochar is a concentrated, energy rich compound. Combustion of 1kg biochar would produce approximately 30 to 35MJ of energy (Ryu *et al.* 2007), which is higher than that produced during combustion of the same mass of dry biomass (14-21 MJ/kg; Demirbas 2004). In theory, this implies that biochar can provide soil organisms with an energy source; however, this is entirely dependent on what proportion of the biochar carbon is in a form that is available to soil organisms (i.e. labile), and whether or not these organisms have the proper enzymes that can degrade the carbon. The proportion of labile and recalcitrant biochar carbon is closely related to both production conditions and feedstock (discussed in more detail in Chapters 2 and 7). Although biochar may indeed provide a source of labile carbon for microbial decomposition, the majority will remain in the soil over the long term (hundreds of years), potentially acting as a 'slow-release' energy source over time.

When and if biochar is consumed by soil microorganisms, the subsequent decomposition will result in a release of CO_2 (commonly known as soil respiration). There is evidence that CO_2 emissions (i.e. respiration) increase following biochar application, although this increase is usually very small in relative terms, with Zimmerman et al. (2011) finding that the largest proportion of mineralised biochar carbon was always less than 4 per cent over one year. This increase in CO_2 emissions is usually associated with the degradation of the labile carbon fraction of biochar, and emissions are generally higher from low temperature biochars (with a higher content of volatile organic matter). The increase in CO_2 emissions following biochar application could also be associated with a 'priming' effect of native soil organic matter (see Chapter 7 for more detail).

Addition of biochar is also reported to lead to an increase in microbial biomass (e.g. Zhang et al. 2014), which suggests that soil microorganisms are able to utilise a fraction of biochar carbon for growth. This is supported by incubation studies in sterile sand, which have shown the degradation of biochar carbon, in the absence of any other source of carbon for mineralisation (e.g. Cross and Sohi 2011; Nelissen et al. 2013), although studies have shown that this fraction can also be degraded abiotically (Zimmerman 2010). Decreases in microbial biomass have also been observed following biochar application, and it has been suggested that microbial community structure, rather than microbial biomass, change following application of biochar (Dempster et al. 2012). Evidence from studies on soil organic matter also suggests that microbial communities are very diverse and adaptable (Hirsch et al. 2009), and could therefore be expected to respond quickly to changes in both substrate and environmental conditions. These contrasting results suggest that alternate mechanisms could be an explanation, with changes in microbial biomass increasing the presence of organisms involved in lignocellulosic degradation (e.g. fungi), or alternatively microbial biomass might be suppressed by the presence of large amounts of volatile organic substances (more common in low temperature biochar). The latter is highlighted by the increasing amount of evidence suggesting that a range of chemicals present in biochar are responsible for disease suppression in plants (Jaiswal et al. 2014).

Increases in CO_2 mineralisation and microbial biomass following biochar addition to soils strongly suggest that microbial degradation and utilisation of biochar occurs, which raises the question of whether or not these effects are direct or indirect, and due to changes in factors including soil moisture, soil temperature and aeration, and whether or not this is comparable to other organic amendments such as compost, particularly regarding support of biological functions.

Changes to soil pH, effects on toxins and the soil and protection against desiccation are among the properties which may influence soil organisms (Lehmann et al. 2011), and the comparison of biochar with compost has been discussed in detail by Fischer and Glaser (2012). Their general conclusion was that compost is quite different from biochar in relation to its bio-decomposition and capacity to provide nutrients for crops. In particular, compost is less stable and provides more nutrients. As compost is essentially produced through microbial degradation of organic materials, it is clear that composts support microbial growth. Fischer and Glaser (2012) ascribed the increase in microbial biomass following biochar application as a result of supporting microbial growth due to the following factors:

- effect of water and nutrient retention;
- formation of 'active' surfaces providing an optimal habitat for microorganisms;
- weak alkalinity; and
- partial inhibition of 'destructive' and simultaneous support of 'beneficial' microorganisms (e.g. Jaiswal *et al.* 2014).

All of these factors tend to support the theory that modification of environmental conditions, rather than direct consumption of biochar, is the most important factor controlling soil microbial growth following biochar application. It is important to consider that this might not necessarily be true for every combination of soil/biochar type, particularly when working with biochars at the more extreme range of labile carbon/volatile organic carbon content.

Refuge for microorganisms

Biochar's ability to act as a refuge for soil microorganisms is widely purported to be one of the potential positive effects of biochar. This has been supported by electron microscopic images which show bacteria or fungi on the surface or inside biochar pores (Verheijen *et al.* 2010). The explanation for this is that microorganisms can enter pores and use them as a refuge, where they are protected from predation. An indirect effect of this is that fungal hyphae could potentially form conduits to move solution and solutes to other parts of the soil and out of the biochar.

The larger biochar pores (most likely the structural remnants of wood xylem, phloem vessels and other larger features) have diameters greater than 10μm. Organisms that consume soil bacteria, including protozoa and nematodes, have diameters less than 10 μm, making access to these larger pores relatively easy. Biochar micro and nanopores are far too small for bacteria to enter, and the presence of microorganisms on the surface and inside biochar pores is insufficient evidence to indicate protection against predation. Another possible mechanism through which soil microorganisms are protected could be hydrophobic adsorption. Biochar is generally (though not in all cases) hydrophobic by nature. This is largely determined by the presence of functional groups, and this hydrophobicity can decrease over time. Some microorganisms can be strongly attached to hydrophobic surfaces, which creates biofilms several bacterial layers thick (Mussati *et al.* 2005), which would provide for some degree of protection from predation.

Other factors affecting soil organisms

The effect of biochar on soil pH is primarily dependent on biochar properties, especially ash content. Soil microorganisms generally respond to changes in soil pH, and so it is not surprising that addition of biochar to soils can affect the microbial population. Fungi mostly favour a low soil pH, while most bacteria prefer higher soil pH values. In the North Sea ring trial field trials (see Chapter 5), some changes in the bacterial community structure were found; however, no effects on fungi were observed. These observations are not direct proof of an increase in pH and may be transient, and could be affected by a range of factors including leaching of biochar ash (which would be the main determinant of pH change) from the soil, and the fact that fungi might not respond quickly to changes in soil pH.

The effects of biochar on the sorption and desorption of pesticides and other organic compounds has been reviewed in detail by Kookana (2010). The sorption capacity of biochar is related to the aromatic molecular structure of biochar. Activated carbon is well known for its adsorption capacity of organic molecules, mainly owing to its hydrophobic nature combined with an extremely high relative surface area. Biochar with a high hydrophobicity and large relative surface areas (properties largely determined by pyrolysis conditions or in post-pyrolysis mixing or co-composting) would therefore also be expected to adsorb organic molecules, including pesticides (see Chapter 11). This might have both positive and negative effects on microbial activity in the soil (negative if biochar becomes effectively saturated by organic molecules (e.g. pesticides) and can no longer absorb them, or positive if the pesticides are inactivated through adsorption).

Conclusions

Soil organic matter plays a crucial role in agricultural soils. Biochar can also have similar beneficial effects on soil functions, depending on the type and amount of biochar applied. The properties of both biochar and soil organic matter have *some* comparable effects on several soil functions, although the magnitude and dynamic of these effects is often different:

1   *Provision of nutrients:* Soil organic matter performs an important function regarding provision of nutrients (especially nitrogen and phosphorus) to plants, via direct nutrient supply, pH buffering capacity and specific CEC. Although biochar can also perform an important nutrient delivery function, this is related more to nutrient delivery via mineral ash and liming effects on soil. The nitrogen delivery function, however, is small, owing to low nitrogen concentrations in most types of biochar and high C:Ns. Effects of biochar on CEC will probably become apparent in the longer term (in specific soils), but will remain small at practical biochar application rates.

2   *Provision of water:* Soil organic matter has an important effect on soil structure, which in turn affects how much water a soil can hold, how readily available it is, and the ease with which it can travel through the soil (which has an important effect on the delivery of nutrients). Biochar can potentially

improve provision of water by improving soil structure; however, these effects will only be minor (unless applied in large amounts), and highly dependent on the receiving soil (e.g. increasing water holding capacity on sandy soils, and hydraulic conductivity on heavy soils).

3   *Habitat for microorganisms:* The amount and quality of soil organic matter will have a large effect on the abundance and diversity of microbial communities, through delivery of nutrients and provision of energy. Biochar could have similar effects by providing a short-term energy source for microbes (and even potentially physical protection within biochar pores). These effects will again be highly dependent on the type and quantity of biochar applied.

The potentially large difference between the properties of biochar and soil organic matter implies that in some situations the effects of biochar application will not be the same as that of compost, animal manure or crop residues. Given these differences, the potential benefits of biochar applications need to be carefully communicated to farmers, particularly when considering comparisons with soil organic matter. Application of animal manure, compost and crop residues are the classical methods for increasing soil organic matter, and although biochar addition will also increase soil organic matter, the residual effects may be very different to the effects of the more traditional methods.

The large variations in biochar properties, and their effects on crop productivity (see Chapter 5), make it difficult (at this moment in time), to give advice on the effects of biochar with anywhere near the same level of accuracy associated with, for example, fertiliser applications. Apart from the fact that biochar is inherently stable and will increase the carbon content of soils, there are very few general effects that will be the same under all circumstances (although the evidence from *terra preta* soils suggests that biochar together with nutrients and microorganisms produce synergistic benefits which will have a potentially marked effect on improving soil fertility in the long term).

Box 4.7 What are the implications for farming practice?

The information presented in this chapter is still rather technical, and therefore we will try to answer some very practical questions relevant for farming practice. Will biochar application result in:

I Higher soil organic matter content?

Yes, application of 20 tonnes of biochar per ha will increase the soil organic carbon content by approximately 0.5 per cent, depending on the biochar quality and the depth of the soil layer through which the biochar is mixed. Biochar with a low ash content (e.g. produced from wood) will have a better effect than biochar with a high ash content (e.g. produced from manure or straw). Biochar in general is very resistant to biological or chemical decomposition. For that reason, biochar organic carbon will remain for much longer in soils than, for instance, compost carbon (although its effects will not necessarily be the same as soil organic matter). However, the composition of biochar is very different from the composition of 'traditional' soil organic matter.

2 Lower fertiliser application?

In most instances no, as biochar in general has a low nutrient content, although lower fertiliser application may be realised through less fertiliser loss. However, an exception will be biochars rich in ash (e.g. produced from manure or rice husk). Nutrients present in biochar ash may also be easily available to crops, although this effect will only be short term (as these nutrients are either taken up by plants or leached). The large proportion of nitrogen in feedstock is usually volatilised during conversion to biochar, meaning that it will have little value to farmers as an additional source of nitrogen. In some cases, more fertilisation could be needed along with biochar additions, owing to temporary nitrogen immobilisation processes.

3 Less need for liming?

Yes, most biochar has a liming effect by itself, which may replace liming with other compounds. This effect is also only short term and liming will again be needed after a few years, which is similar to the use of other liming agents. The extent of any liming effect will be determined primarily by biochar ash content.

4 Lower fuel use to manage (plough) the soil

Potentially yes. Biochar is a light material, and so, application through soil will lead to a lower density of the soil (mass of soil per unit of volume). However, quite large amounts (> 20 tonnes per ha) are probably needed in order to obtain a measurable effect.

5 Better water management?

This is highly dependent on the receiving soil. There can be a positive effect on coarse sandy soils, and biochar could also help retain water during periods of drought, and conversely by helping drain heavy soils, and thus preventing water-logging. This will to a large extent be determined by application rate.

6 Improved soil structure?

Yes, owing to its low specific gravity, application of biochar may result in a lower soil bulk density. This may be beneficial in providing water and nutrients to the crop. Improved aggregation can also help water and gas move to and from the plant roots, and improve soil structure in the longer term.

7 A better soil life?

This is again dependent on the type of biochar and receiving soil. There is potentially a short-term benefit via the labile carbon fraction, providing a source of energy and nutrients to soil organisms. Longer term benefits could be due to indirect effects related to water transport and soil structure, while benefits associated with changes in pH might be both short to longer term.

8 A higher crop yield

In a European (North–West) context, most likely not (for more details see Chapter 5); however, in extremely poor, nutrient and water limited soils (predominantly in the tropics and sub-tropics) there is potential, and evidence for, improved crop yields.

Bibliography

Boekel, P. (1962). Betekenis van organische stof voor de vocht-en luchthuishouding van zandgronden. *Landbouwkundig Tijdschrift* 74: 128–9135 (in Dutch).

Bruun, E.W., Haugaard-Nielsen, H., Ibrahim, N., Egsgaard, H., Ambus, P., Jensen, P.A. and Dam-Johansen, K. (2011). Influence of fast pyrolysis temperature on biochar labile fraction and short-term carbon loss in a loamy soil. *Biomass and Bioenergy* 25: 1182–1189.

Cross, A. and Sohi, S. (2011). The priming potential of biochar products in relation to labile carbon contents and soil organic matter status. *Soil Biology and Biochemistry* 43: 2127–2134.

Demirbas, A. (2004). Combustion characteristics of different biomass fuels. *Progress in Energy and Combustion Science* 30: 219–230.

Dempster, D.N., Gleeson, D.B., Solaiman, Z.M., Jones, D.L. and Murphy, D.V. (2012). Decreased soil microbial biomass and nitrogen mineralisation with Eucalyptus biochar addition to a course textured soil. *Plant Soil* 354: 311–324.

Fischer, D. and Glaser, B. (2012). Synergisms between compost and biochar for sustainable soil amelioration. In S. Kumar (ed.), *Management of Organic Waste*, 167–198. Rijeka: InTech. Available at www.intechopen.com/books/management-of-organic-waste/synergism-between-biochar-and-compost-for-sustainable-soil-amelioration.

Genesio, L., Miglietta, F., Lugato, E., Baronti, S., Pieri, M. and Vaccari, F.P. (2012). Surface albedo following biochar application in durum wheat. *Environmental Research Letters* 7: article 014025 (http://iopscience.iop.org/1748-9326/7/1/014025).

Glaser, B., Haumaier, L., Guggenberger, G. and Zech, W. (2001). The 'Terra Preta' phenomenon: a model for sustainable agriculture in the humid tropics. *Naturwissenschaften* 88: 37–41.

Hirsch, P.R., Gilliam. L.M., Sohi, S.P., Williams, J.K., Clark, I.M. and Murray, P.J. (2009). Starving the soil of plant inputs for 50 years reduces abundance but not diversity of soil bacterial communities. *Soil Biology and Biochemistry* 41: 2021–2024.

Jaiswal, A.K., Elad, Y., Graber, E.R. and Frenkel, I. (2014) *Rhizoctonia solani* suppression and plant growth promotion in cucumber as affected by biochar pyrolysis temperature, feedstock and concentration. *Soil Biology and Biochemistry* 69: 110–118.

Kinney, T.J., Masiello, C.A., Dugan, B., Hockaday, W.C., Dean, M.R., Zygourakis, K. and Barnes, R.T. (2012). Hydrologic properties of biochars produced at different temperatures. *Biomass and Bioenergy* 41: 34–43.

Kookana, R.S. (2010). The role of biochar in modifying the environmental fate, bioavailability, and efficacy of pesticides in soils: a review. *Australian Journal of Soil Research* 48: 627–637.

Lehmann, J. and Joseph, S. (2009). Biochar for environmental management: an introduction. In J. Lehmann and S. Joseph (eds), *Biochar for Environmental Management: Science and Technology*, 1–12. London: Earthscan.

Lehmann, J., Rillig, M.C., Thies, J., Masiello, C.A., Hockaday, W.C. and Crowley, D. (2011). Biochar effects on soil biota – a review. *Soil Biology and Biochemistry* 43: 1812–1836.

Liu X., Han, F. and Zhang, C. (2012). Effect of biochar on soil aggregates in the Loess plateau: results from incubation experiments. *International Journal of Agriculture and Biology* 14: 975–979.

Mukherjee, A. and Lal, R. (2013). Biochar impacts on soil physical properties and greenhouse gas emissions. *Agronomy* 3, 313-339.

Mussati, M.C., Fuentes, M., Aguirre, P. and Scenna, N. (2005). A steady-state module for modeling anaerobic biofilm reactors. *Latin American Applied Research* 35: 255–263.

Nelissen, V., Rutting, T., Huygens, D., Staelens, J., Ruysschaert, G. and Boeckx, P. (2013). Maize biochars accelerate short-term soil nitrogen dynamics in a loamy sand soil. *Soil Biology and Biochemistry* 55: 20–27.

Nelissen, V., Ruysschaert, G., Manka'Abusi, D., D'Hose, T., De Beuf, K., Al-Barri, B., Cornelis, W. and Boeckx, P. (2015). Impact of a woody biochar on properties of a sandy loam soil and spring barley during a two-year field experiment. *European Journal of Agronomy* 62: 65–78.

Okoroigwe, E., Li, Z., Stuecken, T., Saffron, C. and Onyegegbu, S. (2012). Pyrolysis of *Gmelina arborea* wood for bio-oil/bio-char production: physical and chemical characterisation of products. *Journal of Applied Sciences* 12: 369–374.

Ryu, C., Sharifi, V.N. and Swithenbank, J. (2007). Waste pyrolysis and generation of storable char. *International Journal of Energy Research* 31: 177–191.

Shaaban, S.M. (2012). Improvement of peat hydro physical properties by bark and filter mud additions for seedlings production of fodder beet. *Journal of Applied Sciences Research* 8: 4434–4439.

Steinbeiss, S., Gleixner, G. and Antonietti, M. (2009). Effect of biochar amendment on soil carbon balance and soil microbial activity. *Soil Biology and Biochemistry* 41: 1301–1310.

Streubel, J.D., Collins, H.P., Garcia-Perez, M., Tarara, J., Granatstein, D. and Kruger, C.E. (2011). Influence of biochar on soil pH, water holding capacity, nitrogen and carbon dynamics. *Soil Science Society of America Journal* 75: 1402–1413.

Ventura, M., Sorrenti, G., Panzacchi, P., George, E. and Tonon, G. (2013). Biochar reduces short-term nitrate leaching from a horizon in an apple orchard. *Journal of Environmental Quality* 42: 76–82.

Verheijen, F.G.A., Jeffery, S., Bastos, A.C., van der Velde, M. and Diafas, I. (2010). *Biochar Application to Soils: A Critical Scientific Review of Effects on Soil Properties, Processes and Functions*. EUR 24099 EN. Luxembourg: Office for the Official Publications of the European Communities.

Verheijen, F.G.A., Jeffery, S., van der Velde, M., Penížek, V., Beland, M., Bastos, A.C. and Keizer, J.J. (2013). Reductions in soil surface albedo as a function of biochar application rate: implications for global radiative forcing. *Environmental Research Letters* 8: 044008.

Zhang, Q., Dijkstra, F. A., Liu, X., Wang, Y., Huang, J. and Lu, N. (2014). Effects of biochar on soil microbial biomass after four years of consecutive application in the North China Plain. *PLoS ONE* 9(7): e102062.

Zimmerman, A.R. (2010). Abiotic and microbial oxidation of laboratory-produced black carbon (biochar). *Environmental Science and Technology* 44: 1295–1301.

Zimmerman, A.R., Gao, B. and Ahn, M. (2011). Positive and negative carbon mineralization priming effects among a variety of biochar-amended soils. *Soil Biology and Biochemistry* 43: 1169–1179.

Chapter 5

Field applications of pure biochar in the North Sea region and across Europe

Greet Ruysschaert, Victoria Nelissen, Romke Postma, Esben Bruun,
Adam O'Toole, Jim Hammond, Jan-Markus Rödger, Lars Hylander, Tor
Kihlberg, Kor Zwart, Henrik Hauggaard-Nielsen and
Simon Shackley

Introduction

Biochar application can increase carbon storage in soils and is a promising strategy to mitigate climate change (Chapter 7). However, in the absence of substantial subsidies or carbon credits (Chapter 9), larger-scale biochar application on agricultural fields is only to be expected if it leads to improved soil functioning (Chapter 4) and, as a result, improved gross margins. The gross margins can be accomplished if the yield or market value of the harvested product increases and/or if the variable costs of production (including inputs such as fertilisers, machinery and labour) decreases.

Chapter 4 outlines how biochar can influence soil functioning and as such affect crop yield. Depending on the type of biochar and soil, biochar could potentially affect nutrient delivery and water and air provision, for example. However, other effects can also play a role, such as the fact that some biochars can be phytotoxic or can adsorb organic compounds such as pesticides. Owing to its black colour, soils with biochar might heat more rapidly, which can be an advantage for crop emergence in cold springs.

Scientific reports have demonstrated that, on average, the overall short-term biochar effect (less than three years) on above ground biomass and crop yield can be positive (Jeffery et al. 2011; Biederman and Harpole 2013). An average effect across a wide range of climate, soils, crops, crop management practices, use of chemical and organic inputs, etc., should not be over-interpreted, however, especially given the relatively small number of cases. Moreover, most (Jeffery et al. 2011) to many (Biederman and Harpole 2013) of the studies are conducted in tropical or subtropical regions and include both pot and field experiments.

Climate conditions, soil types and initial soil fertility are expected to have a large impact on biochar effects. Jeffery et al. (2011) demonstrated that there is a positive trend between pH increase in the soil after biochar application and crop productivity. The reason is most probably that weathered (sub)tropical soils are often quite acidic and increasing soil pH is beneficial for the availability of nutrients (such as phosphorus) and the reduction of aluminium toxicity. Moreover, the same authors highlight that increases in crop productivity were significant in medium and coarse textured soils but not in fine-textured soils. Coarse and medium textured soils have a lower capacity to retain water compared with fine textured soils, which is a disadvantage during dry periods. Biochar soil application is expected to increase the water holding capacity of soil, especially of lighter-textured soils, although it remains unclear if, in general, the amount of water available for plants increases after biochar is added (Chapter 4).

Given differences in soil and climate conditions and how agriculture is managed, results from (sub)tropical field trials can only provide an idea of potential effects of biochar applied to European soils, which can only be proven under European field conditions. The first European biochar field trials were established in 2007 and 2008 (Baronti *et al.* 2010; Bell and Worrall 2011; Vaccari *et al.* 2011; Borchard *et al.* 2014). Since then, the number of European biochar field trials has increased rapidly. In this chapter, we included 32 trials with pure biochar additions. However, results from many other trials are expected soon, as it usually takes a few years between field trial establishment and publication of the first results.

The central objectives of this chapter are to assess firstly, whether pure biochar addition to agricultural fields can be beneficial for soil functioning and crop productivity under European agricultural management conditions and secondly, to what extent factors such as soil texture, biochar dose and time since biochar addition influence the biochar effect.

In order to answer these questions, firstly initial results of a biochar ring trial across the North Sea region are presented. This ring trial was established since autumn 2011 in seven countries of the North Sea region within the framework of the 'Biochar: Climate saving soils' Interreg IVB North Sea Region Project (Figure 5.1, Table 5.1). In these field trials, we tested the same biochar type generally following the same experimental protocols. Secondly, recommendations for conducting biochar field trials will be given based upon practical experiences. Thirdly, we highlight the results of other European biochar field trials for comparison. This chapter focuses on pure biochar additions, while Chapter 6 deals with biochar additions in combination with other organic amendments. Furthermore, the focus is on crop productivity and less on soil functioning as this is dealt with in Chapter 4.

Biochar ring trial in the North Sea region

Biochar characteristics

The biochar used in the North Sea ring trials, is called RomChar and is produced by the company Carbon Terra using a mix of woody feedstocks (Norwegian spruce, silver fir, Scots pine, beech and oak) and a pyrolysis temperature of 450–480°C. The biochar was produced by a slow pyrolysis technology which, over a few days, leads the char to a small fire-front at the end of the system that helps to destruct potential harmful substances such as polycyclic aromatic hydrocarbons (PAHs) and dioxins (For further details about pyrolysis, see Chapter 2). Characteristics of the RomChar are described in Chapter 3. The RomChar contained 68 per cent carbon, 0.4 per cent nitrogen (Nelissen *et al.* 2014c), 0.0034 per cent phosphorus and 0.31 per cent potassium. Carbon-to-nitrogen (Box 5.1), hydrogen-to-carbon and oxygen-to-carbon (Chapter 3) ratios were 169 (mass ratio), 0.02 (atomic ratio) and 0.13 (atomic ratio), respectively. Volatile matter and ash content were 12.0 per cent and 8.3 per cent, respectively (Nelissen *et al.* 2014c). Biochar's labile carbon (C) fraction, as assessed through microbial carbon mineralisation over time,[1] amounted to 3.95mg C g^{-1} biochar-C (0.4 per cent) 381 days after the start of the incubation, but not all labile carbon had been mineralised when the experiment was stopped (Nelissen *et al.* 2014a). The pH (measured in a water extract) was 8.8, which was in the range typical for comparable biochars. The surface area was 295m^2 g^{-1}, which is high compared with biochars produced from non-woody feedstock, but is relatively low compared with other biochars produced from woody feedstocks, as these biochars can have surface areas of 400–600m^2 g^{-1}. Total pore volume of the RomChar was 0.163cm^3 g^{-1} and the average pore size was very small (i.e. 4.2 nanometres, or 10^{-9} m; Chapter 3). The cation exchange capacity is a measure for the capacity to retain cations and was 46.3cmol$_c$ kg^{-1} (Nelissen *et al.* 2014c). This is rather high for biochar but still much lower than for soil organic matter that can have a cation exchange capacity of a few hundred cmol$_c$ kg^{-1} (Chapter 4). The cation exchange capacity of biochar can, however, increase over time due to oxidation. 80 per cent of the particles had a size ranging from 0.5 to 8mm, 5 per cent was >8mm and 14 per cent <0.5mm. The biochar was not pretreated before application.

Biochar testing sites

The seven field trial locations are indicated on the map in Figure 5.1 and additional information on the soil characteristics for the respective locations can be found in Table 5.1. There was a large site variability in initial soil carbon contents ranging from 0.9 to 9.8 per cent (Table 5.1). The carbon-to-nitrogen ratios (Box 5.1) are highly variable between the field trial sites and range between 7.7 and 24.7. Soil pH was similar in UK, Norway, Sweden, Germany and the Netherlands (little less than 5.0), but higher in Denmark (5.9–7.1) and Belgium (6.4).

Box 5.1 Carbon-to-nitrogen ratio: does it matter?

The carbon-to-nitrogen ratio indicates whether nitrogen will be mineralised or immobilised during the decomposition by microorganisms of soil organic matter or organic amendments (such as biochar). Decomposition of soil organic matter or organic amendments with a carbon-to-nitrogen ratio of below 20 to 24 will, in fact, temporarily release nitrogen (mineralisation). If the carbon-to-nitrogen ratio is above 20 to 24, microorganisms consume mineral nitrogen in the soil to be able to decompose soil organic matter or organic amendments and the nitrogen is then incorporated in microbial tissue. This process is called microbial immobilisation and reduces the mineral nitrogen that's available for the plants. As a consequence, plants may suffer from nitrogen shortage and retarded crop growth. Biochar usually has a high carbon-to-nitrogen ratio causing microbial immobilisation during decomposition of its labile hydrocarbon fractions. Whether this process is important or not is, however, dependent on how large the labile fraction of biochar is. In the case of the North Sea ring trial, the biochar (RomChar) used is so stable that microbial immobilisation is expected to be negligible (only few kg N per ha) (see also Chapter 4).

Establishment of biochar North Sea ring trials and used protocol

The field trial protocol we used for the biochar North Sea ring trial is described in this section. The use of a common protocol allows better comparison between sites.

Biochar application and field trial establishment

At each site of the biochar North Sea ring trial, biochar plots are compared to control plots without biochar amendment. Thus, biochar amendment is the only

Figure 5.1 Map with the seven locations of the biochar North Sea ring trial.

Table 5.1 Characteristics of the biochar North Sea ring trial field sites

Country	Field trial location	Application date (MM/YYYY)	Incorporation depth* (cm)	Sampling depth (cm)	Soil texture[†]	%clay <2μm	%silt 2–50 μm	%sand 50–2000 μm	organic carbon %***	C:N***	pH***
UK	Boghall	11/2011	0–25**	0–30	Loam	NA	NA	NA	5.3	15.2	4.9
Norway	Ås	05/2012	0–15	0–<23	Sand	NA	NA	NA	2.7	10.3	4.9
Sweden	Jönköping	05/2012	0–5	0–20	Loamy sand	4	17	79	1.5	10.4	4.9
Denmark (autumn)	Roskilde	08/2011	0–20	0–25	Sandy loam	11	22	66	1.2	9.0	5.9
Denmark (spring)	Roskilde	03/2012	0–10	0–25	Sandy loam	11	22	66	1.2	7.7	7.1
Germany	Lathen	10/2011	0–25	0–25	Sandy	NA	NA	NA	3.4	18.3	5.0
Netherlands	Valthermond	11/2011	0–25	0–25	Sand	3	7	90	9.8	24.7	5.0
Belgium	Merelbeke	10/2011	0–25	0–25	Sandy loam	5	35	60	0.9	11.4	6.4
Minimum									0.9	7.7	4.9
Maximum									9.8	24.7	7.1

Notes:

† Based on USDA soil texture triangle (Box 5.2), except for the data from Germany.

* At time of sowing; autumn in 2011 (Germany) or spring 2012 (other sites).

** Field was ploughed and biochar was mainly in the 20–25 cm layer in rich veins.

*** Initial values before biochar application.

Box 5.2 Soil texture

The soil contains mineral particles of different sizes. A particle smaller than 2 μm is usually called a clay particle; a particle between 2 and 50 μm is called a silt particle; and a particle between 50 and 2000 μm is a sand particle. A soil with a large share of sand particles will have different properties from a soil with a large share of clay particles. Sandy soils, for example, perform less well with respect to retention of water and nutrients, but are better drained than clayey soils, at least under European climate conditions. Clayey soils have a higher capacity to retain positively charged plant nutrients such as ammonium (NH_4^+) or potassium (K^+).

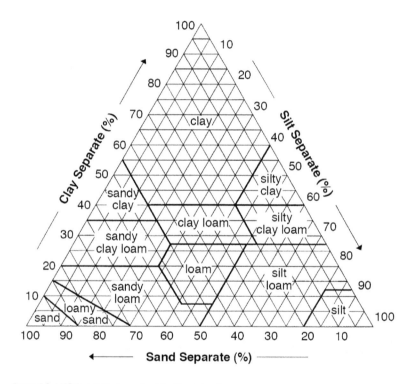

Figure 5.2 USDA soil texture classification triangle.

Depending on the relative share of clay, silt and sand particles, soils are divided in different soil texture classes. An international soil texture classification system that's often used is from the United States Department of Agriculture (USDA; Figure 5.2). By plotting the clay, silt and sand content on the soil texture triangle above, one can read the soil texture class. One should be aware that the definitions of clay, silt and sand particles (size boundaries) and soil texture classes can differ between various soil classification systems.

difference between treatments, except in the Dutch trial, where fertiliser addition was determined per treatment based on standard fertiliser recommendations derived from soil samples at the start of the season. This resulted, in the Dutch trial, in a fertiliser addition of 85kg nitrogen per ha for the control and 93kg nitrogen per ha for the biochar treatment in 2012. The small difference in nitrogen application between the control and biochar plots is probably due to in-field variability. The field trials were organised by a completely randomised (block) design with four replicates (three replicates in the Netherlands). A completely randomised block design means that treatment and control plots are randomly located in a specific area of land over which soil conditions are similar and without pronounced sloping or other obvious topographical differences.

Plot size varied between 36 and $120\,m^2$ (width 3.0–7.5 m, length 8.0–20.0 m) at the different locations. Except for winter wheat in Germany, spring barley was chosen in 2012 as this crop can be grown throughout the North Sea region. In 2013, spring barley was grown in Belgium, Sweden and the UK, winter barley in Germany, winter wheat in Denmark, oat in Norway and potato in the Netherlands. The choice for cover crops (if any) was according to local practices. Mineral fertiliser was used at all sites and doses were chosen according to local recommended practices.

It was assumed that it is more beneficial to apply biochar in autumn, some months before the main cropping season starts, as several incubation (Novak et al. 2010; Bruun et al. 2011; Knowles et al. 2011; Ippolito et al. 2012; Nelissen et al. 2012, 2014b) and pot trials (Deenik et al. 2010; Nelissen et al. 2014b) have shown that biochar can reduce soil mineral nitrogen contents in the short term. This could be by sorption of nitrogen to the biochar surface, for example (i.e. a binding of nitrogen probably in the forms of the ammonium ion (NH_4^+) onto the external and internal surfaces of the biochar) or by biotic immobilisation (Box 5.1) during decomposition of the labile biochar fraction (i.e. the nitrogen is used by microorganisms and this reduces its availability to plants; it only becomes available to the plant once the microorganisms die). By applying biochar some months before the main crop requires nutrients and water (as was the case at the Belgian, Dutch and Scottish sites), such potential disadvantages will be less severe as most of the labile carbon fraction is probably decomposed (mineralised) or sorption sites for nitrogen are saturated. At one site (Denmark), there were two neighbouring trials with different application dates (Table 5.1) in order to test if biochar application date may have an effect on the first year results.

The biochar dose applied in the biochar North Sea ring trial was 20 tonnes ha^{-1}, which corresponds to a total carbon application of 13.5 tonnes ha^{-1} and represents 4 to 46 per cent (depending on the site) of the soil organic carbon in the 0–25cm layer. This dose was applied on a dry matter base of the biochar, except at the German and Swedish sites where biochar was applied on fresh matter base. In Germany, biochar moisture content was 15.6 per cent, which means that of the 20 tonnes ha^{-1} of fresh biochar applied, 3.1 tonnes were water and 16.9 tonnes were dry biochar.[2] The concentration of biochar over the depth in which it was incorporated at the start of the experiment (Table 5.1) ranged between 0.5 and

2.6 per cent (mass percentage). Soil sampling depth was at some sites deeper than incorporation depth (Table 5.1). Biochar concentration over the soil sampling depth, which is relevant for the interpretation of the measured soil characteristics, ranged between 0.5 and 0.7 per cent.

The even distribution of biochar within the plots was achieved by applying the biochar by hand and by dividing the plots into several subplots with stakes or wooden frames (Figure 5.3).

Figure 5.3 Top: Equal division of biochar into subplots. Right: raking biochar evenly over subplots.

Sources: Bioforsk (top); Hawk (right).

Monitoring biochar effects

In order to increase comparison between sites, a standard protocol for soil and crop sampling was developed for the biochar North Sea ring trials. Soil samples were taken just after biochar incorporation, before tillage/sowing and after harvest for organic carbon content,[3] total nitrogen,[4] plant available nutrients,[5] pH[6] and mineral nitrogen[7] analysis. Soil mineral nitrogen (i.e. ammonium, NH_4^+; and nitrate, NO_3^-) was also measured about one week after fertilisation to assess effects on nitrogen availability from fertiliser addition, as we thought that biochar could reduce the nitrogen availability caused by processes described above. To obtain representative soil samples, 10 subsamples were taken across the plot and homogenised to obtain one composite sample per plot. Sampling depths were between 20 and 30cm depending on tillage practices (Table 5.1). Where possible, mineral nitrogen, which is mobile in soil, was measured at three depths: 0–30, 30–60 and 60–90cm. The samples were analysed at one lab to secure appropriate comparability. Owing to local circumstances, not all samples could be taken at all sites. Crop samples were taken at harvest to determine 1000 grain weight (fresh)[8], dry matter[9] yields (grain and straw for cereals) and nitrogen[10] and phosphorus[11] uptake.

For establishing soil water retention curves (Chapter 4) in the laboratory, which display the relation between the force needed to extract water (pF) and soil water content, undisturbed soil cores (height of 5cm) were taken from the middle of the soil layers where biochar was incorporated in the spring of 2013.

Soil samples for measuring the soil microbial community structure (relative abundance of certain types of bacteria and fungi) were undertaken and earthworm counts conducted around the time of flowering of the cereals unless it was too dry to sample (in which case it was postponed). Different microorganisms have different amounts and types of phospholipid fatty acids (PLFA) in their cell membranes. We measured the type and relative abundance of these different fatty acids with gas chromatography and mass spectrometry (Nelissen 2013) in order to determine the importance of different groups of soil microorganisms and thus to assess the soil microbial community structure.

Earthworms are important organisms in soil as they glue together soil particles, making soil aggregates and a stronger soil structure, and because they leave channels behind that can be used for root growth and for water drainage. Earthworms were counted and weighed after adding mustard powder solution to the soil and excavating twice 20 × 20 × 20cm of soil in each plot[12] (Figure 5.4). The mustard powder forces deep-burrowing earthworms to the surface.

Results of the biochar North Sea ring trials

Impacts on soil properties

Soil organic carbon content increased after biochar application at most field sites, due to the large application of biochar carbon (13.5 tonnes ha of carbon) and this was significant at all sites with less than 3 per cent soil organic carbon (Figure 5.5).

Figure 5.4 Sampling earthworms. Top: pouring mustard powder solution onto soil surface to make deep-burrowing earthworms move to the surface. Bottom: collecting earthworms from an excavated 20 × 20 × 20cm frame.

Source: ILVO.

For sites with higher soil organic carbon content, the effect of biochar is more difficult to detect as the relative increase of carbon in the soil is smaller. Mean soil organic carbon content of the control plots in the UK was higher than the soil organic carbon content of the biochar plots, which can be explained by large in-field variability. Consequently, results of other measurements for the UK site need to be considered with care. The data as shown in Figure 5.5 show that the increases in organic carbon are still detected after more than one to two years (harvest 2013), although longer-term monitoring is needed to prove long-term carbon sequestration in the soils. The Dutch site had very high initial carbon content as the trial was located on a reclaimed peat site.

Box 5.3 Statistical data interpretation

P-values and statistically significant differences

When interpreting data, it is important to discern whether differences measured between control and biochar treatments are true differences and not due to chance and measurement variability. In order to do this, statisticians have developed methods for analysing data to see whether differences can be better explained by chance or whether there is a good likelihood (usually 95 per cent) that the difference is due to the different treatments (such as addition of biochar compared to no biochar addition). The computed 'p-value' indicates how likely it is that a detected difference is not a real difference between the treatments. For example, a p-value of 0.05 indicates that there is a probability of 5 per cent that a difference is not true but due to measurement variability or other arbitrary factors. Therefore, the lower the p-value, the more certain we can be that a measured difference is true and explained by the treatment versus the control, if the experimental setup was appropriate. In general, differences are considered to be statistically significant when the p-value is below 0.05. In the remainder of the text, we say that differences are significant when differences are statistically significant at $p < 0.05$. Replication of the treatment and control conditions is necessary in order to calculate the variation in response, allowing calculation of the p-value. Without replication of plots with the same treatment, it is not possible to calculate whether differences are statistically significant or just down to chance.

Error bars

In graphs, variability between the data measured in each replicate of a treatment is indicated by error bars. The smaller the spread of these bars, the less variability between the measurements and the easier it is to prove that differences found are statistically significant. Typically standard deviations or standard errors of the mean are displayed. The standard deviation is a way of measuring how much the separate measurements deviate around the mean value. Standard errors of the mean are equal to standard deviation divided by the square root of the sample size (= number of replicates).

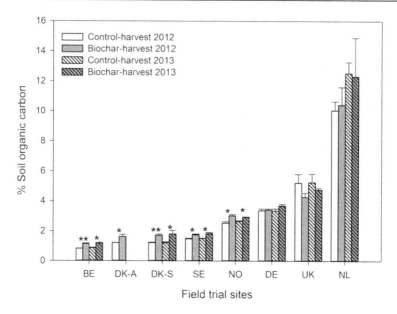

Figure 5.5 Soil organic carbon contents at the different sites of the biochar North Sea ring trial, determined around the harvesting period of 2012 and 2013.

Note: BE: Belgium; DE: Germany; DK-A: Denmark – autumn application; DK-S: Denmark – spring application; NL: the Netherlands; NO: Norway; SE: Sweden; UK: United Kingdom. Error bars are standard errors. Significant higher carbon contents in the biochar treatments compared to the control are indicated by * if p <0.05 and ** if p <0.001. Application dates and sampling depths are indicated in Table 5.1.

The biochar's total nitrogen content was 0.4 per cent. 80kg of nitrogen is thus added to the soil with 20 tonnes of biochar per ha. On the one hand, 80kg of nitrogen is important if biochar is to be considered as a fertiliser and if the nitrogen content needs to be accounted for when calculating total nitrogen inputs in soil. Adding nitrogen to soil is controlled in several regions under the European Nitrates Directive because of the risk of run-off and leaching of nitrates from the land into streams, rivers, lakes and other water bodies causing pollution. For many biochars, however, a large part of nitrogen is locked up in aromatic rings (Chapter 3), making it hard for microorganisms to access it and turn it into nitrate or ammonium, so that it is thought to be unavailable to life forms, except in the long term. On the other hand, the addition of 80kg of nitrogen from biochar is not much, considering the total nitrogen mass that is already available in the soil. The total soil nitrogen over the sampling depth (ranging between 20 and 30cm, Table 5.1) at the ring trial field sites ranged from 2,405 to 13,650kg per hectare. As a result, the nitrogen that was added with biochar ranged between 0.6 and 3.3 per cent of the total soil nitrogen inventory. Therefore, total nitrogen content was only affected to a limited extent and significant increase in total nitrogen due to biochar applications could mostly not be detected in the first two years of monitoring.

Biochar is usually alkaline. Therefore, it has a capacity to neutralise acidic soils and is thus expected to increase the soil pH. Biochar can, therefore, act as liming agent and as such improve acidic soils (e.g. by increasing nutrient availability and improving soil structure). The pH of the biochar used at the ring trial sites was 8.8, which is higher than the soil pH of all field sites[13] (Figure 5.6). Soil pH in the biochar treatments was indeed mostly higher than in the control treatments (Figure 5.6), but significant only at the Norwegian site, which had a low initial pH and only just after biochar application. At later sampling dates, this difference could not be found at this site. Despite the limited effect of biochar on the pH of the bulk soil, the pH surrounding the biochar particles can be higher, locally influencing soil processes such as nitrification (conversion of ammonium to nitrate) or denitrification (conversion of nitrate to gaseous nitrogen forms).

Biochar can have effects on soil nutrient dynamics (Nelissen *et al.* 2012) through, for example, pH increases, adsorption of nutrients by the biochar particles, the nutrient content of biochar itself, or biochar's high carbon-to-nitrogen ratio, leading to microbial nitrogen immobilisation during decomposition of the labile fraction (Box 5.1). The latter is especially important if biochar has a high labile

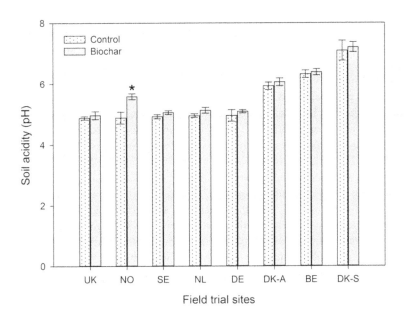

Figure 5.6 Soil pH (determined in 1 M KCl extract) at the different sites of the biochar North Sea ring trial, determined in spring 2012 a few months after biochar application (except for DE and SE at harvest in 2012).

Note: BE: Belgium; DE: Germany; DK-A: Denmark – autumn application; DK-S: Denmark – spring application; NL: the Netherlands; NO: Norway; SE: Sweden; UK: United Kingdom. Error bars are standard errors. Significantly higher pH values in the biochar treatments compared to the control are indicated by * ($p < 0.05$).

carbon fraction (Bruun *et al.* 2012), which is usually not the case for woody biochars and for biochars produced by slow pyrolysis at higher temperatures. Also, the RomChar used in the biochar North Sea ring trial appeared to have a small labile carbon fraction, as less than 0.4 per cent of the RomChar-carbon was emitted in an incubation experiment that lasted 381 days (Nelissen *et al.* 2014a). This is comparable to other biochar types (Chapter 7).

In general, the biochar applied in the North Sea ring trial did not affect the availability of soil mineral nitrogen (nitrate and/or ammonium). This was concluded from the North Sea ring trial results, regular sampling starting from biochar application (every one or two months) at the Belgian site (Nelissen *et al.* 2015), and from samples taken before and one or two weeks after fertilisation at two sites in 2012 and five sites in 2013. As a consequence, our results do not support the hypothesis that less fertiliser would be needed in biochar amended soil because the biochar would retain ammonium nitrogen better, increasing its use efficiency. The earlier mentioned hypothesis that in the first months after biochar application there is a risk that less nitrogen is available due to the decomposition of the labile biochar fraction, is also not supported by our results. We can thus conclude that for the biochar used, there is no need to change nitrogen fertiliser doses in the first two years after application.

Whether biochar has an effect on the availability of other nutrients to the plant was also investigated in the first growing season. This was measured with an acid ammonium lactate extract. There was no effect on the availability of iron, calcium, phosphorus, magnesium or sodium. For potassium and manganese, some statistical effects were found. Except for in the UK, potassium availability was always higher in the biochar treatments and this was significant in three cases. This is perhaps not surprising as the biochar used contained quite high potassium content (0.31 per cent; Chapter 3). Along with 20 tonnes of biochar, we also applied 62 kg of potassium, which could have been mainly concentrated in the ash fraction and as such available.

The water retention curves (also see Chapter 4), measured in the laboratory using soil cores of the field sites (Figure 5.7), reveal that there were no differences for the Dutch site (p > 0.05). For the other sites, some differences for some pF-values were measured but the results were variable. In Belgium, Denmark and Sweden, moisture contents were higher in the biochar plots when the same force was applied to the soil (at equal pF-values). In Norway and the UK, the opposite was the case. The plant-available water[14] was slightly higher in the biochar treatment in Denmark and Belgium. If biochar was incorporated into the 0–25 cm layer, this meant a difference of maximum 5 mm of water. No statistical differences were found at the other locations for plant-available water.

In Denmark, soil moisture content was monitored during the 2012 growing season, but no differences were observed between control and biochar plots. At the Belgian site, soil moisture contents were measured with continuous soil moisture sensors that were calibrated per treatment. Soil moisture contents were generally higher in the biochar plots, but only significantly at certain times and only when soil moisture contents were higher, which is consistent with the pF-curves (Figure 5.7). Thus, there is no evidence so far that the biochar applied can retain more moisture during drought periods at the Belgian site (Nelissen *et al.* 2015). In Norway, however, soil moisture contents were systematically higher in the biochar plots at 5, 15 and 20cm depths and the difference between the biochar and control treatments increased during dry periods. This was in contrast with the results of the water retention curves that revealed lower moisture contents for the biochar plots, especially at the lower soil moisture content range.

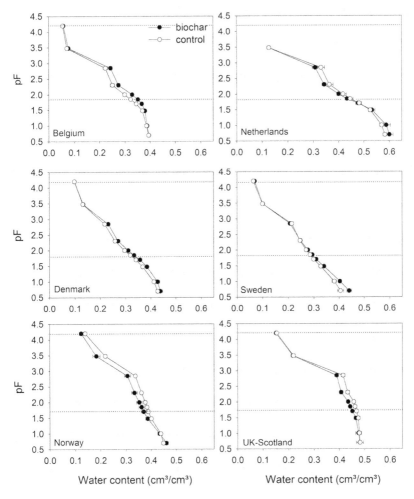

Figure 5.7 Relation between the force needed to extract water (pF) and the fraction of water in soil at the different sites of the biochar North Sea ring trial. Below a pF of 1.8 (field capacity), water is removed by gravity, above a pF of 4.2, water cannot be extracted by plants anymore (permanent wilting point). Plant-available water is therefore between field capacity and wilting point.

Earthworms were sampled at four sites of the biochar North Sea ring trial. It was the intention to sample around barley flowering (June–July), but weather conditions were too dry in Denmark and too wet in the UK at that time. At those sites, sampling was postponed until August–September. Earthworm masses and numbers were very variable (Figure 5.8). A statistical difference could only be detected in Norway, resulting in higher numbers and mass per m^2 of earthworms in the biochar plots compared to the control. There was no significant difference for the mean mass per worm between control and biochar plots.

In each country (except for Sweden), soil microbial community structure in the field trial was investigated by means of phospholipid fatty acid (PLFA) profiles.

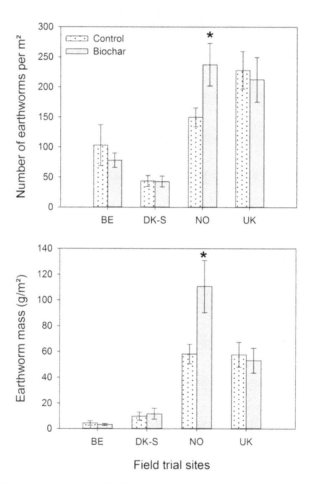

Figure 5.8 Earthworm numbers (top) and earthworm mass (bottom) per square metre at four sites of the biochar North Sea ring trial.

Note: BE: Belgium; DK-S: Denmark – spring application; NO: Norway; UK: United Kingdom. Significantly higher results (p<0.05) are indicated by *. Error bars are standard errors.

Phospholipid fatty acids are essential structural components of the cell membranes of all living cells. They are synthesised during microbial growth and, in general, rapidly decompose after cell death. Therefore, the concentration of total PLFAs is a measure of the microbial biomass, and the individual fatty acids provide details about the groups of organisms or community structure (Burns 2011). In this way, whether biochar influences the presence of microorganisms such as fungi and bacteria in soil can be investigated.[15]

Biochar could change the microbial community structure by inducing an altered soil environment through changing the resource base (e.g. available carbon, nutrients, water) or abiotic factors (e.g. pH, toxic elements), or through locally providing a different habitat (Lehmann *et al.* 2011). At four sites of the biochar North Sea ring trial, there were some differences in relative abundance of soil microorganism groups between control and biochar plots, but no consistent differences could be found. These differences were also only detected for bacteria, but not for fungi. The sum of the absolute PLFA concentrations, a measure for microbial biomass, was not affected by biochar addition at any of the sites (Nelissen 2013).

Impacts on crop yield and nutrient uptake

At most sites, no crop yield differences were detected in the first year after biochar application. However, in Sweden and the UK, grain yields were significantly higher in the biochar plots and in Norway there was a significant increase in the straw yields. In contrast, at the German site, where winter wheat was grown, grain yields were significantly lower in the biochar plots (Figure 5.9). From the soil mineral nitrogen data obtained from all sites, we cannot conclude that a lower nitrogen availability was the cause of these yield differences. The yield increase in the UK was high but it should be noted that a large in-field variability of different soil parameters was detected and that the crop yield of the control was exceptionally low. In addition, the year was anomalously wet and soils on parts of the field site were water-logged, hence yields were only a third of what would normally be expected. The farmer believes that the biochar may have helped to reduce water logging in parts of the soil. In Sweden, higher yields can potentially be attributed to effects on soil water content in a sandy soil with little precipitation, although this hypothesis is not supported by pF-curve measurements in the laboratory (Figure 5.7). Fresh thousand grain weight was determined at four sites, but no differences were detected. Also, in the Netherlands, no difference in grain size distribution was noticed. For both treatments, grain sizes were sufficiently high.

In 2013, spring barley was again grown in Belgium, Sweden and the UK, while the crop choice was winter barley in Germany, winter wheat in Denmark, oats in Norway and potato in the Netherlands. At none of the sites, yield differences (dry matter) were detected between the biochar and control plots.

Nitrogen content in grains is an important grain quality parameter as it indicates protein content. High grain protein contents are for instance required for flour. If barley is used for breweries, protein contents should neither be too high nor too low. No differences in nitrogen content were, however, detected between biochar and control treatments, except for the UK site where nitrogen contents were lower in the biochar plots (p = 0.05). The same conclusion could be drawn for phosphorus (p = 0.03).

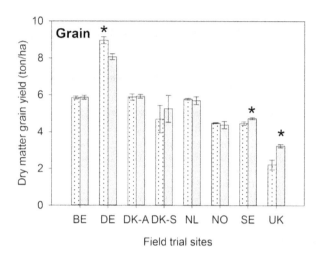

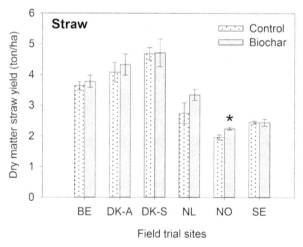

Figure 5.9 Dry matter yields at the different sites of the biochar North Sea ring trial in 2012. Top: grain yields. Bottom: straw yields.

Note: BE: Belgium; DE: Germany; DK-A: Denmark – autumn application; DK-S: Denmark – spring application; NL: the Netherlands; NO: Norway; SE: Sweden; UK: United Kingdom. Spring barley was grown at all sites except for in Germany where winter wheat was grown. Significant higher yields (p<0.05) are indicated by *. Error bars are standard errors. Straw yield data were not available for all sites.

Neither nitrogen nor phosphorus contents were different for straw, except for the lower nitrogen content from the biochar treatments at the Swedish site (p <0.01). In 2013, there were no differences in nitrogen or phosphorus uptake. The nitrogen content of the grains was a little higher in the control plots compared to the biochar plots in Sweden only.

Conclusions from the biochar North Sea ring trials

Biochar field trials were established in seven countries of the North Sea region, using the same biochar applied at a rate of 20 tonnes ha^{-1} and according to a common protocol. Soil properties of the different sites varied widely (Table 5.1). Except for an increase in soil carbon contents at those sites, which had a soil organic carbon content of less than 3 per cent, the biochar used (RomChar) led to little effects on chemical, physical and biological soil properties in the first growing seasons. Effects on grain and straw yields were also limited. The biochar was stable in soil, which resulted in a situation where short-term effects caused by mineralisation (decomposition) of the labile carbon fraction were not detectable. It is likely that biochars with a larger labile carbon fraction may temporarily reduce nitrogen availability and increase (the activity of) microbial soil life.

Based on the results of the ring trials, little impacts are to be expected of biochar application to already fertile soils or in optimised farming systems. In the past century, farming systems in northwest Europe have continuously increased yields by improving soil fertility, crop varieties, tillage, fertilisation and crop protection practices. It is therefore difficult to make new major improvements. It is important to note, however, that except for the German site, the biochar applied did not reduce crop yields, proving that carbon storage is possible without major negative impacts on the farmers' traditional evaluation tool. It is possible that biochar can help in water logged soils, which could be investigated further. Further monitoring is also still needed to understand the longer-term effects of biochar on soil and crop properties as biochar will weather over time.

Tips and tricks for establishing biochar field trials based on practical experience from the biochar North Sea ring trials

Based on the practical experience obtained from the biochar North Sea ring trials, we provide here some general recommendations for conducting biochar field experiments. Please note that these are meant mostly for research and development more than for direct practical purposes. The recommendations are partly also useful to actual farm practice, but the practical application of biochar also depends on several other factors including farmer practice, equipment, economic viability, labour supply, national legislations, etc. (Chapters 9 and 10).

Field trial objectives and hypotheses

Before starting a biochar field trial, the research objectives and hypotheses should be clearly defined because these have implications, for example on field site choice and preparation and on choosing control treatments. If the objective is to investigate whether biochar is beneficial for crop growth in acidic soils, a location with a low-pH soil should be selected. If, on the contrary, the objective is to assess the effect of biochar on crop growth and nutrient availability, exclusive of the potential pH effects, the pH of the site should already be optimal at the start of the experiment. This can be accomplished by liming before the experiment starts. Alternatively, a control without biochar addition but with a pH adjusted to bring the soil pH in line with the biochar treatments should be considered (Jeffery *et al.* 2015). If the objective is to investigate the effect of biochar other than the direct availability of nutrients in the ash fraction of the biochar, a control can be added without biochar but in which the same amounts of direct available nutrients are added. Since biochar may contain many different elements, next to carbon, the experimental layout may become quite complex, especially if the effect of each individual component along with their interaction is part of the study.

Biochar characterisation

Biochar characteristics are highly variable, depending on the feedstock and pyrolysis conditions (Chapter 2). Biochar characteristics might also deviate between batches and change over time in storage due to weathering (Hammond *et al.* 2013). Therefore, it is recommended to always take representative subsamples of the biochar prior to application. Bucheli *et al.* (2014) pointed out that it is important to take representative biochar samples to establish the variance of its properties within a batch. If different batches of the same type of biochar are used at different locations, it is advised to take samples at each location. More information on biochar characterisation is provided in Chapter 3.

Biochar application time

Some pot experiments show reduced crop growth, possibly due to reduced nutrient availability (e.g. Nelissen *et al.* 2014b; Deenik *et al.* 2010). This effect is expected to be limited over time, especially if it is caused by the decomposition (mineralisation) of the biochar labile carbon fraction. But if this is due to strong sorption to the biochar surface, a maximum sorption capacity can be assumed, so the amount of mineral nitrogen that is reduced in soil is expected to stabilise and then diminish over time. Although there was no evidence from the biochar North Sea ring trial of these effects, it might be worthwhile for other biochar project developers to apply biochar some months before the main crop is sown as a precautionary measure.

Recommended doses of biochar

There is no clear answer to the question of what the optimal biochar application dose is, as the effect of biochar dose is dependent on the soil type, biochar type and other local conditions. The dose could be deduced from results of other field experiments with similar biochars in the region or it could be derived from pot experiments that used biochar dose as a factor. These experiments have limitations as they only provide indications for the short-term effect of biochar. Another approach could be to regularly apply small amounts instead of applying a larger amount (10 tonnes of C per ha or more) once (Graves 2013). Applying smaller doses could be a precautionary measure so as not to influence the local soil and plant interrelationship too drastically. Obviously, when planning a biochar dose, it is important to know the moisture content, as only the mass of biochar is important, not the amount of water applied. Biochar moisture contents can vary greatly between biochars and even within large bags owing to gravity effects. Moisture content is, for instance, dependent on the post-production treatment (such as use of water quenching for cooling and removal). Soil moisture contents can be determined by drying biochar for at least 24 h at 105°C and determining moist and dry mass before and after, respectively. For an intended dry biochar dose, how much moist biochar needs to be applied can then be calculated.

Field trial layout

To conduct statistical analysis, it is necessary to have at least three replicate plots randomly distributed over the field site. It is also important to choose a homo- geneous field site to avoid high variability between plots, which makes it difficult to prove the effects of the treatment. Homogeneity could be assessed by asking the field plot owner to assess the homogeneity based on his or her experiences with, for instance, crop development and water infiltration. Taking soil samples and measuring in-field variability for a number of key soil parameters, such as soil organic carbon, total nitrogen and availability of nutrients, is an additional option to assess in-field homogeneity. Otherwise, special statistical layouts such as Latin Square or Latin Rectangles should be used (Glaser *et al.* 2014).

Biochar application

It is important to distribute the intended biochar dose evenly on each plot. This can, for instance, be achieved by applying the biochar by hand and subdividing the plots into subplots (Figure 5.3). Determining the right amount of biochar per subplot can be done by using a field balance or by determining the volume of the required biochar mass in advance (put a mark on a bucket) and applying these volumes in the field. The advantage is that biochar volume is independent of its moisture content.

As dry biochar can be quite dusty and explosive, it should never be applied dry, and those applying biochar should be well protected with breathing masks and eye goggles. Furthermore, dusty biochar can be lost by wind, contaminating control plots and reducing the actual amount of biochar applied to the biochar treatments. In one experiment, Husk and Major (2010) estimated 30 per cent of biochar losses by wind during loading, transport and application of biochar with an agricultural lime spreader. These losses can (partly) be prevented by avoiding using a biochar with a too high dust fraction and/or by pre-wetting the biochar. The challenge is then to make sure that the water is equally distributed by mixing it, for instance, with a concrete mixer. Alternatively, per subplot, one could apply water to the batch and mix the water with the biochar by hand (the operative must wear gloves). In this way, the oven-dry biochar mass is equal for every subplot. In hilly areas, biochar losses by water erosion can also be a problem. Major *et al.* (2010) estimated that 20 to 53 per cent of applied biochar, incorporated to a depth of 10cm, was lost by surface runoff in a field experiment in Colombia. These high losses are, among others, attributed to the hydrophobic nature of biochar, the lack of association with soil minerals just after application and its light nature.

Biochar incorporation

Incorporating biochar immediately after application is strongly recommended. This reduces wind losses and biochar spread onto control plots. If ploughing is used as the first incorporation method, biochar gets concentrated in layers or 'lenses' as was experienced in a Miscanthus biochar field trial in Norway (Figure 5.10). After the first growing season, the field was ploughed a second time, which caused the biochar to resurface. Therefore, applying non-inversion tillage as a first incorporation method so that the biochar is evenly distributed throughout the top layer is advisable. Once the biochar is well distributed vertically, ploughing should no longer be a problem. Graves (2013), however, suggests that intensive mixing of biochar in the soil is suitable if the purpose of biochar is to remediate contaminated soils or if the purpose is to increase water holding capacity, decrease soil bulk density and increase aeration or drainage of physically degraded soils. When biochar is used for promoting plant health, he recommends applying biochar close to root systems using no-tillage or minimum tillage techniques. Other possibilities for biochar application and incorporation include using strip-till techniques and slurry injection or a seed drill for pelletised biochar. These techniques can also be suitable in no-tillage farming systems as intensive mixing of biochar with the soil could trade off against the benefits that no-till farming brings (Jeffery *et al.* 2015).

Horizontal spreading of biochar during tillage should be avoided. This is achieved by tilling with hand machines or driving slowly. Belgium tests showed the disadvantage of applying biochar on (maize) stubble. The cultivator moves the stubble several metres horizontally from the point of growth and, in so doing, some of the biochar is also moved along with the stubble. It is, therefore, better to cultivate stubble before biochar application. Horizontal spreading was also observed at field trials in Angus and Nottinghamshire (UK), where biochar was incorporated during bed formation and in Midlothian (legume trial) during ploughing (Hammond *et al.* 2013).

Besides expressing biochar dose in kg ha^{-1}, it is also appropriate to express biochar dose as a concentration (mass or volume percentage) because incorporation depth is influenced by local machinery used, soil type and pre-cropping, among other factors. At each sampling date, incorporation depth that can vary in time should be assessed and the bulk density of the soil (tonnes m^{-3}) should be measured so that the biochar concentration in soil can be calculated.

Monitoring biochar effects

First, for a correct interpretation and analysis, the vertical and horizontal distribution of the biochar particles should be documented after each tillage or at each sampling date. This can be done by visually inspecting some holes made in the field and assessing if the biochar concentration is similar to what was intended. If biochar is incorporated deeper than sampling depth it means that biochar is diluted and biochar concentration is smaller than intended. Second, avoid taking soil and crop samples close to the plot borders and keep a distance of one to a few metres to avoid border effects (e.g. by horizontal spreading of biochar). If plots are disturbed during the growing season by frequent sampling, you may define a harvest strip which remains untouched in advance, and take other soil and crop

Figure 5.10 Accumulation of biochar in layers or lenses after ploughing as the biochar incorporation method. After the second ploughing operation at the end of the first growing season, the biochar resurfaced.

Note: Biochar field site in Ås, Norway (Bioforsk).

samples during the growing season outside this strip. Third, biochar is very light and might segregate physically from soil particles (especially in dry soil). Therefore, the soil and biochar should be very well mixed just before analysis.

Reporting

Comprehensive reviews and meta-data analyses allow results to be aggregated, enabling more robust extrapolation, which is necessary for policy guidance and directing future research. Therefore, the experimental site and setup should be thoroughly described and as many auxiliary variables as possible – such as pH of the soil before and after application, soil texture and particle size distribution, climate and weather information, crops grown, nutrient content of the biochar, pyrolysis conditions (temperature) and feedstock – should be reported (Jeffery *et al.* 2015). There is a tendency for under-reporting of null results, potentially biasing conclusions from reviews and meta-data analyses. It is hence important that all experimental results are reported and made available in the public domain.

Biochar effects on crop yield in Europe

In this section, an overview is provided of yield responses upon biochar application in European field trials with pure biochar additions (Table 5.2). Unpublished and literature data from 32 field trials were gathered, leading to 75 location–biochar type–biochar dose combinations and 171 location–biochar type–biochar dose–year combinations. 26 trials were established in north and northwest Europe, of which seven field trials were located in the UK. Two trials were conducted in central Europe (Austria) and four in southern Europe (Italy).

In every field trial, crop yield from a plot with biochar addition is compared with crop yield from an unamended plot. In Figure 5.11, the crop yields of the biochar-amended plots are expressed relative to the crop yields of the control plots (i.e. crop yield biochar plots/crop yield control plots × 100). Above the 100-line, crop yields from the biochar plots are higher than the control plots, below the 100-line, crop yields are lower. In 50 per cent of the cases, crop yields in the biochar plots were higher than in the control plots, which also means that in 50 per cent of the cases crop yields were lower in the biochar plots. It must be stressed that this does not mean that crop yields from both treatments are significantly different owing to chance and measurement variability (Box 5.3). Considering a statistical certainty of 95 per cent, crop yields from control and biochar treatments did not significantly differ from each other in 80 per cent of the cases; in 12 per cent of the cases there was a significant positive biochar effect and in 6 per cent of the cases there was a significant negative biochar effect. In 2 per cent of the cases it was not known whether the measured differences were significant or not.

We could question what factors lead to positive or negative crop yield responses in order to learn from previous experience. However, as many interactions occur, such as with the chemical and physical characteristics of each biochar type used, biochar dose, soil and weather conditions and crop type, and as such data are limited, it is difficult to make statements for separate factors. Conclusions are therefore indicative and tentative only and should be interpreted with care.

Biochar feedstock and production technology

In European field trials, almost all biochars tested derived from wood or wood residues (Table 5.2). Most were produced by slow pyrolysis originating mainly from the traditional charcoal industry (such as RomChar). Smålandskol charcoal, used in the Swedish trials, and the charcoal applied in Italy, seemed to perform well, although it is unclear whether this can be attributed to a better biochar type or the soil type and/or climatic conditions. It would be useful to test these biochars at other locations and to characterise them well in order to get a better understanding of their effects under different soil, crop and climate conditions and management regimes. Other slow pyrolysis wood-based biochars (e.g. Dalkeith and RomChar) performed positively or negatively in various cases, but results were not consistent. This suggests that other factors than the physical and chemical characteristics of the biochar employed play a more important role.

Some alternative biochars that were tested include activated carbon, gasification char (at >700°C), flash pyrolysis char and hydrochar produced by hydrothermal carbonisation (see Chapter 2 on production technologies). Activated carbon and

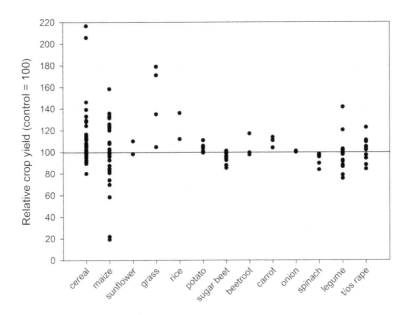

Figure 5.11 Crop yield measured in biochar plots relative to crop yield in control plots shown per crop type, based on European field trials (Table 5.2).

Note: t/os rape = turnip or oilseed rape.

Table 5.2 European field trials with pure biochar additions (not mixed with organic materials)

Trial location	Soil texture[a]	Application date[b]	Biochar name/type[c]	Biochar dose (t/ha)	Y	Source
NO, Ås	Sand[a]	s/2012	Romchar (W)	20[d]	2	Unpublished, Bioforsk; this chapter
NO, Sel	Silty sand	s/2011	Skogenskol (W)	18	1	Unpublished, Bioforsk
SE, Jönköping	Loamy sand[a]	s/2012	Romchar (W)	20	2	Hylander et al., pers. comm. (2013); this chapter
SE, Jönköping	Silty loam	s/2010	Smalandskol charcoal (W)	10/20[d]	2	Kihlberg et al. (2013); Hylander L., pers. comm. (2013)
SE, Jönköping	Silty loam	s/2011	Smalandskol charcoal (W)	10[d] so	2	Kihlberg et al. (2013); Hylander L., pers. comm. (2013)
SE, Jönköping		s/2011		10[d]	2	
FI, south	Silt loam	s/2009	Charcoal (W)	9	1	Karhu et al. (2011)
FI, Viiki	Sandy clay loam[a]	s/2010	Slow pyrolysis biochar (W)	5/10[d]	3	Tammeorg et al. (2014a)
FI, Viiki	Loamy sand	s/2011	Debarked spruce chips (W)	5/10/20/30[d]	2	Tammeorg et al. (2014b)
DK, Roskilde	Sandy loam[a]	a/2011	Romchar (W)	20[d]	1	Unpublished, Riso, DTU; this chapter
DK, Roskilde	Sandy loam[a]	s/2012	Romchar (W)	20[d]	2	
DK, Roskilde	Sandy loam[a]	s/2011	Skogenskol (W)	20[d]	2	Unpublished, Riso, DTU; Kumari et al. (2014)
NL, Kollumerwaard	Clay loam[a]	a/2011	Carbo charcoal (W)/ Activated carbon	20[d]	1	Ros et al. (2013)
NL, Kollumerwaard		s/2010 + yearly		5[f]	3	Ros et al. (2013)
NL, Lelystad	Loam[a]	s/2010 + yearly	Carbo Charcoal (W)	2.5/5[f]	3	Ros et al. (2013)
NL, Valtermond	Sand[a]	s/2010 + yearly	Carbo charcoal (W)/ Activated carbon	5[f]	3	Ros et al. (2013)
BE, Merelbeke	Sandy loam[a]	a/2011	Romchar	20[d]	2	Nelissen et al. (2015); this chapter
BE, Merelbeke		a/2011	Romchar (W)	20[d]	2	
DE, Lathen	Sandy	a/2011	Romchar (W)	20[f]	2	Unpublished, HAWK; this chapter
DE, Braunschweig	Sandy silt	s/2009	Pine wood chips(W)	2.5/5	4	Panten K., pers. comm. (2013)
DE, Bonn	Sandy	s/2008	Charcoal (W)/ Gasification coke (W)/ Flash pyrolysis char (W)	45[d]	3	Borchard et al. (2014); Borchard N., pers. comm. (2013)

Table 5.2 Continued

Trial location	Soil texture[a]	Application date[b]	Biochar name/type[c]	Biochar dose (t/ha)	Y	Source
DE, Göttingen	Silty	s/2008	Charcoal (W)	45/300[d]	3	Gajic and Koch (2012)
			Gasification char (W)/ Flash pyrolysis char (W)	45[d]	3	
	Silt (loam)[a]	s/2010	Hydrochar (sugar beet pulp)/ Hydrochar (beer draff)	10[d]	1	Gajic and Koch (2012)
DE, Gorleben	Sandy	s/2012	Biochar-650°C (W)	1/40[d]	1	Glaser et al. (2014)
AT, Kaindorf	Clay loam	s/2011	Romchar (W)	24/72[d]	2	Karer et al. (2013)
AT, Traismauer	Silt loam	s/2011	Romchar (W)	24/72[d]	2	Karer et al. (2013)
UK, Boghall	Loam	a/2011	Romchar (W)	20[d]	2	Unpublished, UKBRC; this chapter
UK, East Lothian	Silty clay loam	s/2009	Dalkeith charcoal(W)	10	3	Hammond et al. (2013)
		a/2009		10/20/40	2	
UK, Angus	Sandy loam	s/2009	Dalkeith charcoal (W)	10/20/40	1	Hammond et al. (2013)
UK, Auchtermuchty	Clay loam	s/2010	Dalkeith charcoal (W)	10/20/40	1	Hammond et al. (2013)
UK, Nottinghamshire	Loamy sand	s/2010	Dalkeith charcoal (W)	6/12/24/40/80	1	Hammond et al. (2013)
UK, Midlothian	Loam	a/2010	Dalkeith charcoal (W)	10/30	1	Hammond et al. (2013)
		s/2011		10/30	1	
UK Abergwyngregyn	Sandy clay loam	s/2009	BioRegional HomeGrown® (W)	25/50	3	Jones et al. (2012)
IT, Pistoia	Silty loam[a]	a/2008	Charcoal (W)	30/60	2	Vaccari et al. (2011)
		a/2009	Charcoal (W)	30/60	1	
IT, Empoli	Clay loam[a]	a/2007	Charcoal (W)	10	1	Baronti et al. (2010)
IT, Beano	Silt loam[a]	NA	Charcoal (W)	10	1	Baronti et al. (2010)
IT, North	Sandy loam	s/2010	Gasification char (W)/ Slow pyrolysis char (W)	40	1	Lugato et al. (2013)

Note: [a] based on USDA soil texture classification; [b] a = autumn application, s = spring application; [c] W = wood based; [d] expressed on dry-base; [e] expressed on fresh biochar base; without indication of d and f it was unclear if the doses were calculated on fresh or dry base; so = biochar soaked with fertiliser for few hours before application; Y = number of monitoring years; NA = data not available.

gasification char did not lead to positive or negative yield effects. Fresh flash pyrolysis char suppressed germination of maize kernels, probably by phytotoxic compounds in the biochar (Borchard *et al.* 2014). Crop yields after addition of hydrochar made from sugar beet pulp resulted in yield decrease at low nitrogen fertilisation levels probably due to the high carbon-to-nitrogen ratio of the hydrochar and resulting microbial nitrogen immobilisation (Box 5.1). It was also observed that after emergence, 1–10 per cent of the plants ceased to grow and, in the case of another 10–40 per cent, there was a substantial reduction in growth. Fresh hydrochar could contain highly phytotoxic compounds (Gajic and Koch 2012). More research is needed on the correlation between biochar properties, in combination with other factors, such as soil properties and crop yield responses.

One possible serious drawback of biochars with high specific surface area (Chapter 3) is the adsorption of herbicides and pesticides making these products inactive and potentially able to accumulate in soil where they can cause herbicide injury (Graber *et al.* 2012). The specific surface area, and thus adsorption capacity, usually increases with increasing pyrolysis temperature. On the other hand, a slower herbicide and pesticide transport downwards in the soil profile may provide longer retention time for microbial degradation of unwanted chemical compounds and also reduce the risk for pollution of water streams.

Biochar dose

Results regarding biochar application rates and effects on crop yields or biomass production are contradictory. Neither Jeffery *et al.* (2011) nor Biederman and Harpole (2013) found a correlation between biochar application rate and crop productivity. Biochar doses applied in European field trials range from 2.5 to 300 tonnes of biochar per hectare (Table 5.2; Figure 5.12a). In some cases, significant positive crop responses are already possible from a biochar dose of 10 tonnes ha^{-1} (Kihlberg *et al.* 2013; Hammond *et al.* 2013). There may be a risk that high biochar doses can be harmful for crop growth due to nutrient immobilisation or toxicity effects for example, but biochar doses of more than 40 to almost 90 tonnes ha^{-1} have been tested without negative crop yield effects (Figure 5.12a). Some biochars resulted in positive crop responses even at high doses of 60 (Vaccari *et al.* 2011) or 72 tonnes ha^{-1} (Karer *et al.* 2013). However, care is needed as high doses may certainly involve risks. Charcoal application of 300 tonnes ha^{-1} to a German silty soil, for instance, resulted in serious negative crop responses even three years after application. This was attributed to imbalances of nutrients and nitrogen limitations (Borchard *et al.* 2014; Figure 5.12a). A pot trial by Baronti *et al.* (2010) found a positive crop response up to a biochar application rate of 60 tonnes ha^{-1}, but again, resulted in lower crop productivity if rates were 100 or 120 tonnes ha^{-1}. In a Danish farmer's field, Skogens kol biochar did not affect yield except negatively at the highest dose of 50 tonnes ha^{-1} (unpublished results, DTU). The optimal biochar application dose should be derived for every biochar separately using e.g. a dose-response strategy, preferably under different soil conditions. For flash pyrolysis char

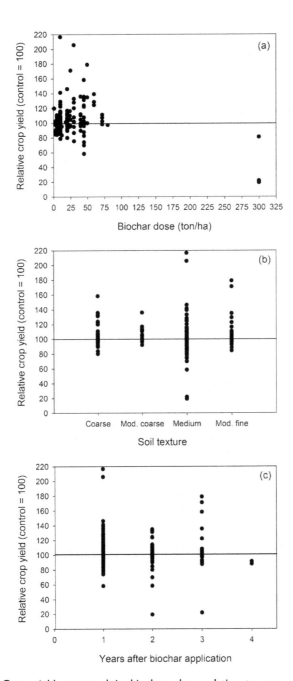

Figure 5.12 Crop yield measured in biochar plots relative to crop yield in control treatments shown per (a) biochar dose, (b) soil texture type and (c) number of years after biochar application.

Note: Mod. = moderately. Based on European field trials; see Table 5.2.

applied at a dose of 45 tonnes ha^{-1}, for instance, there was a negative yield effect in a silty soil, but not in a sandy soil (Borchard 2012). Thus, this specific biochar was either not suited for a silty soil, or the dose was already too high for this kind of soil.

In most field trials, biochar was only applied once, after which the effects were investigated in the consecutive years. Exceptions are the field trials in the Netherlands, where doses of five tonnes per ha of charcoal and activated carbon have been applied each year. So far, there is no indication that spreading biochar application over several years would be more beneficial than applying larger doses at one time. On the other hand, it could be speculated that adding smaller amounts of biochar as part of the yearly tillage operations might improve the longer-term response without short-term negative impacts on soil–crop interactions, as compared to one major application which might imply risks (see higher).

Time after biochar application

Short-term effects of biochar on soil and crops are expected to be different from long-term effects, as the effect of biochar can change over time due to the incorporation of biochar particles in the soil matrix and weathering and mineralisation processes.

Temporary short-term effects could be positive owing to the fact that biochar can act as a liming agent. In (sub)tropical acidic soils, rising soil pH can be beneficial to reduce aluminium toxicity. Raising soil pH can also be beneficial for the availability of nutrients to plants. Nutrient availability can also be increased by easily available nutrients in the ash fraction of biochar. Effects on pH and nutrient availability are expected to be less important in European agriculture where soils are younger and usually sufficiently limed and fertilised. pH effects have been demonstrated to decrease over time as was observed in several European field trials ranging from one cropping season to three years (Castaldi et al. 2011; Vaccari et al. 2011; Jones et al. 2012; Glaser et al. 2014). Likewise, effects of direct additions of easily available nutrients in the ash fraction of the biochar are expected to be short-lived as nutrients are utilised or leached from the soil (Jeffery et al. 2015).

One of the potential negative short-term effects is reduced soil mineral nitrogen contents as is indicated by some incubation (Novak et al. 2010; Bruun et al. 2011; Knowles et al. 2011; Ippolito et al. 2012; Nelissen et al. 2012, 2014b) and pot trials (Deenik et al. 2010; Nelissen et al. 2014b). There can be multiple causes, including adsorption of nutrients to the biochar surface and biotic nitrogen immobilisation caused by decomposition (mineralisation) of the labile fraction of the biochar (see Box 5.1; Chapter 4). Owing to the high carbon-to-nitrogen ratio of biochar, microorganisms will consume nitrogen in soil while mineralising the carbon in the unstable hydrocarbon fraction of biochar. This can be important for those biochars which have a large labile fraction. These reductions in soil mineral nitrogen could cause a reduction in crop growth or the need for more fertiliser.

Short-term effects could also be attributed to phytotoxic effects because of some chemical components that inhibit germination and growth or due to high salt levels. These effects might decrease over time. However, some chemical components in biochar could also cause positive effects. Graber *et al.* (2010) hypothesised that biochar compounds could either have a positive effect on microorganisms that improve plant growth or have a direct stimulating plant growth effect when used in small doses. In addition, phytotoxic compounds in soil can be inactivated by adsorption to biochar.

To avoid negative short-term effects, it could be more beneficial to apply biochar some months before crop sowing or planting. Application time was investigated in Denmark (unpublished results, Technical University of Denmark, DTU; Table 5.2), where four field trials were established side-by-side at the same location. Two were with RomChar and two with Skogens kol. For each of these biochar types, there was one field trial with an autumn application and one with a spring application. There were no crop yield effects; therefore no effect of the application date of biochar upon crop yield could be identified. However, it was observed that less soil mineral nitrogen was available in the Skogens kol plots in the weeks after the spring application, which resulted in a slightly lower plant height. This effect was not observed in the following year on the same plots suggesting that the biochar was already fully 'charged' with nitrogen or that the labile carbon fraction of the biochar was already decomposed so that no soil nitrogen was used anymore for this decomposition process.

From the other European field experiments, it can also be concluded that crop yields in the first year after spring application appear to be mostly equal to or higher than those in the controls. There are certain risks reported in field trials with flash pyrolysis biochar and hydrochar; however, these could (partly) be attributed to phytotoxic effects (Borchard *et al.* 2014; Gajic and Koch 2012).

On the other hand, partial oxidation over time can result in beneficial properties such as an increased capacity to retain nutrients. In this case, crop yield increases are expected some years after application. This weathering can also occur during storage as was experienced by Hammond *et al.* (2013) with Dalkeith biochar, leading to a 10 times higher capacity to exchange nutrients.

The maximum reported monitoring period for European field trials was four years (Table 5.2, Figure 5.12c). Overall, time effects appeared to be variable and inconsistent (Figure 5.12c), so that general conclusions cannot be drawn. More monitoring years are needed to see a consistent time effect and to exclude interaction with weather conditions and crop types grown.

Soil texture

From the non-European field and pot trials, mostly in (sub)tropical soils, one would expect a positive crop response in coarse and medium textured soils (Jeffery *et al.* 2011). The hypothesis is that this is attributed to better water retention in these soils after biochar addition, also resulting in improved nutrient retention and plant availability. Improved water holding capacity could be one of the most promising effects of biochar in temperate regions, especially in sandy soils and during prolonged periods of drought. This might also influence root depth and density, which in turn can lead to increased water and nitrogen uptake from the subsoil, reducing nitrogen leaching and possibly increasing crop yields. In some European trials, higher soil moisture contents and/or higher water holding capacity in biochar amended plots were indeed recorded (e.g. Karhu *et al.* 2011; Liu *et al.* 2012; Hammond *et al.* 2013; Karer *et al.* 2013, Bruun *et al.* 2014, Glaser *et al.* 2014; unpublished results DTU and Bioforsk). In other field experiments, however, such effects were not observed (e.g. Jones *et al.* 2012; Case *et al.* 2014) or not found to be statistically significant (Nelissen *et al.* 2015). A better understanding of the mechanisms behind the water holding capacity of biochar is needed to predict if higher water holding capacity is resulting in higher water and nutrient availability for plants and in which soil types this is the case.

Soil texture is not always easy to compare owing to different classification systems used and lack of data to convert soil texture from one classification system to another (Box 5.2). Nevertheless, we have tried to classify the soil texture of the European field trials in five broad classes: coarse, moderately coarse, medium, moderately fine and fine-textured soils (Soil Survey Division Staff 1993). Coarse-textured soils have a high sand content, whereas fine textured soils have a high clay content. Except for fine-textured soils, observations were made in all four of the other soil texture classes (Figure 5.12b). From the European data, displayed in Figure 5.12b, there was no effect of soil texture on crop response to biochar, and similarly not from the field trials in which the same biochar was used at different locations.

Crop type

A large variety of crop types have been tested in European biochar field trials, of which cereals (43 per cent of the cases) and maize (18 per cent of the cases) were most frequently cultivated (Figure 5.11). For 23 per cent of the observations with cereals, the crop yield was (statistically) significantly higher in the biochar plots compared to the control plots. In only 3 per cent of the cases, the effect was significantly negative. In 17 per cent of the observations with maize, a significant negative crop yield effect and no significant positive effects were observed. Some of the cases with decreased maize yields can be explained by the high biochar doses applied (45–300 tonnes ha^{-1}) and in one case the biochar type, as it was derived from flash pyrolysis (Borchard 2012). For the other crop types, there are not enough data to draw any conclusions. At the spinach site in Nottinghamshire, the crop growth was already highly optimised and biochar application may have disrupted a well-tuned growing system, with negative crop yields as a result (Hammond *et al.* 2013). Niggli and Schmidt (2010) showed 62 crop yield results of the same biochar application in vegetable gardens across Switzerland. Results were very variable. For most crops, negative (−10 per cent), neutral and positive (+10 per cent) crop yield responses were recorded. The most striking effect was seen in the cabbage family (broccoli, cauliflower, cabbage turnip, cabbage); of the nine observations, six were positive and no negative effects were seen. More research is needed to get better insights into whether some crop types react better to biochar addition or not or if other factors such as soil, weather conditions and farm management play a more important role.

Conclusions

In this chapter, we have tried to assess the potential of pure biochar additions to soils to improve crop productivity in Europe. The data gathered were mostly from north and northwest Europe. In most field experiments, there were no effects of pure biochar addition on crop productivity in the first years after application (effects beyond three or four years have not yet been recorded). In 12 per cent of the cases, there were statistically significant positive crop yield effects, and in 6 per cent of the cases, significant negative crop yield effects were reported. Cereals were the most studied crop, and in nearly one fourth of the observations, the grain yield was significantly higher in the biochar plots than in the control plots and almost no detrimental effects were noted. Positive crop yield effects may already take place at the relatively low application rate of 10 tonnes ha^{-1}. Very high biochar doses may lead to detrimental effects on crop productivity. The optimal application dose should be determined for every biochar type separately, ideally in different soil types and with respect to particular crops. In the short term, it seems that biochar has less potential to improve yields on already fertile European soils and in optimised farming systems than it does on the much less fertile soils in the (sub)tropics, where fewer inputs such as fertilisers and liming agents are employed. Carbon sequestration and (uncertain) longer term improvement of soil fertility seems, therefore, to be the main argument for biochar soil application in northwestern and northern Europe. However, this raises the question of whether the aim of carbon storage in soil is sufficient to persuade farmers to integrate biochar soil application on their farms. At the minimum, this would require an economic compensation for the investment made (Chapter 9). Therefore, further research needs to target site-specific problems such as drought and to work with biochars specifically tailored to these problems and with biochar in combination with (organic) fertilisers using regional resources (Glaser *et al.* 2014; Chapter 6).

It is not known if, how, and at what pace biochar characteristics will change over time and if this change would have any effect on soil 'health' and crop productivity. Long-term field experiments (a decade or longer) are therefore urgently needed and they can contribute to decision making processes on whether or not biochar is a legal and valuable soil amendment within European farming systems. Future research should not only focus on crop productivity but also on other effects such as the fate and effectiveness of herbicides and pesticides, effects on biodiversity and the general impact on the biological ecosystem, soil functions and services. So far, most research has focused on wood-based biochars produced in modified traditional kilns and retorts. Given the high demand and value of wood, biochar for wide-scale application is more likely to be made from other feedstock materials such as straw, slurry, residues from fermentation and digestates from biological processes, fibres, dung, sewage sludge, etc. (Glaser *et al.* 2014). Field experiments with these biochars, especially in southern and more drought prone countries in Europe, are needed. However, without a better fundamental knowledge of biochar mechanisms, the outcomes of such experiments remain to be based on trial and error. Presently, the variations in biochar feedstock composition, biochar production conditions, climate, soils and crop types and all the possible interactions that may occur between them are still so large that it is as yet impossible to provide solid evidence-based information to farmers regarding biochar application.

Acknowledgements

We would like to thank the following persons for their kind contributions to the European field trial overview: Jasmin Karer and Gerhard Soja (Austrian Institute of Technology), Gerard Ros (Nutrient Management Institute, the Netherlands), Nils Borchard (Rheinischen Friedrich-Wilhelms-Universität Bonn, Germany), Claudia Kammann (Geisenheim University, Germany), Kerstin Panten (Julius Kühn-Institut, Germany), Priit Tammeorg (University of Helsinki, Finland), and Sara Buckingham (Scotland's Rural College, Edinburgh, UK). The Interreg IVB North Sea Region Project 'Biochar: Climate saving soils' is acknowledged for financial support for the biochar ring trials.

Bibliography

Baronti, S., Alberti, G., Delle Vedove, G., Di Gennaro, F., Fellet, G., Genesio, L., Miglietta, F., Peressotti, A. and Vaccari, F.P. (2010). The biochar option to improve plant yields: first results from some field and pot experiments in Italy. *Italian Journal of Agronomy* 5: 3–11.

Bell, M.J. and Worrall, F. (2011). Charcoal addition to soils in NE England: a carbon sink with environmental co-benefits? *Science of Total Environment* 409: 1704–1714.

Biederman, L.A. and Harpole, W.S. (2013). Biochar and its effects on plant productivity and nutrient cycling: a meta-analysis. *Global Change Biology Bioenergy* 5: 202–214.

Borchard, N. (2012). Interaction of biochar (black carbon) with the soil matrix and its influence on soil functions. Dissertation, Institut für Nutzpflanzenwissenschaften und Ressourcenschutz (INRES), Bonn.

Borchard, N., Siemens, J., Ladd, B., Möller, A. and Amelung, W. (2014). Application of biochars to sandy and silty soil failed to increase maize yield under common agricultural practice. *Soil and Tillage Research* 144: 184–194.

Bruun, E.W., Müller-Stöver, D., Ambus, P. and Hauggaard-Nielsen, H. (2011). Application of biochar to soil and N_2O emissions: potential effects of blending fast-pyrolysis biochar with anaerobically digested slurry. *European Journal of Soil Science* 62: 581–589.

Bruun, E.W., Ambus, P., Egsgaard, H. and Hauggaard-Nielsen, H. (2012). Effects of slow and fast pyrolysis biochar on soil C and N turnover dynamics. *Soil Biology and Biochemistry* 46: 73–79.

Bruun, E.W., Petersen, C.T., Hansen, E., Holm, J.K. and Hauggaard-Nielsen, H. (2014).

Biochar amendment to coarse sandy subsoil improves root growth and increases water retention. *Soil Use and Management* 30(1): 109–118.

Bucheli, T.D., Bachmann, H.J., Blum, F., Bürge, D., Giger, R., Hilber, I., Keita, J., Leifeld, J. and Schmidt, H.-P. (2014). On the heterogeneity of biochar and consequences for its representative sampling. *Journal of Analytical and Applied Pyrolysis* 107: 25–30.

Burns, R.G. (2011). Soil biology and biochemistry citation classic X. *Soil Biology and Biochemistry* 43: 1619–1620.

Case, S.D.C., McNamara, N.P., Reay, D.S. and Whitaker, J. (2014). Can biochar reduce soil greenhouse gas emissions from a *Miscanthus* bioenergy crop? *Global Change Biology Bioenergy* 6(1): 76–89.

Castaldi, S., Riondino, M., Baronti, S., Esposito, F.R., Marzaioli, R., Rutigliano, F.A., Vaccari, F.P. and Miglietta, F. (2011). Impact of biochar application to a Mediterranean wheat crop on soil microbial activity and greenhouse gas fluxes. *Chemosphere* 85(9): 1464–1471.

Deenik, J.L., McClellan, T., Uehara, G., Antal, M.J. and Campbell, S (2010). Charcoal volatile matter content influences plant growth and soil nitrogen transformations. *Soil Science Society of America Journal* 74: 1259–1270.

Gajic, A. and Koch, H.J. (2012). Sugar beet (*Beta vulgaris* L.) growth reduction caused by hydrochar is related to nitrogen supply. *Journal of environmental quality* 41: 1–9.

Glaser, B., Wiedner, K., Seelig, S., Schmidt, H.-P. and Gerber, H. (2014). Biochar organic fertilizers from natural resources as substitute for mineral fertilizers. *Agronomy for Sustainable Development* 35(2):667–678.

Graber, E.R., Harel, Y.M., Kolton, M., Cytryn, E., Silber, A., David, D.R., Tsechansky, L., Borenshtein, M. and Elad, Y. (2010). Biochar impact on development and productivity of pepper and tomato grown in fertigated soilless media. *Plant and Soil* 337: 481–496.

Graber, E.R., Tsechansky, L., Gerstl, Z. and Lew, B. (2012). High surface area biochar negatively impacts herbicide efficacy. *Plant and Soil* 353: 95–106.

Graves, D. (2013). A comparison of methods to apply biochar into temperate soils. In N. Ladygina and F. Rineau (eds), *Biochar and Soil Biota*, 202–260. Boca Raton, FL: CRC Press.

Hammond, J., Shackley, S., Prendergast-Miller, M., Cook, J., Buckingham, S. and Pappa, V. (2013). Biochar field testing in the UK: outcomes and implications for use. *Carbon Management* 4(2): 159–170.

Husk, B. and Major, J. (2010). Commercial scale agricultural biochar field trial in Québec, Canada, over two years: effects of biochar on soil fertility, biology, crop productivity and quality. Available at www.researchgate.net/publication/273467636_Commercial_Scale_Agricultural_Biochar_Field_Trial_in_Quebec_Canada_Over_two_Years_Effects_of_Biochar_on_Soil_Fertility_Biology_and_Crop_Productivity_and_Quality.

Ippolito, J.A, Novak, J.M., Busscher, W.J., Ahmedna, M., Rehrah, D. and Watts, D.W. (2012). Switchgrass biochar effects two aridisols. *Journal of Environmental Quality* 41: 1123–1130.

Jeffery, S., Verheijen, F.G.A., van der Velde, M. and Bastos, A.C. (2011). A quantitative review of the effects of biochar application to soils on crop productivity using meta-analysis. *Agriculture, Ecosystems and Environment* 144: 175–187.

Jeffery, S., Bezemer, T.M, Cornelissen, G., Kuyper, T.W., Lehmann, J., Mommer, L., Sohi S., Van De Voorde, T.F.J., Wardle, D.A. and Van Groeningen, J.W. (2015). The way forward in biochar research: targeting trade-offs between the potential wins. *Global Change Biology Bioenergy* 7(1): 1–13.

Jones, D.L., Rousk, J., Edwards-Jones, G., DeLuca, T.H. and Murphy, D.V. (2012). Biochar-mediated changes in soil quality and plant growth in a three year field trial. *Soil Biology and Biochemistry* 45: 113–124.

Karer, J., Wimmer, B., Zehetner, F., Kloss, S. and Soja, G. (2013). Biochar application to temperate soils: effects on nutrient uptake and crop yield under field conditions. *Agricultural and Food Science* 22: 390–403.

Karhu, K., Mattila, T., Bergström, I. and Regina, K. (2011). Biochar addition to agricultural soil increased CH$_4$ uptake and water holding capacity: results from a short-term pilot field study. *Agriculture, Ecosystems and Environment* 140: 309–313.

Kihlberg, T., Storm, J.-O., Abrahamsson, E. and Hylander, L.D. (2013). A three year biochar field trial on a fertile soil in Sweden. In A. O'Toole and P. Tammeorg (eds), *Book of Abstracts: 2nd Nordic Biochar Seminar 'Towards a Carbon Negative Agriculture'*, 41. Helsinki: University of Helsinki.

Knowles, O.A., Robinson, B.H., Contangelo, A. and Clucas, L. (2011). Biochar for the mitigation of nitrate leaching from soil amended with biosolids. *Science of the Total Environment* 409: 3206–3210.

Kumari, K.G.I.D., Moldrup, P., Paradelo, M., Elsgaard, L., Hauggaard-Nielsen, H. and de Jonge, L.W. (2014). Effects of biochar on air and water permeability and colloid and phosphorus leaching in soils from a natural calcium carbonate gradient. *Journal of Environmental Quality* 43: 647–657.

Lehmann, J., Rillig, M.C., Thies, J., Masiello, C.A., Hockaday, W.C. and Crowley, D. (2011). Biochar effects on soil biota – a review. *Soil Biology and Biochemistry* 43: 1812–1836.

Leroy, B.L.M., Van den Bossche, A., De Neve, S., Reheul, D. and Moens, M. (2007). The quality of exogenous organic matter: Short-term influence on earthworm abundance. *European Journal of Soil Biology* 43: S196–S200.

Liu, J., Schulz, H., Brandl, S., Miehtke, H., Huwe, B. and Glaser, B. (2012). Short-term effect of biochar and compost on soil fertility and water status of a Dystric Cambisol in NE Germany under field conditions. *Journal of Plant Nutrition and Soil Science* 175(5): 698–707.

Lugato, E., Vaccari, F.P., Genesio, L., Baronti, S., Pozzi, A., Rack, M., Woods, J., Simonetti, G., Montanarella, L. and Miglietta, F. (2013). An energy-biochar chain involving biomass gasification and rice cultivation in Northern Italy. *Global Change Biology Bioenergy* 5: 192–201.

Major, J., Lehmann, J., Rondon, M. and Goodale, C. (2010). Fate of soil-applied black carbon: downward migration, leaching and soil respiration. *Global Change Biology* 16: 1366–1379.

Nelissen, V. (2013). Effect of biochar on soil functions and crop growth under northwestern European soil conditions. PhD thesis, Ghent University, Belgium.

Nelissen, V., Rütting, T., Huygens, D., Staelens, J., Ruysschaert, G. and Boeckx, P. (2012). Maize biochars accelerate short-term soil nitrogen dynamics in a loamy sand soil. *Soil Biology and Biochemistry* 55:20–27.

Nelissen, V., Rütting, T., Huygens, D., Ruysschaert, G. and Boeckx, P. (2014a). Temporal evolution of biochar's impact on soil nitrogen processes – a ^{15}N tracing study. *Global Change Biology Bioenergy*.

Nelissen, V., Ruysschaert, G., Müller-Stöver, D., Bodé, S., Cook, J., Ronsse, F., Shackley, S., Boeckx, P. and Hauggaard-Nielsen, H. (2014b). Short-term effect of feedstock and pyrolysis temperature on biochar characteristics, soil and crop response in temperate soils. *Agronomy* 4: 52–73 (doi:10.3390/agronomy4010052).

Nelissen, V., Saha, B.K., Ruysschaert, G. and Boeckx, P. (2014c). Effect of different biochar and fertilizer types on N$_2$O and NO emissions. *Soil Biology and Biochemistry* 70: 244–255.

Nelissen, V., Ruysschaert, G., Manka'Abusi, D., D'Hose, T., De Beuf, K., Al-Barri, B., Cornelis, W. and Boeckx, P. (2015). Impact of a woody biochar on properties of a sandy loam soil and spring barely during a two-year field experiment. *European Journal of Agronomy* 62: 65–78.

Niggli, C. and Schmidt, H.P. (2010). Hobby gardeners' experiments on biochar: preliminary evaluation. *Ithaka Journal* 2010(1): 339–344.

Novak, J.M., Busscher, W.J., Watts, D.W., Laird, D.A., Ahmedna, M.A. and Niandou, M.A.S. (2010). Short-term CO_2 mineralization after additions of biochar and switchgrass to a Typic Kandiudult. *Geoderma* 154: 281–288.

Ros, G.H., van Balen, D., de Haan, J.J., de Haas, M.J.G., Bussink, D.W. and Postma, R. (2013). *Biochar Application in Agriculture: Results from Field Trials in the Netherlands from 2010 to 2012.* Report 1348.N.13. Wageningen: Nutrient Management Institute.

Soil Survey Division Staff (1993). *Soil Survey Manual.* Washington, DC: US Government Printing Office.

Tammeorg, P., Simojoki, A., Mäkelä, P., Stoddard, F.L., Alakukku, L. and Helenius, J. (2014a). Biochar application to a fertile sandy clay loam in boreal conditions: effects on soil properties and yield formation of wheat, turnip rape and faba bean. *Plant and Soil* 374: 89–107.

Tammeorg, P., Simojoki, A., Mäkelä, P., Stoddard, F.L., Alakukku, L. and Helenius, J. (2014b). Short-term effects of biochar on soil properties and wheat yield formation with meat bone meal and inorganic fertilizer on a boreal loamy sand. *Agriculture, Ecosystems and Environment* 191: 108–116.

Vaccari, F.P., Baronti, S., Lugato, E., Genesio, L., Castaldi, S., Fornasier, F. and Miglietta, F. (2011). Biochar as a strategy to sequester carbon and increase yield in durum wheat. *European Journal of Agronomy* 34: 231–238.

Notes

1   Incubation of biochar with pure sand and a microbial inoculum in the lab with regular measurements of the emitted CO_2 to estimate the easily decomposable fraction of biochar.

2   Under the assumption of a soil bulk density of 1.3 tonnes of soil per m^3.

3   Using ISO 10694:1995.

4   Using the Dumas method.

5   Using ammonium lactate extract.

6   Using 1M KCl extract; ISO 10390.

7   Using 1M KCl extract; ISO 14256-2.

8   The mass of 1000 grains from a subsample.

9   Dry matter is obtained by putting the plant material in an oven at 65°C for at least 48 hours.

10   Using Dumas method; ISO 16634-1.

11   ICP-OES measurement; after destruction with HNO_3 and H_2O_2.

12   Method based on Leroy et al. (2007).

13   Note that the pH of biochar and soil pH was measured with different methods (water versus KCl extract). Nelissen (2013), however, measured the pH of the biochar with the same extract as for the soil samples and obtained a biochar pH of 8.6.

14   Here measured as the difference in water content between a pF of 1.8 and 4.2.

15   This includes co-called arbuscular mycorrhizal, actinomycetes, Gram-positive and Gram-negative bacteria.

Chapter 6

Combining biochar and organic amendments

*Claudia Kammann, Bruno Glaser
and Hans-Peter Schmidt*

The ancient enigma of 'terra preta': the cradle of biochar concepts

More than four centuries have passed since the Dominican pater Gaspar de Carvajal described grandiose and splendid, densely populated Indian garden cities and settlements in the Amazon basin. Now, four centuries later, findings of straight broad street residues and settlement places (Heckenberger *et al.* 2003, 2008), and in particular the patches of fertile Amazonian dark earths (ADE) or *terra preta* (Sombroek 1966; Glaser *et al.* 2004; Kern *et al.* 2004; Neves *et al.* 2003), finally reveal a true core in the Orellana's 450-year-old 'legend'. Thus, the Amazon basin was, up to 450 years before today, likely not the native, untouched green wilderness with just a few 'primitive' hunter-gatherer cultures that our textbooks taught us. In the 1960s, the famous Netherlands soil scientist Wim Sombroek published his stimulating book on soils in the Amazon basin, their properties, geology, genesis and associated vegetation (Sombroek 1966). He described the unusually fertile *terra preta* patches and islands that existed in an ocean of infertile, highly weathered soils (mostly Ferralsols, according to World Reference Base) within the Amazon basin (Sombroek 1966). For a long time, it was hotly debated whether ADE soils were of anthropogenic origin or the result of volcanic deposits, nutrient-rich sediments from the rivers of the Andes etc. (Glaser *et al.* 2004).

As Glaser *et al.* summarised in 2001:

> The enhanced fertility of 'terra preta' soils is expressed by higher levels of soil organic matter (SOM), nutrient holding capacity, and nutrients such as nitrogen, phosphorus, calcium and potassium, higher pH values and higher moisture-holding capacity than in the surrounding soils. According to local farmers, productivity on the 'terra preta' sites is much higher than on the surrounding poor soils.
>
> (Glaser *et al.* 2001: 37)

In all ADE soils (i.e. the continuum from *terra preta* to *terra mulata*, which is lighter, with enhanced fertile properties), large amounts of charred organic residues of up to 130 tonnes per hectare and one metre soil depth were found (Glaser *et al.* 2001). However, Glaser *et al.* (2001) have already suggested that these ADE soils were not formed by just putting pure charcoal/biochar into (or onto) agricultural soils but must have been amended repeatedly with fertilising organic material. In the meantime, it was scientifically proven that organic additions were involved from the start in *terra preta* genesis, either as nutrient-rich human kitchen and human/animal faecal waste, as forest litter debris (Glaser and Birk 2012; Glaser *et al.* 2001) or other forms. Although under dispute, it is believed that garbage (compost) piles surrounding the central village areas were occasionally burned to get rid of the waste; the smouldering likely created low-temperature biochars of all kinds (Kern *et al.* 2004; Neves *et al.* 2003), in addition to kitchen and ceramic making fire charcoal waste.

After the garden–city cultures vanished, ADE patches were quickly overgrown by rainforest. Thus, the *terra preta* sites would on average have received dry leaf litter amounts of 8.6Mg ha^{-1} every year for the last 450 years (Chave *et al.* 2010: meta-analysis of over 81 Amazon basin study sites). Despite of the lack of anthropogenic litter inputs, the natural litter input over centuries may have either increased or sustained its original fertility (how fertile the soils were during garden–city times is unknown, but 'demanding' crop remains have been unearthed). Either way, the close association of biochar with organics in the enigmatic *terra preta* soils, as we know them today, suggests a beneficial marriage of biochar and organic litter in general, or in particular with anthropogenic nutrient-rich organic waste streams which warrants a closer look.

Black carbon-rich soils: soil organic matter build-up beyond biochar?

Soils containing condensed aromatic black carbon (BC) are often richer in total organic carbon (OC) than soils without BC (i.e. in OC *besides* BC; Glaser and Amelung 2003). The close correlation of OC content with increased BC content has been observed in a variety of different soils of completely different texture and genesis, climate and with different organic matter inputs. For example, this close correlation occurs in Amazonian dark earths (Glaser *et al.* 2001), in temperate chernozem or chernozemic soils (Glaser and Amelung 2003; Rodionov *et al.* 2010), or in old charcoal-making sites in Germany compared to the surrounding calcareous or acidic forest soils (Borchard *et al.* 2014).

(Glaser *et al.* 2001) investigated five different *terra preta* sites (two sandy and three clayey) and found mean BC contents of 11g kg^{-1}, corresponding to 50Mg ha^{-1}, which is 70 times higher compared to surrounding soils, while the mean soil organic carbon (SOC) stocks were 'only' 2.7 times higher than in the respective adjacent soils. In the five *terra preta* sites that were investigated, a significantly larger percentage of the organic matter was aromatic BC (on average 20 per cent but up to 35 per cent versus 9 per cent in the topsoil; Glaser *et al.* 2001). Chernozem (Russian for 'black earth') in temperate climates is thought to have developed without direct human influence (i.e. waste nutrient inputs) under steppe vegetation; these BC-rich soils are among the most fertile agricultural soils in, for example, the Hildesheimer and Magdeburger Börde in Germany. In chernozem, BC also comprises the oldest and most recalcitrant carbon fraction (Rodionov et al. 2010; Schmidt et al. 1999). It can amount to 45 per cent of the total OC fraction, with up to 8g kg^{-1} of BC (Schmidt et al. 1999).

In old charcoal-making places, the biochar application over time was not associated with nutrient-rich human waste inputs. However, decade- to century-old former kiln sites with regrown forest, or surrounded by forest, have also received continuous leaf litter deposition. European forests are spotted with such ancient charcoal-making (kiln) sites that they are more readily found now via satellite imaging. These places have mostly been investigated for archaeological purposes rather than for effects of the BC contents on SOC accumulation. Borchard et al. (2014) investigated five former charcoal-making sites that had been unused for decades, located on acidic soils (in the German Siegerland region) or base-rich carbonaceous soils (in the German Eifel region) in western Germany, respectively. Both reported mean BC concentrations of 25g kg^{-1} and 10g kg^{-1} in the top 0–20cm of the charcoal-making sites, respectively, compared to nearly zero in the adjacent sites surrounding the kiln places. In both instances, increased BC contents significantly correlated with higher mean non-BC OC stocks at the kiln sites, which were 4.0 and 2.9 times larger than those of the corresponding adjacent non-kiln locations in the acidic or carbonaceous places in the Siegerland and Eifel regions.

A striking feature of the terra preta soils is their significantly elevated nutrient content, which is significantly correlated to the SOC content (Sombroek 1966). In terra preta, tremendous inputs of kitchen waste including fish and chicken bones, partly composted biomass residues, including forest litter and human excrement, would have contributed to the increase in nutrient stocks, in particular phosphorus (P) (Glaser and Birk 2012). However, the acidic German charcoal-making soil site revealed increased P concentrations as well as P stocks, despite the significantly lower bulk densities (Borchard et al. 2014). Here, far away from settlements, kitchen waste inputs are highly unlikely; leaf litter could have contributed over decades to this P stock build-up. The mechanisms behind the phenomenon clearly need further research.

OC stocks in soils are the result of two carbon (C) flows, namely the magnitude of the net input (gross primary production or GPP) and the net output by ecosystem respiration (Chapin et al. 2006). About 50 per cent of GPP is autotrophic plant respiration; the remainder, NPP (net primary productivity), is in mature ecosystems in balance with decomposition processes. Microbial decomposition of organic material by aerobic and anaerobic heterotrophic processes is the result of soil, climate, vegetation and biological soil activity (based on physico-chemical redox soil properties (Briones 2012). The increased OC stocks (C pool size) in terra preta soils besides the BC may thus be the result of changes in the C input–output flows; that is, theoretically increased C inputs (NPP increase due to more fertile soils), combined with reduced litter decomposition (reduced output), or reduced output only. A lower rate of C loss may be caused by a more efficient SOC use by soil microorganisms (Briones 2012; Chen *et al.* 2014). This may be due to external electron transport by biochar-mineral humic complexes or by an increased partitioning of catabolic C flows during decomposition into stable humic compound formation.

The natural reforestation that took place after the decline of the Amazon civilisation may have enhanced the NPP input in the ADE and non-ADE soils alike, up to the stage where litter C inputs balanced outputs under the respective soil conditions (equilibrium). Therefore, it is not likely that forming biochar-mineral-organic complexes (Lin *et al.* 2012) enhanced the retention of forest litter C in the long run, generating a higher SOC level in the ADE soils. However, during the transition from human inhabitation to secondary forest, increased NPP litter input through trees and understory plants may have contributed to the higher equilibrium SOC content we find today. For chernozem (prior to human cultivation), it is unknown if a positive input feedback loop existed, but we can assume that it did. Continued organic inputs are always necessary to prevent depletion of SOC stocks in agricultural soils. In Russian chernozem, more than 30 per cent of the non-BC OC compounds were lost during 55 years of bare fallow soil cultivation, compared to the non-BC OC content of mown steppe which followed a suppressed fire regime (Vasilyeva *et al.* 2011). The BC, however, was hardly diminished, and its quality (aromaticity) remained unchanged. Thus, the aromatic backbone of the biochar, its black carbon content, seems to be quite stable.

However, biochar itself is not completely inert to decomposition over decades (Abiven *et al.* 2011); in particular the heterocyclic nitrogen (N) of grassy biochar can be prone to solubilisation (de la Rosa and Knicker 2011). Lately, an extensive study on the export of the soluble fractions of charcoal, dissolved black carbon (DBC), within major waterways around the globe, indicated that DBC represents a major pathway of C export to the oceans (Jaffé *et al.* 2013). However, the functioning and effects of the soluble aromatic DBC fractions derived from charcoal (or biochar) are not yet fully understood. Mechanistic understanding will be crucial for the safe implementation of biochar as a tool for environmental management (Jaffé *et al.* 2013; Masiello and Louchouarn 2013), either to avoid unwanted negative effects, or to understand if the soluble DBC compounds may even be a *prerequisite* for the genesis of fertile black earth soils. Thus, the fate and functioning of DCB in soils will be a hot topic for future research.

In the two studied charcoal-making site regions in Germany, increased NPP could be excluded as a cause for the SOC build-up because the kiln places were small, and sparsely covered by the same vegetation as the surrounding forest (if covered at all; Borchard *et al.* 2014). However, the mechanisms of reduced decomposition (i.e. reduced C output) by more efficient microbial metabolism or increased partitioning in stable fractions, or a role of the soluble DBC in litter-C cycling, may have contributed to the non-BC stock increase. In *terra preta*, such mechanisms are suggested by the results of Liang *et al.* (2010). The authors observed that in *terra preta* soils in three different sites, fresh organic litter inputs decomposed 25 per cent slower. This resulted in a larger fraction of the fresh litter being incorporated into more stable fractions in the *terra preta*, as compared with the adjacent soils amended with the same litter; moreover, the pre-existing SOC was less severely primed[i] in *terra preta* (i.e. less native SOC was decomposed by the 'C-fuel' addition of fresh litter; Liang *et al.* 2010). A similar observation was made earlier by Glaser (1999) who compared five *terra preta* topsoils with the respective adjacent soils in a laboratory incubation. Steiner *et al.* (2008b) observed a higher respiratory efficiency in ADE soils; it is possible that the decomposer microbiology of BC soils shifts towards a higher fungi:bacteria ratio (Glaser and Birk 2012). Recent investigations in one of the kiln site soils also revealed a significant shift towards fungi via phospholipid fatty acid analyses (C. Bamminger and S. Marhan, personal communication; Kammann *et al.* 2015).

Taken together, there are indications of an improved stabilisation of SOM and plant nutrients that serve soil fertility in ancient biochar-containing soils. One large unknown remains: the size of the original biochar/charcoal inputs and their decomposability over centuries. How much of today's non-BC SOC has its origin in the biochar inputs? Is their transformed decomposition product probably DBC that was incorporated into complexes with humus-forming litter? And how much is actually derived from subsequently humified (increased) litter inputs? In other words, the crucial question is if biochar will, in the long term or under certain conditions, aid in the build-up of additional stable organic matter fractions besides those derived from biochar? In the best case, biochar-C 'investments' may enlarge beneficial SOC pool 'returns' via increased NPP, reduced decomposition, and larger partitioning into stable humus fractions during decomposition. However, this may vary greatly in different types of soil, litter inputs, biochars, climate etc. In the worst case, native SOC would be faster mineralised as observed with needle litter rich in phenolics over 10 years in boreal forest (Wardle *et al.* 2008).

To sequester C in SOM, for each quantity of carbon, stoichiometric quantities of nitrogen and other nutrients are required, which are bound in humic fractions (C:N ratio of 10–15). Therefore, to balance an increase in the SOC content in the top 30cm per hectare by 1 per cent, about 2000kg N is necessary. The C:N ratio of BC-rich soils is in most cases higher than that of adjacent soils (Glaser 2007; Glaser *et al.* 2004; Oguntunde *et al.* 2004). Although the exact cultural practices that led to *terra preta* genesis are not completely understood, it is evident that charcoal was not the sole ingredient. Rather, it was applied alongside organic wastes (Kern *et al.* 2004; Neves *et al.* 2003) rich in calcium (Ca) and P, for example, from fish bones and human faeces (Glaser and Birk 2012; Lehmann *et al.* 2004). It is therefore imperative that biochar should be co-applied together with nutrient-rich organic amendments if we do not want to wait for decades, and under forest, for the build-up of nutrient pools to balance the recalcitrant C inputs. In particular, composting, with its thermophilic heat rotting phase with up to 65–70°C, may thus accelerate the desired changes of properties in biochar (Fischer and Glaser 2012).

Biochar as composting additive: a way forward?

Aerobic composting

Composting of organic wastes is a bio-oxidative process involving the mineralisation and partial humification of organic matter, leading to a stabilised final product with certain humic properties which should be free of phytotoxicity and pathogens (Bernal *et al.* 2009). Composting releases part of the carbon fixed during photosynthesis back into the atmosphere, while part of the input carbon is transferred into more stable humic compounds (Amlinger *et al.* 2005, 2008; Bernal *et al.* 2009; Fischer and Glaser 2012). Composting techniques may vary from simple approaches such as waste collection in rotting heaps around settlements to sophisticated industrial-scale facilities, and from small-scale household or open windrow composting to large encapsulated composting systems, including waste air treatment (NH_3 stripping) and mechanical biological waste treatment. Input feedstock materials may be green waste, biowaste, sewage sludge or animal manures. New livestock production systems, based on intensification in large farms, produce huge amounts of manures and slurries, without enough agricultural land for their direct application as fertilisers (Bernal *et al.* 2009). The same is true for sludge from wastewater treatment plants close to large cities in China, for example (Chen *et al.* 2010; Hua *et al.* 2009, 2012; Park *et al.* 2011). Here, composting is increasingly considered a beneficial pathway for recycling surplus manures as a stabilised and sanitised end product for agriculture. However, this end product needs to be a high quality substrate or soil conditioner/fertiliser to overcome the costs associated with its production (Bernal *et al.* 2009).

Biochar use during composting can be a strategy to achieve high quality compost end products for agricultural or horticultural use with added economic value. However, simply adding biochar to compost does not guarantee that badly made composts (comparable to 'raw humus') will turn into a miraculous fertile substrate. The main principles of quality composting are:

- To start with a balanced C/N of the feedstock blend of around 35.
- To add sufficient amounts of clay (~5 per cent vol.) and rock powder (>1 per cent weight) to promote aggregate formation.
- To avoid anaerobic conditions for more than one to three days to minimise putrescence and prevent detrimental shifts in the decomposing and humifying microbial community.
- To achieve this, compost windrows should not have diameters of more than 3m (see Figure 6.1).
- Windrows should be turned at least every three days during the heat rotting phase for oxygenation (Figure 6.1).
- Low bulk density materials such as lignin-rich green clippings, sawdust or biochar (also known as 'bulking agents') should be added for aeration and liquid absorption.
- The water content should be controlled and remain at about 40 per cent.

Figure 6.1 Example of optimal windrow size permitting highly aerobic composting with regular turning.

Source: Ithaka Institute.

Actually only few composts are produced following these guidelines. The majority of composts are piled on huge heaps (up to 5 m high, covering hundreds of m²), are turned only two or three times and rot partly or completely anaerobically for half a year or longer. Odours of butyric acid and ammonia are typical for these types of compost piles, with high losses of carbon, nitrogen and nitrous oxide and methane (greenhouse gas or GHG) emissions.

To illustrate the importance of the intrinsic compost quality we report an experimental example: Kammann *et al.* (2010), grew radish in 500ml pots where a poor coarse sand was mixed 1:1 (vol/vol) with either one of two different commercial horticultural peat substrates (ED73 and Compo Sana), or a German commercial green waste compost or a high quality Austrian compost produced for organic farming (Bionica), and compared to a control where the sandy substrate was not blended with organics. Within the different organic amendments, woody biochar (produced at about 600°C by Pyreg GmbH, Germany) was added at different rates or blended with the greenwaste compost before (BC compost) or after composting (compost plus BC; Bionica compost always plus BC). The pots were wick-watered from water reservoirs containing either pure water or a full fertiliser nutrient solution. The differences between the German commercial greenwaste compost and the Austrian Bionica compost were considerable, and it was only to some extent ameliorated by providing nutrients (Figure 6.2, top and bottom). Adding biochar to the low-quality compost during composting increased the (very low) radish yield by about 20 per cent compared to the compost alone (Figure 6.2 top), but yields were tiny compared to a high-quality compost without biochar (Bionica). In the latter, just mixing in biochar did not further boost radish growth.

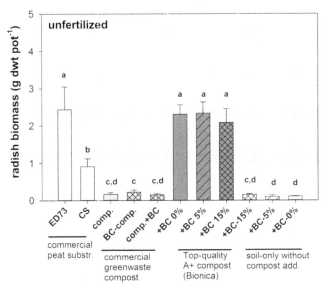

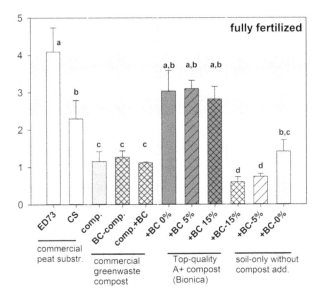

Figure 6.2 Mean radish yield + standard deviation (n = 4) in poor sandy, wick-watered substrate mixed 1:1 with different organic amendments. The grey shaded area indicates the range of responses obtained with commercial peat substrates. Hatched patterns indicate biochar addition. Different letters within a figure indicate significant differences between treatments (One-way ANOVA; $p < 0.05$).

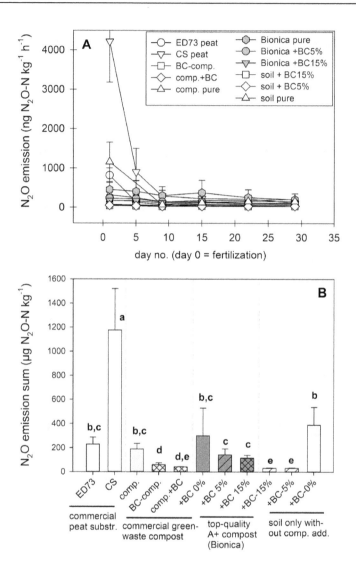

Figure 6.3 Mean N$_2$O emissions + standard deviation (n = 4) from fertilised soil mixtures used for radish growth (Figure 6.2), adjusted to 65 per cent of their maximum WHC in an incubation study over time (A) or as cumulative emissions (B).

However, the addition of the woody biochar to the growth media tremendously reduced N$_2$O (and also CO$_2$) emissions, compared to the peat substrates or respective controls (Figure 6.3). Therefore, biochar addition significantly reduced the GHG 'costs' of radish production, when used together with compost (Kammann *et al.* 2010). The reduction of GHG emissions may easily be achieved with a large number of suitable woody biochars by adding small amounts (<5 per cent) to horticultural substrates. Meanwhile a first study was published reporting reduced N_2O emissions during composting by biochar addition (Wang *et al.* 2013). In a meta-analysis, Cayuela *et al.* (2014; see also Van Zwieten *et al.* 2015) showed that particularly woody biochars can significantly reduce N_2O emissions when added to soils; the same effect may be expected in compost.

Biochar as a composting additive

In *terra preta*, charcoal particles are in contact with decomposable organic substances, dissolved OC, N and other nutrients and microbial residues, and with living plant roots, their symbionts and other microorganisms. These turnover processes involving organics and minerals may be a prerequisite for biochar's beneficial effects. Therefore, the combined use of biochar and compost is a logical step forward (Fischer and Glaser 2012; Liu *et al.* 2012; Prost *et al.* 2013; Steiner *et al.* 2008a, 2010, 2011). There are principally two options: first, the combination of mature compost or manure mixed with biochar, as described by Liu *et al.* (2012) in a field experiment in eastern Germany on sandy soil, or in a greenhouse experiment by Schulz and Glaser in 2012; and second, the option to co-compost biochar by adding it as a bulking agent (Dias *et al.* 2010; Steiner *et al.* 2011), for heavy metal remediation (Hua *et al.* 2012) or N retention (Hua *et al.* 2009; Prost *et al.* 2013; Steiner *et al.* 2010). Adding biochar during composting or pre-composting treatments (e.g. bokashi or potpourri in Japan) has been practised in the Asian-pacific region since pre-historic times (Wiedner and Glaser 2015). Therefore, unsurprisingly, some of the first systematic investigations of biochar as an additive for composting originate in Asia (Chen *et al.* 2010; Hua *et al.* 2009, 2012; Jindo *et al.* 2012). These techniques are very well known, easy to use and have likely been in use in Asian cultures for thousands of years. This contradicts current approved patents (Boettcher *et al.* 2009; Wolf and Wedig 2007 in the US) on 'biological activation' of biochar or the generation of *terra preta*-like planting substrates.

Using biochar during composting is a special case of using biochar blends with organic materials. Others may include the post production mixture of biochar to composts, co-application of biochar with animal manures and slurries (Schmidt 2011), or the use of biochar in animal housing with subsequent land application. There is only a small body of literature regarding biochar and composting (see also Fischer and Glaser 2012; Schulz *et al.* 2013). 'Best recipes' can be found in some practical guides (Schmidt 2011; Taylor 2010) and are used by a growing number of commercial enterprises that sell biochar-based fertilisers, soil substrates, animal feed, liquid manure additives (e.g. Carbon Gold, Swiss Biochar, Sonnenerde, Cool Planet, Seek Fertiliser). Interestingly, not much scientific research (with the exception of Joseph *et al.* 2013) has been done to investigate the function and performance of these new biochar-based products already on the market. This is clearly a future challenge that needs to be jointly addressed by scientists and commercial enterprises alike.

The following questions will be addressed in this section:

- From a 'biochar-centred' mechanistic point of view, how do biochar surfaces and properties change with composting as a post-treatment? And do composted biochars subsequently have plant-growth promoting properties that are superior to those of non-composted biochar?
- Can biochar addition provide benefits for the composting process as such to produce 'better' (economically and ecologically more valuable) composts?
- Can biochar addition during composting reduce GHG emissions associated with composting, namely N_2O and NH_3, improving N retention?
- Will a biochar-compost have properties that are superior to the pure compost without biochar addition, in particular for plant growth when biochar-compost is used as soil additive?
- What concentrations of what kinds of biochars have been used that resulted in significant, intended effects? That is, how low can the biochar dosage be to achieve certain goals? The latter question has hardly been addressed at all so far.

All of these questions are important to assess and understand the effects observed in field studies using biochar-organic blends or biochar composts; the first evidence in literature is provided below.

First, how do biochar properties change with composting? Not many studies have addressed this issue directly (see Prost *et al.* 2013; Schmidt *et al.* 2014; Kammann *et al.* 2015; Fischer and Glaser 2012), but findings from biochar ageing experiments, retrieval of aged biochar from soils, or comparative physicochemical and biological investigations on biochar-mineral complexes provide first insights.

Biochar ageing (understood here to be the development of more functional groups on the biochar surfaces by oxidative processes) is likely accelerated during composting, in particular due to the high temperature during the heat rotting phase (~60–70°C, mostly over two to four weeks), the active oxygenation and the highly biologically active environment (Briones 2012; Cheng *et al.* 2006; Joseph *et al.* 2013; Prost *et al.* 2013). However, the BC content (i.e. the aromatic backbone) remained quantitatively unaltered in commercial charcoal and gasifier biochar in a 130-day composting experiment (Prost *et al.* 2013). Instead, the biochar started to carry considerable amounts of dissolved OC and various nutrients during composting (Prost *et al.* 2013). The same was observed by Kammann *et al.* (2015) where the composted woody biochar contained up to 2.2g kg^{-1} N (retrieved by electro-ultra-filtration) and up to 5.3g kg^{-1} N when retrieved by repeated washings, mostly in the form of nitrate (compared to initially zero); most of it was strongly bound to the biochar. Therefore, not all was removed by normal analytical procedures such as KCl extraction. This strong adsorption of N may even lead to the wrong conclusion that there was only low N retention by biochar addition.

It is unclear if the nutrient enrichment observed in biochar during composting is simply due to the uptake of liquids loaded with soluble nutrients into smaller micro and larger nanopores of biochar, or if the nutrients are precipitated at, or react with, the biochars' surfaces. It is likely that all these mechanisms play a role (as assumed by Prost *et al.* 2013). Sarkhot *et al.* (2012) found that 8.3 per cent more N than initially present was found per gram of hardwood chip biochar produced at 300°C, after shaking it only 24 hours with 100 ml of filtered dairy manure effluent, centrifuging, filtering and oven drying the biochar. Another potential mechanism is the bounding of water shelled ions via weak hydrogen bonding to the biochar surfaces within the porous system as described by Conte *et al.* (2014), discussed in Kammann *et al.* (2015).

Surface reactions and interactions with minerals are likely to become more important over time in contact with composting materials. It has been demonstrated that biochar organo-mineral complexes (BMCs) can form during biochar ageing in the soil in the presence of organic matter (decomposition) and minerals, but also by post treatment with minerals and organic matter (Chia *et al.* 2012; Joseph *et al.* 2013; Lin *et al.* 2012, 2013). S. Joseph *et al.* (2013) use the term 'enhanced biochars' to describe these post-treated biochars, and they recently reported that significant surface chemistry changes can occur when biochar reacts with minerals and nutrients, even without composting. After four weeks of incubation of a wheat straw biochar (~25 per cent) with clay minerals (bentonite, 5 per cent), urea, KCl and mono-ammonium phosphate in sealed containers under ambient conditions, they undertook examinations of the micro- and nano-structure of the original and the incubated biochar via infrared spectroscopy (FTIR), X-ray photoelectron spectroscopy, scanning microscopy and transmission electron microscopy (TEM). It was apparent that the carbon lattice of the biochar had reacted with the clay minerals and chemicals, and that there was a migration of soluble cations and anions into the biochar pores. The most interesting feature, however, was that this enhanced biochar provided a better rice growth. In a field study near Nanjing, China, Joseph *et al.* (2013) with his Chinese colleagues, G. Pan *et al.*, observed, with an economically interesting 'tiny' dose of biochar (enhanced-biochar fertiliser), significantly promoted grain yields by +39 per cent at overall reduced rates of N fertilisation (168kg N ha^{-1}) compared to conventional chemical fertilisation (210kg N ha^{-1}, i.e. −20 per cent). Thus, the N use efficiency was significantly improved, as was the N uptake into grains over that into shoots in the rice plants grown with the biochar fertiliser (Joseph *et al.* 2013).

BMCs are thought to act as electron shuttles, allowing external electron transfer between microbes for a multitude of microbial physiological processes including anaerobic respiration, acting as redox reactors in soils together with metals and clay minerals (Briones 2012; Joseph *et al.* 2013). It is likely that such BMCs will form in the compost, in particular when high quality composts are produced where rock powder and clay have been added. Since high quality composts contain con-siderable amounts of stable humic substances (Bernal *et al.* 2009; Fischer and Glaser 2012), the BMCs may be embedded in, or connected to, stable humic substances, thus functioning in soils comparable to *terra preta* (Briones 2012; Joseph *et al.* 2013). Theoretically, these enhanced biochars should provide their best effects in promoting plant growth in highly weathered or sandy soils, and this seems to be the case (see Schulz *et al.* 2013; Figure 6.4).

Do composted biochars have growth-promoting features that the non-composted biochars do not have? There is evidence for such a positive change, besides the one mentioned above. To investigate the effect of composting ('ageing') on biochar performance, a completely randomised full-factorial pot experiment was carried out in the greenhouse using the pseudo-cereal *Chenopodium quinoa* (Kammann *et al.* 2015). Two factors in the study were (I) type of biochar added (composted or untreated biochar, or no biochar as a control) and (II) addition of compost. The growth medium was poor loamy sand. Biochars and compost were added at a rate of 2 per cent (w/w) to the soil. From the start, there was a considerable difference between the growth of quinoa plants with untreated versus composted biochar. The untreated biochar produced the often-observed reduction in plant growth, compared to the unamended control. The reduction was alleviated to a certain extent by the addition of compost. In contrast, the co-composted biochar always significantly stimulated quinoa growth (roots, shoots, inflorescences). This relative stimulation was greatest when the growth conditions were the most unfavourable (no compost addition: 305 per cent versus − 61 per cent with composted versus untreated biochar, compared to the control). The stimulation was least pronounced when the growth conditions were most favourable (compost addition and high fertilisation). It turned out that the composted biochar was strongly enriched with a broad spectrum of nutrients that were partly fixed to the biochar with greater strength, in particular N, P, K and Ca, but not Mg (Kammann *et al.* 2015). Besides the nutrient-loading effects, microbial colonisation with beneficial plant-growth promoting microbes (as observed by Anderson *et al.* 2011) may also have provided beneficial effects on plant growth, though this was not investigated in the above-mentioned trial.

With respect to optimising the biochar amount in biochar composts or *terra preta* substrates, Schulz *et al.* (2013) conducted a greenhouse study in which both total application amount of biochar compost and the biochar content of the applied products were varied (Figure 6.4). Their results clearly indicate that oat yields were greater the higher the amount of applied biochar-compost and the higher the biochar content in the biochar compost was (up to 50 per cent in the final biochar compost mixture); this can be summarised as 'the more biochar in the compost the better' (Figure 6.4). In loamy soil, the effect was not as clear, but the tendency was the same as in the sandy substrate (Figure 6.4, top and bottom).

However, the growth-promoting effects of a biochar-compost mixture over that of the same weight of pure compost will largely depend on the question if, per unit of weight, the biochar will acquire more nutrients and offer them in a more suitable plant-available form than delivered by the same amount of slowly mineralising compost. If the properties of the biochar-compost mix are essentially the same than those of the pure compost no difference in plant growth will be visible (compare Figure 2 in Kammann *et al.* 2015). Here, longer-term effects of, for example, nitrate retention against leaching may play the largest role, if the co-composted biochar within the compost may help to intercept the (compost-mineralised) nitrate and keep it in the plant rooting zone (Mengel 2013).

Second, what benefits can biochar provide for the composting process itself? Adding biochar to compost influences the composting process in various ways. Owing to its high water holding capacity, the humidity of the compost can be

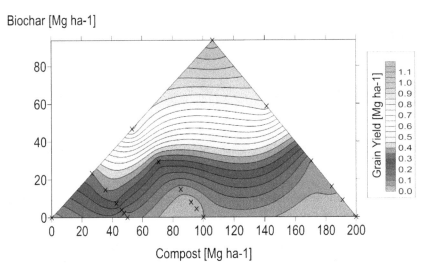

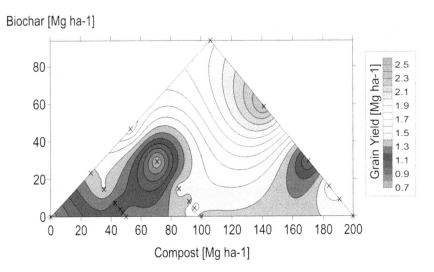

Figure 6.4 Plant biomass (oats, *Avena sativa*, greyshade code) in a pot experiment on sandy (top) and loamy (bottom) substrate as function of the amount of applied biochar compost (x-axis) and the biochar ratio of total applied amount (y-axis).

Source: Schulz *et al.* (2013).

better kept in an optimum range if regularly watered. Biochar's high porosity helps to keep the oxygen level higher for a longer time, enabling an aerobic microbial milieu. Biochar can adsorb toxins that may be detrimental to decomposing and aggregating bacteria and fungi. It can reduce ammonia volatilisation and N_2O emissions, help retain N and eventually promote the formation of stable humus compounds, as detailed below. However, the relative contribution of biochar addition during composting to different tasks, such as increasing the N retention (adsorptive capacity) versus the aeration ('bulking agent' capacity), will probably depend on the particle size of the added biochar. To our knowledge, no studies on the effects of different particle size of biochar in composting exist.

Composting of wet nutrient-rich materials inevitably requires the addition of bulking and adsorbing agents with a high C/N ratio such as wood chips, grain husks, cereal straw, cotton waste or other nutrient-poor dry porous materials (Amlinger et al. 2003; Steiner et al. 2011). These are added for aeration purposes as well as for achieving the desired initial C:N ratio of about 35. Biochar proved to be an excellent compost additive in recent studies (Dias et al. 2010; Steiner et al. 2010, 2011; Wang et al. 2013). Moreover, in composted pig sludge, the mobility of heavy metals such as copper and zinc (Cu and Zn), or contaminants such as polycyclic aromatic hydrocarbons (PAH) was significantly reduced with 7 per cent (wt/wt) addition of bamboo charcoal (Hua et al. 2009, 2012). In subsequent plant growth studies with ryegrass (*Lolium perenne*) or festuce (*Festuca arundinacea*) in two different soils, the addition of 40 per cent of the final biochar-compost to the soil (as a control) significantly reduced Cu, Zn and PAH accumulation in plants, compared to the pure compost alone; the BC sludge compost addition increased plant biomass growth by up to 16 per cent and 27 per cent over that with pure compost in fescue and rye, respectively (Hua et al. 2012).

Bamboo biochar addition during composting of sludge and rapeseed mark reduced ammonia volatilisation and hence NH_3 loss (Chen et al. 2010; Hua et al. 2009, 2012); the same was observed with pine chip biochar (Steiner et al. 2010), or with acidified pine chip and peanut hull biochars (Doydora et al. 2011) during poultry litter composting. However, not all biochars are able to adsorb ammonia to the same extent; low temperature (<400°C) grassy or aged biochars (e.g. Taghizadeh-Toosi et al. 2012) usually have larger amounts of functional groups for NH_3 adsorption. Chen et al. (2014) reported that in a biochar that is neutral or slightly acidic, NH_3 sorption is dominated by the biochars' surface properties such as surface area and sorption capacity; but at lower or higher pH values, the pH effect dominates the sorption characteristics (Chen et al. 2013).

In an unreplicated study with one compost windrow with biochar, and one without biochar addition (60kg of bamboo biochar, 600°C, pH_{H2O} 10.4 and BET surface 359m^2 g^{-1}; amended to 1200 kg of pig manure and 800 kg wood chips and sawdust), Wang et al. (2013) observed significantly lower cumulative N_2O emissions during composting by 26 per cent with this comparably tiny biochar addition. Interestingly, the authors reported a significant increase in the gene expression of nitrous oxide reductase (nosZ), coding for the last enzyme in the denitrifier reduction chain (from N_2O to N_2) in the presence of biochar, which they attributed to the pH increase and lowered moisture due to biochar addition (Wang et al. 2013). In addition, Harter et al. (2014) also observed in a fully replicated soil incubation experiment at field capacity that the gene expression and activity of the nosZ gene was significantly enhanced with biochar addition. When untreated hardwood biochar chips were added to an aerobic manure compost to constitute 30 per cent of the volume (vol/vol; = 11 per cent by weight), there were significant reductions in N_2O emissions across a range of water holding capacities of 60 per cent to 90 per cent of the maximum water holding capacity. Where biochar was added to manure before composting (constituting 11 per cent wt/wt in the end product) and co-composting was undertaken, the N_2O emissions were significantly reduced at 60 per cent water holding capacity (Kammann and Schmidt, unpublished data).

Schimmelpfennig et al. (2014) did not compost pig slurry, but added the slurry to clay loam grassland soil, amended with equal amounts of carbon as feedstock (*Miscanthus* straw), as hydrochar or as biochar, respectively. The mixtures were incubated over three months and GHG fluxes were measured weekly. The feedstock and hydrochar decomposed quickly, while biochar did not; N_2O emissions were reduced by 50 per cent with biochar compared to the unamended control, but tremendously increased with the un-carbonised feedstock. Interestingly, the biochar-amended manure-fertilised soil was the only treatment that delivered +30 per cent growth increase in *Lolium perenne* planted subsequently after the incubation; the most likely cause of improvement was increased nitrate retention by reduced gaseous N losses (Schimmelpfennig et al. 2014).

In summary, biochar as composting additive deserves future research and development. Biochars with a large surface area (BET) and high amounts of functional groups (e.g. grassy or bamboo biochar) should be selected and additionally acidified if too alkaline. Using these as bulking agents during composting will likely reduce NH_3 losses and N_2O emissions, providing economic benefits by N retention and improved end product quality. Moreover, many of the above-cited papers report observations of odour reduction when biochar was added during the composting; we noted the same effect when biochar was mixed with pig manure prior to application in a field study (Schimmelpfennig et al. 2014). In addition to economic goals such as N retention, acceptance of conventional compost facilities (usually not producing aerobic high-quality compost) by neighbouring citizens may promote biochar use in such facilities simply due to odour reduction purposes.

From a global change mitigation point of view, accelerated formation of stable humic substances might be the most interesting property of biochar addition during composting. Jindo et al. (2012) added 2 per cent of an oak biochar produced in a traditional Japanese kiln (400–600°C) to a mixture of poultry manure, rice husk and apple pomace during composting. The authors observed that biochar addition reduced the water-soluble carbon content of the product by 30 per cent, increased the microbial diversity of fungi and, most notably, increased the formation of humic acids over fulvic acids (Jindo *et al.* 2012). Increased humic acid formation was also observed by Dias *et al.* (2010). There is the possibility that biochar addition during composting reduces the overall (non-BC) organic-carbon losses, as reported by Fischer and Glaser (2012). Thus, a central future question is if using biochar composts in agriculture could provide a 'carbon build-up return of investment' (i.e. a retention of more organic material), in particular the desirable stable humus fraction in the compost and subsequently in the soil, *beyond* the pure biochar-C deposition. The evidence from old soils that contain charcoal and decomposed organic wastes suggests that such a mechanism may exist as discussed earlier.

However, this C preservation from labile organic substrates can only occur when biochar particles come into close contact with organics and minerals and are exposed to microbial and plant root activity, as likely happened during *terra preta* formation (Briones 2012; Glaser and Birk 2012). Putting pure, untreated biochar into agricultural soils that are often fallow or bare half of the year may, or more likely rather may not, have the same effect – but certainly, much longer time scales will be required.

Biochar composts as growing media

In 2011, a fully replicated (n = 3) biochar-compost trial was undertaken at the Ithaka Institute in Switzerland. A feedstock blend of cow manure (75 per cent vol), horse manure (9 per cent vol), chicken manure (1.5 per cent vol), straw (4.5 per cent vol), clay soil (9 per cent vol) and rock powder (1 per cent vol) was composted with either 20 vol per cent biochar or without biochar addition (pure compost) as control. The study was set up at an industrial scale using 15m³ piles per replicate and was executed with the composting technology as described above for aerobic quality composting.

The biochar compost quickly reached a higher temperature during the intensive heat-rotting phase, showed higher oxygen concentrations during post rotting and completed the entire rotting process earlier than the pure-compost control. The biochar compost had lower ammonia and higher nitrate concentrations during the composting. Total C and N losses were reduced by approximately 20 per cent and 10 per cent in the biochar compost compared to the non-amended compost, respectively.

Both substrates were subjected to multiple further trials to investigate the effect of biochar in the final substrate. An earthworm avoidance test resulted in a non-significant preference of the finished biochar compost substrate compared to the pure compost. Pot studies with seven different plant species at mixing rates of 50 per cent (vol/vol) with coarse sand showed a 10 per cent increase in growth in the biochar compost and 60 per cent increase in fruit yield, compared to pure compost and to commercial peat substrate, respectively. The pure biochar compost emitted significantly less N_2O compared to the pure compost at 60 per cent water holding capacity (which is in the upper range for unimpeded growth for horticultural plants).

Overall, the biochar addition during composting improved the quality of the final product and improved the nutrient retention efficiency. Comparable substrates are currently produced by several companies in Switzerland and Austria, as potting substrates for urban gardening, and as planting substrate for trees and special cultures where the substrate is given in a concentrated form into the plant hole under the roots.

In horticultural studies concerning reduced peat use or peat replacement, Zwart and Kim (2012) amended peat moss potting substrate with 5%, 10% and 20% of pure, untreated (non-composted) biochar. Here, the authors reported significant reductions in stem lesions caused by two host-specific *Phytophthora* species (fungal pathogens) in *Acer* and *Quercus* seedlings grown with 5% biochar. However, adding more biochar did not linearly lead to better results, as was also observed with even lower pure biochar amendments to horticultural substrates (Elad *et al.* 2010; Graber *et al.* 2010; Meller-Harel *et al.* 2012). Rather, with larger amounts of pure, untreated biochar, the opposite effect was found – larger biochar amendment resulted in reduced plant fitness – suggesting that an optimum dosage exists for stimulating plant defences (Elad *et al.* 2010; Graber *et al.* 2010; Meller-Harel *et al.* 2012; Zwart and Kim 2012). The most likely explanation for the U-shaped dose-response curves observed with pure untreated, as opposed to co-composted, biochars ('the more the better characteristic'; see Schulz *et al.* 2013) is the rich cocktail of hundreds of different volatile organic compounds in the production-fresh biochar, which likely provide hormesis-like phytohormonal effects (Graber *et al.* 2010; Meller-Harel *et al.* 2012). These compounds will likely have vanished owing to microbial decomposition after co-composting the biochar, which explains the different concentration-dependent effects. Another explanation might be the lack of beneficial microorganisms in untreated biochar, as the peat moss or coconut fibre turf horticultural substrates were probably less biologically active than in compost. Thus biochar composts may have potential for reducing the use of peat in at least the private sector of horticulture, where customers are more inclined to pay for peat-free planting substrates.

Biochar as additive in soil-less growth media

There are two ways that biochar may serve as a beneficial additive in horticultural growth substrates and potting mixtures. First, added in small doses of only a few percentages or even below 1 per cent, and secondly, used as a peat substitute as pure biochar or biochar compost, or perhaps even composted hydrochar (Busch *et al.* 2013).

It has been demonstrated by researchers at the Volcani Center (Israel) that biochars produced from greenhouse waste and added in small amounts of 1 to 5 per cent to soil-less growth media (a coconut fibre-tuff potting mixture) can significantly improve pepper and tomato growth (Graber *et al.* 2010). Organic solvent extracts of the biochar, which was used in the planting substrate, contained several organic compounds such as n-alkanoic acids, hydroxy and acetoxy acids, benzoic acids, diols, triols, and phenols. They found that the rhizosphere of biochar-amended pepper plants had significantly greater abundances of culturable microbes belonging to groups with known beneficial effects on plant growth (Graber *et al.* 2010). The authors concluded that the observed growth stimulation was either due to the increase in beneficial rhizobiota or a chemical 'hormesis' effect of low doses of otherwise biocidal chemicals on the biochar (Graber *et al.* 2010; Kolton *et al.* 2011). In other studies, researchers described that their biochar, added to the growth medium (1 and 3 per cent), induced and improved systemic resistance in pepper and tomato to foliar fungal pathogens (*Botrytis cinerea*, grey mould; or *Leveillula taurica*, powdery mildew), as well as to the broad mite pest (*Polyphagotarsonemus latus*) in pepper (Elad *et al.* 2010). Moreover, Meller-Harel *et al.* (2012) showed that 1 and 3 per cent biochar amendment to potting substrates induced systemic resistance to three fungal pathogens with different infection strategies in strawberry by up-regulating the defence-related gene expression.

The 'optimum' responses observed in many studies argues for a hormesis-like effect (Elad *et al.* 2010); it will likely be highly biochar-specific and probably also depend on the time that has elapsed after the biochar was produced (oxidation or evaporation of the labile compounds) (Jaiswal *et al.* 2014; Kammann and Graber 2015). Co-composting will most likely reduce this effect as the biochars' labile compounds decompose or become aggregated during composting. In summary, small amounts of biochar addition for accelerating plant growth and suppressing pathogens in horticulture may be economically very interesting, as soon as the mechanisms are better understood and thus reproducible. If designer biochars with the desired properties can be purposefully produced and targeted, this would open up a new pathway of reducing fungicide and insecticide use in horticulture.

Application of biochar and organics in field studies

Positive effects of the combined use of biochar with organics in the field have predominantly been observed in the tropics in highly weathered soils where most of the early studies were carried out, often in the context of aiding rural development programs (Cornelissen *et al.* 2013; Steiner *et al.* 2007). For example, F. Jenny started using biochar composts in Ghana, based on the knowledge reported above that was the result of the pure biochar. The biochar-compost had yield increases on all soils. That was the story we wanted to tell here!!

In 2010, pure biochar was 150 per cent of control in the South and 90 per cent in the North. In 2012, when biochar-compost was applied, yield was 260 per cent in the South and 240 per cent in the North. Jenny reported the results of biochar compost amendments (1:1 mixture) across a transect ranging from fertile coastal areas to highly weathered interior soils in Ghana, at the 74th ANS Symposium in Potsdam. He reported that there was no measurable biochar effect in the fertile coastal soils, when only biochar was applied. Whereas while he found increases of 200 to 250 per cent in maize yield on both, the fertile coastal and the poor weathered interior soils. The same was already found by Steiner *et al.* (2007) in

Manaus (Brazil). This is in line with findings that even the use of pure biochar without organic amendments was shown to have positive effects in poor tropical soils, in particular when they were sandy, acidic or highly weathered, but the effects were smaller or absent when soils were already fertile (Cornelissen et al. 2013; Haefele et al. 2011; Nguyen et al. 2008; Oguntunde et al. 2004). However, in a field trial of pumpkins in fertile tropical soils in Nepal, Schmidt et al. (2015) reported fourfold harvestable yield increases when biochar was used as a carrier for organic fertilizers (urine) and used in combination with compost.

Results of field experiments in temperate regions where combined biochar and organic amendments were applied are scarce, and those that do exist show mixed results so far. Documented scientific field experiments comprising biochar mixed with organic fertilisers under temperate climate conditions have existed since 2009 (Liu et al. 2012). The authors reported yield improvements of up to 80 per cent, due to biochar addition of 20 Mg ha^{-1} together with 30 Mg ha^{-1} compost, compared to 30 Mg ha^{-1} compost application only. In the meantime, biochar-based organic fertilisers are iteratively further optimised (e.g. Fischer and Glaser 2012).

Schmidt et al. (2014) used the same biochar, aerobic quality compost and biochar compost described above for the horticultural trials in a vineyard field experiment in Valais (Switzerland). The amendment of untreated biochar alone (10t ha^{-1}, produced from wood at 500°C), aerobic compost (72 m^3 ha^{-1}) and biochar compost (10t ha^{-1} + 72m^3 ha^{-1}, mixed before the composting process) were compared to an unamended control. The soil of the field trial was a poor stony Haplic Regosol with legume-rich green cover between the vine rows. During the years 2011 and 2012, various vine and green cover growth, vine health and grape quality parameters were monitored. Biochar compost treatments induced only small and mostly non-significant effects over the two years. Some positive trends on grape quality parameters, such as sugar, polyphenols, yeast-available nitrogen and protein content were found only in the first, drier year in the biochar compost treatment compared to pure biochar, but not in the second year when there was above average precipitation. Therefore Schmidt et al. (2014) concluded that topsoil application of higher amounts of biochar has no immediate economic value for vine growing in poor-fertility alkaline-temperate soils when applied homogeneously over the entire surface with subsequent superficial tillage. For further biochar trials with deep-rooting permanent cultures like vine, it seems more advantageous to inject (nutritive enhanced) biochar or biochar-containing substrates into the deeper root sphere, or to incorporate it in trenches of 30 to 50cm on one or both sides of the vines. Higher concentrations of biochar or biochar-containing substrates placed closer to the main roots of the vines may increase the effect on the vines, creating hotspots with higher water retention, nutrient concentration and biological activity in an otherwise poor soil environment (Blackwell et al. 2010; Graves 2013). As biochar compost had a positive rather than negative effect on vine growing, its application might become a suitable tool to improve ecosystem services and decrease the environmental impact of pesticides, herbicides and chemical fertilisers used currently in high quantities in viticulture.

The same biochar and biochar composts produced at the Ithaka Institute and used for horticultural substrates or in the field trial (Schmidt *et al.* 2014) were also used in a study conducted at the University of Geisenheim, Germany. Here, the roots of the young grapevine plants were directly planted into the first 30cm of topsoil, which was homogeneously mixed with biochar, compost, biochar compost and compost plus untreated biochar, respectively, atop of the subsoil. Grapevine (White Riesling, Clone 198-30Gm) was grown under field conditions in large 120 L planting containers filled with poor sandy soil. The top 30cm of the containers were filled with the pure soil, or soil mixed with either 0, 30 or 60Mg ha^{-1} of pure biochar, pure compost, compost plus untreated biochar, or the biochar-compost, respectively (Mengel 2013). The containers were slowly and continuously irrigated and heavily irrigated monthly to promote nutrient leaching, and the leachate was collected and analysed. Over the vegetative season, the unamended sandy control soil lost an amount of nitrate equal to the added N in the fertiliser. The addition of 60t ha^{-1} compost reduced total nitrate leaching (most of the nitrate was leached primarily *before* the fertiliser was added; i.e. it was derived from compost minerali-sation); but the total nitrate leached was still more than half of the added fertiliser. This did not happen in the corresponding biochar compost or compost plus biochar treatments; these lost only about 20 per cent or 10 per cent of the leached amount in the control, respectively. Unsurprisingly, the grape yield increased 2- to 4-fold in the compost and biochar-compost amendments when compared to the poor sandy control soil. This was significant only for the 60t ha^{-1} biochar compost and compost treatments, respectively. Adding pure biochar, however, did not increase grape yields. The biochar compost and pure compost (60t ha^{-1}) yields were not significantly different from each other. This indicates that the composted biochar (although initially nutrient-poor) must have delivered nutrients at considerably reduced nitrate leaching rates (Mengel 2013). The comparison with the previously described vineyard field trial undertaken with the same compost and biochar-compost indicates that the application technique of the biochar substrates in permanent cultures is likely of predominant importance. Only when the roots of the plants come into direct contact with the substrates can we expect optimal effects.

Contrary to permanent cultures, in trials with arable crops, homogenous and superficial tillage of the biochar substrates are preferred. Fischer *et al.* (unpublished) conducted a field experiment in Bayreuth (Bavaria, Germany) in a sandy loam under ecological farming conditions (no pesticides, no synthetic fertilisers) with 10 treatments arranged in a Latin rectangle design, in which each of the four replicates were arranged regularly so that in each row and column, each treatment occurred only once. This design enables a proper statistical evaluation in the case that the field, which was rather large (about 1 hectare in total), was not homogeneous. The 10 treatments were: two different controls (no addition at all); pure biochar (nine and 32 tonnes per hectare); pure compost from greenwaste (20 and 70 tonnes per hectare); biochar (nine and 32 tonnes per hectare, respectively) mixed with compost (20 and 70 tonnes per hectare, respectively); and co-composted biochar (amounts and mixtures were the same as biochar mixed with compost). In the first year, maize yields did not differ significantly among treatments but a tendency was observed that biochar mixed with compost had a lower yield while composted biochar tended to produce a higher yield compared to all other treatments. When looking at plant quality, the C/N ratio of maize is a good indicator for nitrogen nutrition. Here, no significant differences could be observed. However, C/N ratio was in tendency higher when pure biochar or biochar mixed with compost were applied, and in tendency lower when biochar was composted. These results indicate the beneficial effects of composted biochar with respect to plant growth and nutrition. Similar results were obtained in a field experiment on sand in the Wendland (Northern Germany) during the first year. In the second year, significant benefits on plant growth and nutrition compared to conventional mineral fertilisation were observed when composted biochar products were applied (Glaser et al. 2015).

With respect to soil properties, synergistic effects of biochar and compost can theoretically be obtained by the stability of biochar (C sequestration) and nutrient-rich compost (contributing nutrients). As described above, the best synergism can be obtained if biochar is co-composted (biological 'activation'). In the same biochar/compost field experiment in Bayreuth, SOC concentrations significantly increased in the upper 10cm of the soil to which the materials were incorporated, especially after biochar addition. It is clear that these short-term experiments cannot forecast SOC long-term stability, which is crucial for its C sequestration potential and *terra preta* substrates. Therefore, long-term experiments are clearly required.

On the other hand, compost without biochar blending contributed surprisingly little to the C sequestration, especially at the 'lower' application rate of 20 tons per hectare. From a nutrient supply perspective, the situation was reversed; compost significantly increased the total N content in the topsoil while biochar had no effect. The results indicate synergism between biochar and organic amendments with respect to plant growth and plant and soil quality, even when compared to conventionally optimised mineral fertilisation.

Schimmelpfennig et al. (2014) top-dressed C-rich substrates (*Miscanthus* feedstock, hydrochar and biochar) along the respective similar C content onto wet, species-rich grassland near Giessen, Germany, in a randomised block design (n = 4). The substrates were soaked in liquid pig slurry prior to application to enrich them with nutrients and prevent dust formation. As expected, biochar immediately reduced the bad odours from the pig slurry. The grassland soil is biologically a very active clay loam soil (pH 6) with 3.5–4.5 per cent SOC (Kammann et al. 2008). Pig slurry was applied for fertilisation two times per year. After 2.5 years, the biochar, still visible as a black band, had moved down into the top 5cm of the mineral soil of the grassland below the rooting mat at the top, while the two other C additives were hardly discernible anymore. While the hydrochar significantly reduced the grassland yield over two years, the biochar addition did not, but it promoted a shift towards the more nutrient-rich group of forbs. The latter was without any negative impact on fodder quality (Schimmelpfennig et al. 2015).

The 'plant water availability' factor is of particular importance in poor sandy soils with a low water holding capacity. Here, the combination of compost and biochar seems to provide added value. In a field study in the east of Germany where poor sandy soils predominate, Liu *et al.* (2012) set up a field trial where they applied 30 t ha⁻¹ compost or none (control), and added increasing amounts of the sieved remainder of charcoal production (woody biochar), either 0, 5, 10 or 20 t ha⁻¹. They observed that the highest rate of combined compost-charcoal addition significantly doubled the amount of plant-available water holding capacity (i.e. the volumetric amount available per volume of soil between field capacity (pF 1.8) and the permanent wilting point (pF 4.2).

Conclusion: shaping beneficial pathways of biochar use

For centuries, charcoal has been a valuable energy carrier used for heat generation, ore smelting or as a reduction agent in metallurgy. Therefore, charcoal and thus also biochar definitely has its price, ranging from €300 to €1,000 per tonne. If one intends to use it as a soil additive, beneficial biochar effects must repay the investment unless biochar is produced from biomass wastes not used for any other beneficial purpose or for which even payments are necessary (e.g. because of organic contaminants due to intensive pesticide application).

If we think of biochar/charcoal as a valuable end product, why is there such a large area of *terra preta* and other charcoal-enriched ancient settlement soils? It either did repay its price (e.g. labour efforts to make it) by its effects, or it was just the by-product of the removal of something that was unwanted (e.g. rotting waste and garbage piles around villages that were smouldered). Thus, there might have been the intended or, more likely, unintended formation of organo-mineral biochar complexes at the root of *terra preta* genesis.

Scanning the scarce available literature on biochar and organic additives and biochar and composting (with an 'Asian' focus, since the technique has been in use there for centuries) revealed that it may offer a way forward, although not necessarily an easy way. Adding 'black fairy powder' to badly made compost does not turn it into a miracle, but combining C-rich biochar with N-rich organics shows promising benefits, which are rooted in biochars' ancient Amazonian past.

Bibliography

Abiven, S., Hengartner, P., Schneider, M.P.W., Singh, N. and Schmidt, M.W.I. (2011). Pyrogenic carbon soluble fraction is larger and more aromatic in aged charcoal than in fresh charcoal. *Soil Biology and Biochemistry* 43: 1615–1617.

Amlinger, F., Gotz, B., Dreher, P., Geszti, J. and Weissteiner, C. (2003). Nitrogen in biowaste and yard waste compost: dynamics of mobilisation and availability: a review. *European Journal of Soil Biology* 39: 107–116.

Amlinger, F., Peyr, S., Hildebrandt, U., Müsken, J., Cuhls, C. and Clemens, J. (2005). *Stand der Technik der Kompostierung.* Vienna: Richtlinie des Bundesministeriums für Land- und Forstwirtschaft, Umwelt und Wasserwirtschaft.

Amlinger, F., Peyr, S. and Cuhls, C. (2008). Greenhouse gas emissions from composting and mechanical biological treatment. *Waste Management and Research* 26: 47-60.

Anderson, C.R., Condron, L.M., Clough, T.J., Fiers, M., Stewart, A., Hill, R.A. and Sherlock, R.R. (2011). Biochar induced soil microbial community change: Implications for biogeochemical cycling of carbon, nitrogen and phosphorus. *Pedobiologia* 54: 309–320.

Bernal, M.P., Alburquerque, J.A. and Moral, R. (2009). Composting of animal manures and chemical criteria for compost maturity assessment: a review. *Bioresource Technology* 100: 5444–5453.

Blackwell, P., Krull, E., Butler, G., Herbert, A. and Solaiman, Z. (2010). Effect of banded biochar on dryland wheat production and fertiliser use in south-western Australia: an agronomic and economic perspective. *Australian Journal of Soil Research* 48: 531–545.

Boettcher, J., Pieplow, H. and Krieger, A.E. (2009). Verfahren zur Herstellung von humus- und nährstoffreichen sowie wasserspeichernden Böden oder Bodensubstraten für nachhaltige Landnutzungs- und Siedlungssysteme. Available from Google Patents.

Borchard, N., Ladd, B., Eschemann, S., Hegenberg, D., Möseler, B.M. and Amelung, W. (2014). Black carbon and soil properties at historical charcoal production sites in Germany. *Geoderma* 232–234: 236–242.

Briones, A.M. (2012). The secrets of El Dorado viewed through a microbial perspective. *Frontiers in Microbiology* 3: 239–239.

Busch, D., Stark, A., Kammann, C.I. and Glaser, B. (2013). Genotoxic and phytotoxic risk assessment of fresh and treated hydrochar from hydrothermal carbonization compared to biochar from pyrolysis. *Ecotoxicology and Environmental Safety* 97: 59–66.

Cayuela, M.L., van Zwieten, L., Singh, B.P., Jeffery, S., Roig, A. and Sánchez-Monedero, M.A. (2014). Biochar's role in mitigating soil nitrous oxide emissions: a review and meta-analysis. *Agriculture, Ecosystems and Environment* 191: 5–16.

Chapin, F.S., Woodwell, G.M., Randerson, J.T., Rastetter, E.B., Lovett, G.M., Baldocchi, D.D., Clark, D.A., Harmon, M.E., Schimel, D.S., Valentini, R., Wirth, C., Aber, J.D., Cole, J.J., Goulden, M.L., Harden, J.W., Heimann, M., Howarth, R.W., Matson, P.A., McGuire, A.D., Melillo, J.M., Mooney, H.A., Neff, J.C., Houghton, R.A., Pace, M.L., Ryan, M.G., Running, S.W., Sala, O.E., Schlesinger, W.H. and Schulze, E.D. (2006). Reconciling carbon-cycle concepts, terminology, and methods. *Ecosystems* 9: 1041–1050.

Chave, J., Navarrete, D., Almeida, S., Álvarez, E., Aragão, L. E. O. C., Bonal, D., Châtelet, P., Silva-Espejo, J. E., Goret, J. Y., von Hildebrand, P., Jiménez, E., Patiño, S., Peñuela, M. C., Phillips, O. L., Stevenson, P. and Malhi, Y. (2010). Regional and seasonal patterns of litterfall in tropical South America. *Biogeosciences* 7: 43–55.

Chen, C.R., Phillips, I.R., Condron, L.M., Goloran, J., Xu, Z.H. and Chan, K.Y. (2013). Impacts of greenwaste biochar on ammonia volatilisation from bauxite processing residue sand. *Plant and Soil* 367: 301–312.

Chen, S., Rotaru, A.-E., Shrestha, P.M., Malvankar, N.S., Liu, F., Fan, W., Nevin, K.P. and Lovley, D.R. (2014). Promoting interspecies electron transfer with biochar. *Scientific Reports* 44: article 5019 (doi: 10.1038/srep05019).

Chen, Y.-X., Huang, X.-D., Han, Z.-Y., Huang, X., Hu, B., Shi, D.-Z. and Wu, W.-X. (2010). Effects of bamboo charcoal and bamboo vinegar on nitrogen conservation and heavy metals immobility during pig manure composting. *Chemosphere* 78: 1177–1181.

Cheng, C.-H., Lehmann, J., Thies, J.E., Burton, A.J., and Engelhard, M. (2006). Oxidation of black carbon by biotic and abiotic processes. *Organic Geochemistry* 37: 1477–1488.

Chia, C.H., Gong, B., Joseph, S.D., Marjo, C.E., Munroe, P. and Rich, A.M. (2012). Imaging of mineral-enriched biochar by FTIR, Raman and SEM-EDX. *Vibrational Spectroscopy* 62: 248–257.

Conte, P., Hanke, U.M., Marsala, V., Cimò, G., Alonzo, G. and Glaser, B. (2014). Mechanisms of water interaction with pore systems of hydrochar and pyrochar from poplar forestry waste. *Journal of Agricultural and Food Chemistry* 62: 4917–4923.

Cornelissen, G., Martinsen, V., Shitumbanuma, V., Alling, V., Breedveld, G., Rutherford, D., Sparrevik, M., Hale, S., Obia, A. and Mulder, J. (2013). Biochar effect on maize yield and soil characteristics in five conservation farming sites in Zambia. *Agronomy* 3: 256–274.

de la Rosa, J.M. and Knicker, H. (2011). Bioavailability of N released from N-rich pyrogenic organic matter: An incubation study. *Soil Biology and Biochemistry* 43: 2368–2373.

Dias, B.O., Silva, C.A., Higashikawa, F.S., Roig, A. and Sanchez-Monedero, M.A. (2010). Use of biochar as bulking agent for the composting of poultry manure: effect on organic matter degradation and humification. *Bioresource Technology* 101: 1239–1246.

Doydora, S.A., Cabrera, M.L., Das, K.C., Gaskin, J.W., Sonon, L.S. and Miller, W.P. (2011). Release of nitrogen and phosphorus from poultry litter amended with acidified biochar. *International Journal of Environmental Research and Public Health* 8: 1491–1502.

Elad, Y., David, D.R., Harel, Y.M., Borenshtein, M., Kalifa, H.B., Silber, A. and Graber, E.R. (2010). Induction of systemic resistance in plants by biochar, a soil-applied carbon sequestering agent. *Phytopathology* 100: 913–921.

Fischer, D. and Glaser, B. (2012). Synergisms between compost and biochar for sustainable soil amelioration. In K. Sunil and A. Bharti (eds), *Management of Organic Waste*, 167–198. Rijeka: InTech.

Glaser, B. (1999). Eigenschaften und Stabilität des Humuskörpers der Indianerschwarzerden Amazoniens. *Bayreuther Bodenkundliche Berichte* 68, S.

Glaser, B. (2007). Prehistorically modified soils of central Amazonia: a model for sustainable agriculture in the twenty-first century. *Philosophical Transactions of the Royal Society B* 362: 187–196.

Glaser, B. and Amelung, W. (2003). Pyrogenic carbon in native grassland soils along a climosequence in North America. *Global Biogeochemical Cycles* 17: 1064.

Glaser, B. and Birk, J.J. (2012). State of the scientific knowledge on properties and genesis of Anthropogenic Dark Earths in Central Amazonia (terra preta de Índio). *Geochimica et Cosmochimica Acta* 82: 39–51.

Glaser, B., Haumaier, L., Guggenberger, G. and Zech, W. (2001). The 'Terra Preta' phenomenon: a model for sustainable agriculture in the humid tropics. *Naturwissenschaften* 88: 37–41.

Glaser, B., Guggenberger, G. and Zech, W. (2004). Identifying the Pre-Columbian anthropogenic input on present soil properties of Amazonian Dark Earths (terra preta). In B. Glaser and W.I. Woods (eds), *Amazonian Dark Earths: Explorations in Space and Time*, 145–158. Heidelberg: Springer.

Glaser, B., Wiedner, K., Seelig, S., Schmidt, H.-P. and Gerber, H. (2015). Biochar organic fertilizers from natural resources as substitute for mineral fertilizers. *Agronomy for Sustainable Development* 35: 667–678.

Graber, E.R., Harel, Y.M., Kolton, M., Cytryn, E., Silber, A., David, D.R., Tsechansky, L., Borenshtein, M. and Elad, Y. (2010). Biochar impact on development and productivity of pepper and tomato grown in fertigated soilless media. *Plant and Soil* 337: 481–496.

Graves D. (2013). A comparison of methods to apply biochar to temperate soils. In N. Ladygina N and F. Rineua F (eds), *Biochar and Soil Biota*, 202–260. London, CRC Press.

Haefele, S.M., Konboon, Y., Wongboon, W., Amarante, S., Maarifat, A.A., Pfeiffer, E.M. and Knoblauch, C. (2011). Effects and fate of biochar from rice residues in rice-based systems. *Field Crops Research* 121: 430–440.

Harter, J., Krause, H.-M., Schuettler, S., Ruser, R., Fromme, M., Scholten, T., Kappler, A. and Behrens, S. (2014). Linking N_2O emissions from biochar-amended soil to the structure and function of the N-cycling microbial community. *ISME Journal* 8: 660–674.

Heckenberger, M.J., Kuikuro, A., Kuikuro, U.T., Russell, J.C., Schmidt, M., Fausto, C. and Franchetto, B. (2003). Amazonia 1492: pristine forest or cultural parkland? *Science* 301: 1710–1714.

Heckenberger, M.J., Russell, J.C., Fausto, C., Toney, J.R., Schmidt, M.J., Pereira, E., Franchetto, B. and Kuikuro, A. (2008). Pre-Columbian urbanism, anthropogenic landscapes, and the future of the Amazon. *Science* 321: 1214–1217.

Hua, L., Wu, W., Liu, Y., McBride, M. and Chen, Y. (2009). Reduction of nitrogen loss and Cu and Zn mobility during sludge composting with bamboo charcoal amendment. *Environmental Science and Pollution Research* 16: 1–9.

Hua, L., Chen, Y. and Wu, W. (2012). Impacts upon soil quality and plant growth of bamboo charcoal addition to composted sludge. *Environmental Technology* 33: 61–68.

Jaffé, R., Ding, Y., Niggemann, J., Vähätalo, A.V., Stubbins, A., Spencer, R.G.M., Campbell, J. and Dittmar, T. (2013). Global charcoal mobilization from soils via dissolution and riverine transport to the oceans. *Science* 340: 345–347.

Jaiswal, K.A., Elad, Y., Graber, E.R. and Frenkel, O. (2014). *Rhizoctonia solani* suppression and plant growth promotion in cucumber as affected by biochar pyrolysis temperature, feedstock and concentration. *Soil Biology and Biochemistry* 69: 110–118.

Jindo, K., Suto, K., Matsumoto, K., Garcia, C., Sonoki, T. and Sanchez-Monedero, M.A. (2012). Chemical and biochemical characterisation of biochar-blended composts prepared from poultry manure. *Bioresource Technology* 110: 396–404.

Joseph, S., Graber, E.R., Chia, C., Munroe, P., Donne, S., Thomas, T., Nielsen, S., Marjo, C., Rutlidge, H., Pan, G.X., Li, L., Taylor, P., Rawal, A. and Hook, J. (2013). Shifting paradigms: development of high-efficiency biochar fertilizers based on nano-structures and soluble components. *Carbon Management* 4: 323–343.

Kammann, C. and Graber, E.R. (2015). Biochar effects on plant eco-physiology. In J. Lehmann and S. Joseph (eds), *Biochar for Environmental Management: Science, Technology and Implementation*, 2nd edn, 391–420. Abingdon: Routledge.

Kammann, C., Müller, C., Grünhage, L. and Jäger, H.-J. (2008). Elevated CO_2 stimulates N_2O emissions in permanent grassland. *Soil Biology and Biochemistry* 40: 2194–2205.

Kammann, C., Kühnel, Y., von Bredow, C. and Gößling, J. (2010). C-Sequestrierungspotential und Eignung von Torfersatzstoffen, hergestellt aus Produkten der Landschaftspflege und Biochar [C sequestration potential and suitability of peat substitutes, produced from nature-conservation waste products and biochar]. Project Report, Department of Plant Ecology, Justus-Liebig University Gießen, Germany, Gießen.

Kammann, C., Finke, C., Lima, A., Clough, T.J., Tsai, S.M., Teixeira, W.G., Braker, G. and Müller, C. (2012). Will aged biochars continue to reduce N_2O emissions? Abstract Eurosoil Conference 2012, Bari, Italy. Available at www.researchgate.net/profile/Claudia_Kammann2/contributions.

Kammann, C.I., Schmidt, H.-P., Messerschmidt, N., Linsel, S., Steffens, D., Müller, C., Koyro, H.-W., Conte, P. and Joseph, S. (2015). Plant growth improvement mediated by nitrate capture in co-composted biochar. *Scientific Reports* 5 (doi 10.1038/srep11080).

Kern, D.C., da Costa, M.L. and Frazao, F.J.L. (2004). Evolution of the scientific knowledge regarding archaeological black earths of Amazonia. In B. Glaser and W.I. Woods (eds), *Amazonian Dark Earths: Explorations in Space and Time*, 19–28. Heidelberg: Springer.

Kolton, M., Meller Harel, Y., Pasternak, Z., Graber, E.R., Elad, Y. and Cytryn, E. (2011). Impact of biochar application to soil on the root-associated bacterial community structure of fully developed greenhouse pepper plants. *Applied and Environmental Microbiology* 77: 4924–4930.

Kuzyakov, Y., Ehrensberger, H., Stahr, K. (2001). Carbon partioning and below-ground translocation by Lolium perenne. *Soil Biology and Biochemistry* 33: 61–74.

Lehmann, J., Campos, C.V., de Macedo, J.L.V. and German, L. (2004). Sequential P fractionation of relict Anthropogenic Dark Earths of Amazonia. In B. Glaser and W.I. Woods (eds), *Amazonian Dark Earths: Explorations in Space and Time*, 113–123. Heidelberg: Springer.

Liang, B., Lehmann, J., Sohi, S.P., Thies, J.E., O'Neill, B., Trujillo, L., Gaunt, J., Solomon, D., Grossman, J., Neves, E.G. and Luizão, F.J. (2010). Black carbon affects the cycling of non-black carbon in soil. *Organic Geochemistry* 41: 206–213.

Lin, Y., Munroe, P., Joseph, S. and Henderson, R. (2012). Migration of dissolved organic carbon in biochars and biochar-mineral complexes. *Pesquisa Agropecuaria Brasileira* 47: 677–686.

Lin, Y., Munroe, P., Joseph, S., Ziolkowski, A., van Zwieten, L., Kimber, S. and Rust, J. (2013). Chemical and structural analysis of enhanced biochars: thermally treated mixtures of biochar, chicken litter, clay and minerals. *Chemosphere* 91: 35–40.

Liu, J., Schulz, H., Brandl, S., Miehtke, H., Huwe, B. and Glaser, B. (2012). Short-term effect of biochar and compost on soil fertility and water status of a Dystric Cambisol in NE Germany under field conditions. *Journal of Plant Nutrition and Soil Science* 175: 698–707.

Masiello, C.A. and Louchouarn, P. (2013). Fire in the ocean. *Science* 340: 287–288.

Meller-Harel, Y., Elad, Y., Rav-David, D., Borenstein, M., Shulchani, R., Lew, B. and Graber, E.R. (2012). Biochar mediates systemic response of strawberry to foliar fungal pathogens. *Plant and Soil* 357: 245–257.

Mengel, J. (2013). Auswirkungen von Biokohle auf den Boden, die Nährstoffversorgung von Votos vinifera und die Auswaschung von Nährstoffen. Hochschule Geisenheim University, Geisenheim, Germany.

Neves, E.G., Petersen, J.B., Bartone, R.N. and Silva, C.A.D. (2003). Historical and socio-cultural origins of Amazonian Dark Earths. In J. Lehmann, D.C. Kern, B. Glaser and W.I. Woods (eds), *Amazonian Dark Earths: Origin, Properties, Management*, 29–50. Dordrecht: Kluwer Academic Publishers.

Nguyen, B., Lehmann, J., Kinyangi, J., Smernik, R., Riha, S. and Engelhard, M. (2008). Long-term black carbon dynamics in cultivated soil. *Biogeochemistry* 89: 295–308.

Oguntunde, P.G., Fosu, M., Ajayi, A.E. and Giesen, N. (2004). Effects of charcoal production on maize yield, chemical properties and texture of soil. *Biology and Fertility of Soils* 39: 295–299.

Park, J.H., Choppala, G.K., Bolan, N.S., Chung, J.W. and Chuasavathi, T. (2011). Biochar reduces the bioavailability and phytotoxicity of heavy metals. *Plant and Soil* 348: 439–451.

Prost, K., Borchard, N., Siemens, J., Kautz, T., Sequaris, J.-M., Moeller, A. and Amelung, W. (2013). Biochar affected by composting with farmyard manure. *Journal of Environmental Quality* 42: 164–172.

Rodionov, A., Amelung, W., Peinemann, N., Haumaier, L., Zhang, X., Kleber, M., Glaser, B., Urusevskaya, I. and Zech, W. (2010). Black carbon in grassland ecosystems of the world. *Global Biogeochemical Cycles* 24: GB3013.

Sarkhot, D.V., Berhe, A.A. and Ghezzehei, T.A. (2012). Impact of biochar enriched with dairy manure effluent on carbon and nitrogen dynamics. *Journal of Environmental Quality* 41: 1107–1114.

Schimmelpfennig, S., Müller, C., Grünhage, L., Koch, C. and Kammann, C. (2014). Biochar, hydrochar and uncarbonized feedstock application to permanent grassland: effects on greenhouse gas emissions and plant growth. *Agriculture, Ecosystems and Environment* 191: 39–52.

Schimmelpfennig, S., Kammann, C., Moser, G., Grünhage, L. and Müller, C. (2015). Changes in macro- and micronutrient contents of grasses and forbs following *Miscanthus x giganteus* feedstock, hydrochar and biochar application to temperate grassland. *Grass and Forage Science* in press (doi: 10.1111/gfs.12158).

Schmidt, M.W.I., Skjemstad, J.O., Gehrt, E. and Kögel-Knabner, I. (1999). Charred organic carbon in German chernozemic soils. *European Journal of Soil Science* 50: 351–365.

Schmidt, H.-P., Kammann, C., Niggli, C., Evangelou, M.W.H., Mackie, K.A. and Abiven, S. (2014). Biochar and biochar-compost as soil amendments to a vineyard soil: influences on plant growth, nutrient uptake, plant health and grape quality. *Agriculture, Ecosystems and Environment* 191: 117–123.

Schmidt, H.-P., Pandit, B.H., Martinsen, V., Cornelissen, G., Conte, P. and Kammann, C.I. (2015). Fourfold increase in pumpkin yield in response to low-dosage root zone application of urine-enhanced biochar to a fertile tropical soil. *Agriculture* 5: 723–741.

Schulz, H., Dunst, G. and Glaser, B. (2013). Positive effects of composted biochar on plant growth and soil fertility. *Agronomy for Sustainable Development* 33: 817–827.

Schulz, H. and Glaser, B. (2012). Effects of biochar compared to organic and inorganic fertilizers on soil quality and plant growth in a greenhouse experiment. *Journal of Plant Nutrition and Soil Science* 175(3): 410–422 (doi:10.1002/jpln.201100143).

Sombroek, W.G. (ed.) (1966). *Amazon Soils: A Reconnaissance of the Soils of the Brazilian Amazon Region.* Wageningen: Verslagen van Landbouwkundige Onderzoekingen.

Steiner, C., Teixeira, W., Lehmann, J., Nehls, T., de Macêdo, J., Blum, W. and Zech, W. (2007). Long term effects of manure, charcoal and mineral fertilization on crop production and fertility on a highly weathered Central Amazonian upland soil. *Plant and Soil* 291: 275–290.

Steiner, C., Glaser, B., Teixeira, W.G., Lehmann, J., Blum, W.E.H., and Zech, W. (2008a). Nitrogen retention and plant uptake on a highly weathered central Amazonian Ferralsol amended with compost and charcoal. *Journal of Plant Nutrition and Soil Science* 171: 893–899.

Steiner, C., Teixeira, M. and Zech, W. (2008b). Soil respiration curves as soil fertility indicators in perennial central Amazonian plantations treated with charcoal, and mineral or organic fertilisers. *Tropical Science* 47: 218–230.

Steiner, C., Das, K.C., Melear, N. and Lakly, D. (2010). Reducing nitrogen loss during poultry litter composting using biochar. *Journal of Environmental Quality* 39: 1236–1242.

Steiner, C., Melear, N., Harris, K. and Das, K.C. (2011). Biochar as bulking agent for poultry litter composting. *Carbon Management* 2: 227–230.

Taghizadeh-Toosi, A., Clough, T., Sherlock, R. and Condron, L. (2012). Biochar adsorbed ammonia is bioavailable. *Plant and Soil* 350: 57–69.

Taylor, P. (ed.) (2010). *The Biochar Revolution: Transforming Agriculture and Environment.* Victoria, Australia: Global Publishing Group.

Van Zwieten, L., Kammann, C., Cayuela, M. L., Singh, B.P., Joseph, S., Kimber, S., Donne, S., Clough, T., and Spokas, K. (2015). Biochar effects on nitrous oxide and methane emissions from soil. In J. Lehmann and S. Joseph (eds), *Biochar for Environmental Management: Science, Technology and Implementation*, 2nd edn, 487–518. Abingdon: Routledge.

Vasilyeva, N.A., Abiven, S., Milanovskiy, E.Y., Hilf, M., Rizhkov, O.V., and Schmidt, M.W.I. (2011). Pyrogenic carbon quantity and quality unchanged after 55 years of organic matter depletion in a Chernozem. *Soil Biology and Biochemistry* 43: 1985–1988.

Wang, C., Lu, H., Dong, D., Deng, H., Strong, P.J., Wang, H. and Wu, W. (2013). Insight into the effects of biochar on manure composting: Evidence supporting the relationship between N$_2$O emission and denitrifying community. *Environmental Science and Technology* 47: 7341–7349.

Wardle, D.A., Nilsson, M.-C. and Zackrisson, O. (2008). Fire-derived charcoal causes loss of forest humus. *Science* 320: 629.

Wiedner, K. and Glaser, B. (2015). Traditional use of biochar. In J. Lehmann and S. Joseph (eds), Biochar for Environmental Management: Science, Technology and Implementation, 2nd edn, 15–38. Abingdon: Routledge.

Wolf, R. and Wedig, H. (2007). Process for manufacturing a soil conditioner. Available from Google Patents.

Zwart, D.C., and Kim, S.-H. (2012). Biochar amendment increases resistance to stem lesions caused by *Phytophthora* spp. in tree seedlings. *HortScience* 47: 1736–1740.

Note

1   Soil (organic carbon) priming means the stimulation of its decomposition by the addition of fresh organic material with nutrients and labile carbon that fuels microbial life and thus assists in the decomposition of more recalcitrant carbon fractions (Kuzyakov *et al.* 2001).

Chapter 7

Biochar carbon stability and effect on greenhouse gas emissions

*Esben Bruun, Andrew Cross, Jim Hammond,
Victoria Nelissen, Daniel P. Rasse and
Henrik Hauggaard-Nielsen*

Introduction

When organic matter is added to soils, it is used as a source of energy and nutrients by microorganisms. The carbon is thereby unlocked from chain-like molecules from which plants are composed. Microorganisms such as fungi and bacteria get energy by breaking down these often long molecules into smaller units such as sugars, which are in turn broken down to provide a source of energy and carbon. While some of the carbon is used by microorganisms as a building block in multiplication and reproduction, another part of the carbon is oxidised by reaction with oxygen in the soil to create the greenhouse gas (GHG) carbon dioxide (CO_2). As microorganisms reproduce and die rapidly, CO_2 is also produced as a result of microbial decomposition soon after organic matter is added to soil. This process by which carbon locked in organic molecules is converted into the gas CO_2 is called 'mineralisation'. The speed of mineralisation varies greatly depending upon soil temperature – a higher temperature (say, between 15 and 30°C) is more conducive to microbial growth than the lower temperatures in temperate climates (between 20 and 15°C). This is the main reason why soil organic carbon (SOC) levels are generally higher in cooler climates than in the (sub)tropics, though other factors such as water logging (creating very low–oxygen conditions) are very important.

In most soils, the microorganisms are constrained by availability of food and energy – hence, when biomass is added to the soil, it is rapidly exploited. Most of the organic matter added to soils is mineralised to CO_2 in a few years in hotter climates and in 10 or so years in cooler climates. For example, one experiment in the tropics using rice husk, showed that in air-rich soils (aerobic conditions), between 80 and 100 per cent of the carbon in the husk was decomposed to CO_2 within three years (Knoblauch *et al.* 2010). This compares to cooler, temperate climates where roughly 20 per cent of organic matter added to soils was still there 10 years later (Knoblauch *et al.* 2010). Typically a small portion (<1 per cent) of the organic matter introduced to soils turns into a substance called humus, which remains in the soil for decades, and often binds with mineral particles such as clays to form a stable 'aggregate'. For this reason, soil organic carbon (SOC) levels are only sustained if organic matter is continually added to the soil. If the organic inputs stop, SOC declines to the low levels at which it remains.

The proposition behind biochar as a form of carbon storage in soils (also called 'carbon sequestration') relates to a change in the chemical structures in which carbon atoms are bound.

Organic matter, such as straw or wood, is thermally decomposed in an oxygen-depleted atmosphere (pyrolysis; see Chapter 2), and modified to form structures that are much more resistant to biological and chemical degradation as compared to the original feedstock (see Figure 7.1). Pyrolysis of agricultural biomass residues can stabilise up to half of the carbon that is fixed annually in these residues through photosynthesis. Thus, by stabilising plant-captured carbon as biochar and sequestering it on a long-term basis in the soil, application of biochar is a way to withdraw CO_2 from the atmosphere. This moves carbon out of the relatively short natural carbon cycle (i.e. a few years), and into the long carbon cycle (i.e. where it takes hundreds to thousands of years for this carbon to be degraded and returned to the atmosphere). In addition, renewable energy can potentially be generated from pyrolysis in the form of bio-oil and syngas, which can be used to replace fossil fuels (Bruun *et al.* 2011a). Because of the long-term (>100 years) net removal of atmospheric CO_2 carbon (fixed in the biochar), the overall pyrolysis process can be called 'carbon-negative' (Figure 7.1; Woolf and Lehmann 2012), unlike CO_2 emissions from traditional fossil fuel-based energy production, which generate a lot of carbon per unit of useful energy delivered and medium, or close to neutral at best, carbon emissions from wind, solar or biomass based energy production (also see Chapter 8).

This chapter deals with the stability of biochar and the physicochemical parameters and production conditions which may influence its degradation in soil. The question of whether biochar may influence the degradation of native soil organic matter is also touched upon. Moreover, the chapter explores the effect of biochar on other GHG emissions apart from CO_2.

Stability of biochar in soil

The stability of biochar determines how long biochar carbon will remain sequestered in the soil after application (also expressed as the mean residence time (MRT), which is the average time that the biochar remains in the soil). Biochar is primarily characterised by its high organic carbon content, which mainly comprises of joined up or conjugated aromatic compounds of six carbon atoms linked together in rings (see Chapter 3). This condensed aromatic 'backbone' is what makes biochar highly stable in the environment, as microorganisms cannot readily utilise carbon and other elements embedded in the biochar matrix as a source for energy and nutrition (unlike the chain-like molecules which constitute plant life). However, biochar is not just one uniform aromatic material; it has a highly heterogeneous composition determined by the feedstock's physicochemical characteristics and the pyrolysis configurations (see Chapter 2 and 3). In addition to the large, relatively

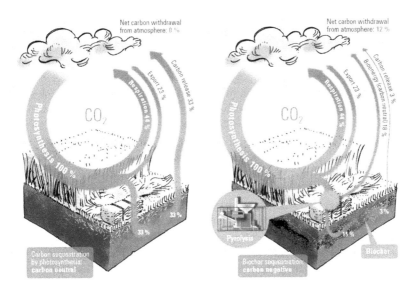

Figure 7.1 Principle behind carbon sequestration with biochar soil application (right) compared to a normal agricultural system (left). In the normal carbon cycle nearly all carbon built into the crops is returned to the atmosphere within a relatively short time scale (typically less than 20 years) through crop respiration and mineralisation processes of the residue. Biochar however will not significantly degrade through microbial activity or chemical reactions in the environment at a climate change relevant timescale (>100 years). In the above example the biochar application results in an overall long-term withdrawal of 14 per cent of the carbon that is photosynthesised from the atmosphere. The pyrolysis products of bio-oil and syngas (represented by the bioenergy arrow in the figure) are assumed to be used to substitute for fossil fuels. The wheat carbon cycle is based on Aubinet *et al.* (2009).

Source: ©E.W. Bruun, Department of Chemical and Biochemical Engineering, DTU.

stable, fractions, biochar contains smaller molecules which are known as 'labile components', and often also minor fractions of volatile carbon compounds, which readily dissolve in water. These labile and volatile compounds can be readily broken down (mineralised) by microorganisms once they become available. The total carbon release from biochar is determined by the decomposition rates of these different fractions.

What is the long-term stability of biochar?

The short (10 years), medium (10–100 years) and long-term (>>100 years) stability of biochar is of fundamental importance for the assessment of potential climate change mitigation, as it influences the quantity and speed at which the indigenous biochar carbon is released again as CO_2. It is well established that biochar and black carbon particles from kilns and prehistoric forest fires are very stable in the environment. The high resistance to microbial degradation is demonstrated by the 1200-year age of charcoal particles from prehistoric pits on Iceland (Church *et al.* 2007; the charcoal was produced in the eighth century AD). The age is measured by radiocarbon dating, a very accurate way in which the age of carbon recovered from charcoal particles in the soil can be dated. Charcoal particles added to the soil several hundred to thousand years ago by indigenous populations of Amazonia which resulted in the dark Amazonian soils or *terra preta* (see Chapter 6) have been radiocarbon dated and found to originate from 500 up to 7000 years BP (Neves *et al.* 2003). However, radiocarbon dating of charcoal merely gives the time elapsed since the carbon was photosynthesised by plants till present time. It does not provide quantitative information on the initial charcoal amount and thus the decomposition rate (i.e. a large proportion of the charcoal added to *terra preta* soil might have been mineralised, and it would not be possible to know the quantity already lost from radiocarbon dating on the remaining charcoal). Precise MRT estimations are therefore difficult to make and as yet, there is no agreed-upon methodology for calculating the long-term stability of biochar. This may be part of the explanation for the large span in biochar residence time estimations reported in literature, with residence times ranging from centennial up to millennial timescales (e.g. Lehmann 2007; Liang *et al.* 2008; Preston and Schmidt 2006; Zimmerman 2010; Spokas *et al.* 2010).

The substantial uncertainties regarding the longer-term longevity of biochar in the environment are also influenced by a number of factors such as pyrolysis reactor characteristics, peak process temperature, heating rate and feedstock quality; all of which can influence the final quality and stability of the biochar produced (discussed in Chapters 2 and 3). Moreover, biochar can be applied to a multitude of soil types under different climates and agricultural systems, which makes general recommendations difficult as one biochar may decompose at a higher rate in one system than in another (Zimmerman *et al.* 2012). However, despite these uncertainties biochar is generally recognised as highly recalcitrant and far more stable than the original feedstock and other soil organic fractions (Masék *et al.* 2013). Thus, regardless of whether biochar remains in soils for hundreds or thousands of years, it may still be considered a long-term sink for the purposes of reducing carbon dioxide emissions. However, in order to qualify for carbon credits, precise longevity estimations (based on standardised methodology) are required (Chapters 9 and 11).

Methods that have been used to determine the long-term stability of biochar have focused on:

- the chemical characterisation of biochar as a predictor of stability, such as hydrogen-to-carbon (H/C) and oxygen-to-carbon (O/C) ratios;
- extrapolations from short-term incubation studies; and
- investigations of charcoal found in the environment (e.g. from wildfire sites) or archaeological sites.

However, there are some fundamental limitations with all of these approaches. Results from chemical characterisation such as ^{13}C nuclear magnetic resonance (NMR), proximate and elemental analysis are very instructive; however, they are only qualitative and do not provide a direct analogue for long-term environmental degradation. Extrapolations from short-term incubation studies are also limited as carbon loss does not always follow simple first-order decay, and will therefore be inherently bound by large errors, whilst investigations of environmental and archaeological charcoal are not able to account for the amount of carbon lost since the charcoal was originally created. It is therefore obvious that a combination of these methods, alongside newly developing techniques based on accelerated ageing analogous to environmental degradation, is needed to definitively predict the long-term stability of biochar. One example hereof is the procedure for determination of the relative long-term stability of biochar (>100 years) samples developed by Cross and Sohi (2013). It is based on oxidative degradation of biochar, analogous to oxidative enzyme-mediated degradation in the environment.

Short-term decomposition of biochar

Long-term studies (>10 years) are required to improve estimations of the decomposition rate of the stable fraction of biochar carbon. However, these studies are expensive and laborious and the number of longer-term studies currently available is limited. Consequently, residence time estimations are typically based on short-term studies (months to a few years) and concurrent use of dynamic models using different decay rates for the labile and stable carbon fractions (Lehmann *et al.* 2009; Zimmerman 2010). These studies have shown how some biochar types are considerably more degradable than others over the short term (Bruun *et al.* 2011a, 2011b; Zimmerman 2010; Zimmerman *et al.* 2012).

The short-term degradability of biochar is commonly determined by comparing the CO_2 respiration from soil amended with biochar to soil without biochar. During the experiment, the soil CO_2 flux originating from respiration is determined repeatedly by gas analysis or through the use of sodium hydroxide. The overall degradation (carbon loss) of the added biochar materials is then calculated using simple difference (i.e. CO_2 emission from the biochar amended soil minus the emission from the control soil – no biochar added; Bruun *et al.* 2012; Cross and Sohi 2011). The difference in CO_2 emissions is then assumed to originate from biochar. Biochar derived CO_2 losses from a number of short-term incubation studies are summarised in Table 7.1, which demonstrates cumulative carbon losses of biochar after a few months/years from almost zero up to about 12 per cent of total carbon in the biochar.

Biochar degrades in the environment owing to both biological (biotic) and physical-chemical (abiotic) decomposition processes. It is well known that abiotic factors such as water, oxygen, soil, temperature and pH influence organic matter decomposition, but the relative magnitude and importance of these factors for the decomposition of biochar is still largely unknown. Zimmerman (2010) showed in

Table 7.1 Short-term carbon losses of a range of biochars when added to soil. All experiments presented in the table are laboratory incubation studies

Biochar feedstock	Pyrolysis type/temp. settings (°C)	Medium type (sand/soil)	Duration of experiment	Cumulative short–term carbon loss (% of added C)	Authors
Aiton sapwood	Slow pyrolysis/250–350°C	Sand	120 days	<2	Baldock and Smernik (2002)
Wheat straw	Slow pyrolysis/ 225°C and 300°C	Not defined	2 years	3.1–9.3	Bruun et al. (2009)
Wheat straw	Fast pyrolysis/ 475–575°C	Sandy loam	115 days	3–12	Bruun et al. (2011a)
Wheat straw	Slow and fast pyrolysis/ 525°C	Sandy loam	65 days	2.9–5.5	Bruun et al. (2012)
Barley straw	Slow pyrolysis (muffle oven)/ 400°C	Sandy soil, (Luvisol)	451	1.8–1.9	Bruun and El-Zehery (2012)
Sugarcane bagasse	Slow Pyrolysis 350, 450 and 550°C	Sand	14 days	0.23–1.08	Cross and Sohi (2011)
Oak wood and maize/rye	Slow pyrolysis/ 800°C and 350°C	Sand	60 days	0.3–0.8	Hamer et al. (2004)
Willow, pine	Slow pyrolysis/450, 550 and 650°C	Sand	357 days	0.04–0.3	Nelissen et al. (2014a)
Mix of woody feedstocks ('RomChar')	Slow pyrolysis/450–480°C	Sand	381 days	0.4	Nelissen et al. (2014b)
Mixture of beech, hazel, oak, birch	Slow pyrolysis/ 500°C	Cambisol	84 days	2.8	Zavalloni et al. (2011)
Six feedstock types incl. hardwoods, grass and sugarcane baggase	Slow pyrolysis/ 400, 525, or 650°C	Sand	2 years	0.4–3 year−1	Zimmerman (2010)

a one-year laboratory experiment using a range of biochar materials mixed in sterilised or non-sterilised soil, that abiotic degradation was equally as important as biotic processes for the short-term decomposition of biochar. Biological minerali-sation of biochar involves mainly microorganisms, but the relative contribution of bacteria and fungi is not clear. It is plausible that specialised fungi are the dominant decomposers of the more stable aromatic components of the biochar structure. For example it has been shown that white root fungi are capable of metabolising very recalcitrant materials like coal (Lehmann *et al.* 2009; Hofrichter *et al.* 1999) and lignin. Earthworms have also been shown to ingest biochar (Topoliantz and Ponge 2005), and may, together with insects and other bigger soil animals, also increase the accessibility for other decomposers by vertical and horizontal mixing of the biochar particles in the soil (Lehmann *et al.* 2011).

The influence of pyrolysis technology on the labile carbon fraction of biochar

When estimating a given biochar's stability, the labile fraction remaining in the biochar after pyrolysis is important to quantify, as this fraction strongly influences the short-term biochar degradability (Lehmann *et al.* 2009; Zimmerman 2010). Freshly produced biochar often contains small,highly labile fractions, which are rapidly mineralised after application to soil (Lehmann *et al.* 2009; Bruun *et al.* 2011a, 2011b). Some of these fractions may dissolve in the soil water, thus representing a highly mobile component providing a readily available source of energy and nutrients for soil microorganisms. Accordingly, increased microbial biomass responses after application of fresh biochar to soils has been reported in a number of studies (Bruun *et al.* 2012; Kolb *et al.* 2009; Kuzyakov *et al.* 2009; Novak *et al.* 2010; Steiner *et al.* 2008). The size of the labile fraction (if any) depends largely on the feedstock and pyrolysis type and settings used. In general, higher pyrolysis temperatures result in lower biochar yields, but which have higher proportions of carbon and stable aromatic structures (Antal and Gronli 2003; Masék *et al.* 2013; Zimmerman 2010; Zimmerman *et al.* 2011). For example, in a laboratory study by Bruun *et al.* (2011a), the degradation rates of biochar produced by fast pyrolysis of wheat straw at 475, 500, 525, 550 and 575°C respectively, were strongly related to the reactor temperature (Figure 7.2). The lowest temperature (475°C) biochar had the largest cumulative carbon loss of 12 per cent in the 115-day experiment, while the highest temperature (575°C) biochar had the lowest carbon loss of 3 per cent (Figure 7.2). The relative amount of labile carbohydrate content (mainly cellulose or hemicellulose) left in the biochar decreased from 36 per cent to 3 per cent with increasing pyrolysis temperatures (from 475 to 575°C). The wide range in decomposition rates in Figure 7.2 illustrates the strong influence even small adjustments of the pyrolysis reactor temperature can have on the biochar quality and degradability and emphasises the importance of having an in-depth understanding of the specific pyrolysis equipment used and its effects on biochar quality (see Chapter 2 and 3). Contrary to slow pyrolysis, fast pyrolysis technology may result in incompletely pyrolysed biomass due to the very fast heating of the biomass particles, which in turn leaves a less stable biochar.

From a carbon mitigation perspective, it is not desirable to apply biochar with large labile fractions in order to avoid emissions of CO_2, and with proper pyrolysis equipment and knowledge it is fairly easy to produce highly stable biochar without labile fractions (see Chapter 2). However, a farmer might consider it more important to apply a biochar that actually provides accessible carbon to the soil microorganisms, providing beneficial soil functions and services, though C/N ratios are a further important consideration (i.e. too much labile C relative to available N can reduce the plants' access to N; see Chapter 5).

Does biochar enhance or reduce the mineralisation of the indigenous soil organic matter? The 'priming effect' controversy

A long-lasting increase in soil carbon content is a major goal of biochar addition to soil, notably in the framework of GHG mitigation strategies. Obviously, the fate and the mineralisation rate of the added biochar carbon is a major question. However, the fate of the indigenous SOC following biochar treatment cannot be neglected. In many biochar experiments, the dominant soil carbon pool remains the indigenous organic matter. Many agricultural soils contain about 1 to 5 per cent of carbon in their plough layer (Brady and Weil 2014). This concentration

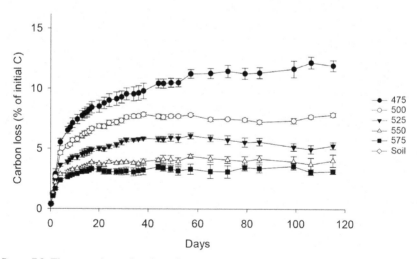

Figure 7.2 The cumulative biochar decomposition (carbon loss) measured as net soil surface CO_2-fluxes after incorporation of 40g soil with 2g biochar produced at different reactor temperatures (475, 500, 525, 550 and 575°C, respectively). Standard error is shown (n = 4).

Source: Bruun *et al.* (2011a).

translates into roughly 30 to 60 tonnes of carbon per hectare. This amount exceeds the biochar incorporation rates of many field experiments, which often range from 10 to 40 tonnes biochar carbon per hectare. In other words, in many soil types, a typical biochar application will at best double the soil carbon content.

An effect known as 'priming' might induce changes in the mineralisation rate of indigenous soil organic carbon when biochar is added to soil. This effect was first reported by Löhnis (1926) and is linked to the fact that the mineralisation of SOC is mediated by soil microorganisms. The respiration of the soil microbes as they consume soil organic matter for energy is a major source of CO_2 flux from soils. When soil microbes are provided with a new food source in the form of any organic carbon material, their overall activity and appetite for the indigenous soil organic matter is affected (Kuzyakov et al. 2000). This can be understood in terms of complementing food sources, known in technical terms as co-metabolism. To prosper, soil microbes need both energy, in the form of easily decomposable organic molecules, and nutrients, especially nitrogen. For example, soil organic matter is often rich in nitrogen but poor in energy-rich molecules that are easily decomposable. Adding a high-energy but low-nitrogen substrate to soil can prompt an enhanced decomposition of the soil organic matter by microorganisms, as the added carbon source gives them the needed energy to mine the indigenous soil organic matter for nitrogen. A typical substrate for this kind of effect would be glucose, which is a highly accessible energy source but contains no nitrogen. In this example, we would be faced with a positive priming effect (i.e. the decomposition of the indigenous soil organic matter is accelerated by the addition of the energy-rich substrate to soil). A negative priming effect can also occur, meaning that the added substrate reduces the mineralisation rate of the indigenous soil organic matter.

Adding biochar to soil can affect microbial access to food sources in many ways. In theory, such modifications of food sources can lead to priming effects. For example, a positive priming effect can potentially be induced when soil microbes utilise the small fraction of easily labile molecules often present in biochar products as an energy source. By contrast, a negative priming effect is expected if biochar decreases the availability of labile molecules in soils. This is thought to be an important mechanism as biochar is generally a high-sorption material (Zimmerman et al. 2011) and will remove some of the molecules making them less accessible to microorganisms. Negative priming effects will lead to a reduction of CO_2 emissions at the field scale.

In 2008, a high-profile study suggested that adding pyrogenic carbon to soils leads to positive priming effects (Wardle et al. 2008), which raised concerns that biochar could induce similar effects. However, this does not seem to be the case, as the majority of recent studies point to an absence of priming effect or even negative priming effect, potentially leading to further increases in C sequestration with biochar (Woolf and Lehmann 2012). The mineralisation rate of the indigenous SOC is often reduced following biochar treatment (Jones et al. 2011; Cross and Sohi 2011; Keith et al. 2011; Zimmerman et al. 2011; Bruun and El-

Zehery 2012). Some studies did report a positive priming effect; however, in these instances, it appears that the positive priming effect was short lived, and actually reversed into negative priming and increased organic matter stability in the longer term (Zimmerman *et al.* 2011). Positive and transient priming effects appear to be associated with low temperature biochar (300–400°C; Zimmerman *et al.* 2011), which is also the type of biochar that is least stable in soil environments and therefore generally not considered for agricultural and soil carbon storage applications. Model simulations based on our current understanding of biochar impacts on microbial food sources suggest that in the long term, biochar will generally contribute to increased stability of the non-pyrogenic organic matter (Woolf and Lehmann 2012). In conclusion, increased stabilisation of the indigenous soil organic matter by biochar appears possible (Soinne *et al.* 2014) and would be a desired trait of biochars produced for agronomic and carbon storage purposes. However, as most research on priming has been conducted in the lab, field studies are needed in order to test priming effects under natural conditions.

Biochar impact on greenhouse gas soil emissions

In addition to long-term carbon sequestration, biochar application to soils has also been shown to reduce emissions of the GHGs, nitrous oxide (N_2O) and methane (CH_4), thereby addressing carbon emissions but also reducing total GHG emissions (Yanai *et al.* 2007; Spokas and Reicosky 2009; Singh *et al.* 2010; Van Zwieten *et al.* 2010; Bruun *et al.* 2011b). Although increases or no effect on N_2O emissions have also been reported (Bruun *et al.* 2012; Case *et al.* 2013) a meta-analysis by Cayuela *et al.* (2014) shows that a reduction in N_2O is much more common than an increase is. N_2O and CH_4 are both potent GHGs that are estimated, respectively, to be 298 and 21 times stronger than CO_2 (Forster *et al.* 2007). On a quantitative basis though, CO_2 is still by far the most significant GHG (Verheijen *et al.* 2010) and the most important GHG to address. The main pathways resulting in production of the different GHGs are illustrated in Figure 7.3.

The effect of biochar additions on soil CO_2 emissions depends on the soil environment and the microbial community present, as well as the physicochemical characteristics of the biochar (e.g. labile fractions stimulating microbial activity and growth, as previously discussed). Usually, increases in CO_2 emissions occur directly after biochar addition to soil in small-scale incubation experiments (e.g. Brodowski *et al.* 2006; Bruun *et al.* 2011a; Hamer *et al.* 2004; Steiner *et al.* 2008), but the effect is typically short-term in nature, lasting no more than a few weeks or months, after which the biochar amended soil will have CO_2 emission rates similar to soil without biochar.

Nitrous oxide is an important GHG, estimated to comprise 8 per cent of global GHG emissions (as CO_2 equivalents) in 2004 (Denman *et al.* 2007). Since the pre-industrial age, the atmospheric N_2O concentration has increased considerably (from 270 to 319 parts per billion in 2005), and the agricultural sector is the largest

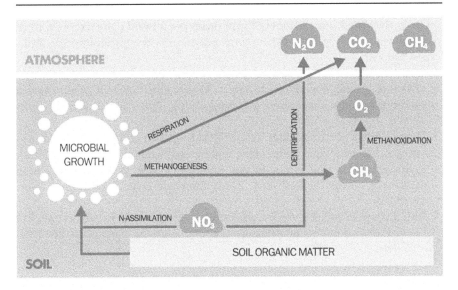

Figure 7.3 Conceptual model of carbon and nitrogen cycling.

Note: NO$_3$ = nitrate; N$_2$O = nitrous oxide; CO$_2$ = carbon dioxide; O$_2$ = oxygen; CH$_4$ = methane.

single emitter (Del Grosso *et al.* 2005; Denman *et al.* 2007). Methods for reducing N$_2$O emissions from agriculture without losing productivity are therefore essential for the mitigation of climate change. Biochar has been shown to decrease N$_2$O emissions from agricultural soils under certain conditions, although other studies have shown no effect, mixed or even negative effects (Bruun *et al.* 2011b; Case *et al.* 2013). The majority of these biochar-N$_2$O studies have used soil with applied nitrogen fertiliser to test if biochar can reduce part of the N$_2$O release usually observed after fertilisation. Ameloot *et al.* (2013) observed, in an incubation experiment, a 50 per cent reduction in N$_2$O emissions when high temperature biochars (10t fresh biochar ha^{-1} at 700°C) were applied to a sandy loam soil. In contrast, the authors applied low temperature biochars (350°C) to the same soil and observed no influence on N$_2$O emissions. Nelissen *et al.* (2014b) conducted a lab experiment in which biochar addition (5g kg^{-1}) reduced N$_2$O emissions by 52 per cent and 84 per cent after urea and nitrate fertiliser application respectively, while no effect was observed after ammonium fertiliser application. Field observations in Europe are limited; Castaldi *et al.* (2011) observed mixed results in a field trial in Italy with 30 and 60t ha^{-1} biochar applications. Generally, N$_2$O emissions were higher in the control compared to the biochar treated plots, but this difference was only statistically significant on two occasions. Karhu *et al.* (2011) observed no significant differences in N$_2$O emissions between birch biochar amended plots (9t ha^{-1}) and control plots in a Finnish field trial. Case *et al.* (2013) did not detect a significant effect of biochar (49t ha^{-1}) on N$_2$O emissions in a UK field trial. The mechanisms behind reduced (or changed) N$_2$O emissions have not yet been elucidated, but a number of hypotheses put forward in literature are listed (see also Figure 7.4; see Box 7.1 for an explanation of the nitrification–denitrification process):

1 Microbial inhibition through increased hormone content with biochar addition reduces nitrification (e.g. Spokas *et al.* 2010; Spokas and Reicosky 2009).
2 Lower nitrate concentrations due to NH_3 formation and subsequent adsorption on biochar particles after urine application (Taghizadeh-Toosi *et al.* 2011).
3 Nitrogen immobilisation reducing substrate availability for denitrification (Bruun *et al.* 2011b; Case *et al.* 2012; Kammann *et al.* 2012; DeLuca *et al.* 2009).
4 An increase in soil pH after biochar addition increases the activity of the N_2O reductase enzyme, forcing denitrification through to N_2 (Singh *et al.* 2010; Van Zwieten *et al.* 2010).

Figure 7.4 Hypotheses for reduced N_2O emissions after biochar addition suggested in literature (see Nelissen *et al.* 2014b and references therein).

5   Restricted oxygen availability is required for denitrification, and through biochar porosity and water absorption, soil aeration is increased (Stewart *et al.* 2012; Yanai *et al.* 2007).

6   The highly porous surfaces of biochar adsorb N_2O (Van Zwieten *et al.* 2009), thus influencing N_2O emissions directly.

It is therefore not possible to currently draw a strong conclusion as both lab and field experiments show complex interactions between soil type, soil water content, biochar type/dose and fertiliser type/dose.

Decreases in soil methane (CH_4) emissions have also been observed following amendment with biochar. Relatively high applications (10 per cent by mass of soil) of 16 different biochars on three North American soils generally showed CH_4 reductions in laboratory incubations (Spokas *et al.* 2009). Field trials using biochar in rice paddy, typically subject to high CH_4 emissions, have shown mixed results. Increases in methane emissions were observed by Zhang *et al.* (2010, 2013) and Feng (2012), while biochar reduced methane emissions in studies by Haefele *et al.* (2011) and Knoblauch *et al.* (2010). Biochar has been used as a feed supplement for ruminants and other livestock, with reduced livestock methane emissions reported under field (Leng 2012a) and laboratory conditions (Leng 2012b, 2013). It seems that biochar has some beneficial effect upon the digestion process as increased weight gain by livestock has also been found (Kana 2011; Leng 2012a). Given the difficulty of finding other ways of reducing methane emissions from

---

Box 7.1 Nitrous oxide formation in soil: the nitrification–denitrification process

The nitrification and denitrification processes are the principal sources of nitrous oxide (N_2O) emission from soil (Figure 7.5). Nitrification is defined as the biological transformation (oxidation) of ammonium (NH_4^+) to nitrite (NO_2^-) and nitrate (NO_3^-). Typically, the bacterial genera *Nitrosomonas* and *Nitrospira* oxidise NH_4^+ to NO_2^-, while *Nitrobacter* converts NO_2^- to NO_3^-. For both processes oxygen is required. NH_4^+ availability is most frequently the factor limiting the nitrification rate (Hutchinson and Davidson 1993). Other important factors influencing this rate are temperature, moisture, pH and oxygen supply (White 2006). Denitrification can be defined as respiratory reduction of NO_3 to NO_2 to gaseous NO (nitric oxide) with N_2O or N_2 (dinitrogen) as gaseous end products. Unlike the narrow species diversity of organisms responsible for nitrification in soil, the process of denitrification can be accomplished by several bacterial groups. Denitrifiers are essentially oxygen requiring (aerobic) bacteria with the alternative capacity to use nitrogen oxides when oxygen becomes limiting. Oxygen availability is the most important factor determining the denitrification process, as restricted oxygen availability is required (Hutchinson and Davidson 1993). Increase in water content hinders aeration and promotes denitrification (Dalal *et al.* 2003). Other important factors influencing the rate of denitrification are temperature and pH (White 2006).

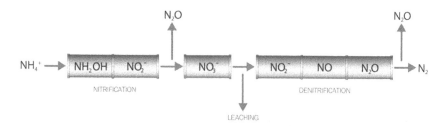

Figure 7.5 The nitrification and denitrification process showing the pathways of N_2O production.

livestock, this potential effect merits further R&D in a European context. Conversely, biochar (charcoal) added to anaerobic digestion tanks has been shown to increase CH_4 emissions (Kumar 1987; Inthapanya *et al.* 2012). The mechanisms behind all these observations remain unclear and warrant further research before general conclusions can be drawn.

Summary

As demonstrated by several scientific studies, there is no doubt that biochar in general is very recalcitrant when compared to other organic matter additions and soil organic matter fractions, and it is possible to sequester carbon at a climate change relevant time scale (~100 years or more) through application of biochar. However, the environmental stability of biochar carbon is strongly correlated with the degree of thermal alteration of the original feedstock (the higher the temperature, the larger the aromatic fraction), and in depth understanding of the technology used and its effect on the biochar quality is necessary in order to produce the most beneficial biochars for soil application. Beside carbon sequestration in soil, biochar may improve the GHG balance by reducing N_2O and CH_4 soil emissions, although contrasting results are found in the literature. The mechanisms behind these reductions remain unclear and more research is required in order to investigate the various hypotheses in more detail, and to unravel the complex interaction between biochar, crop and soil, especially under field conditions. In conclusion, our current knowledge is largely based on short-term lab studies and pot experiments, which have provided detailed insight in certain processes and aspects of biochar application to soils, but suffer from large uncertainties when scaled up to the level of the farmer's field. In order to produce more realistic scenarios of the potential impact of biochar on C sequestration and soil GHG emissions, biochar research needs to be brought up to the field scale with longer-term studies.

Bibliography

Ameloot, N., De Neve, S., Jegajeevagan, K., Yildiz, G., Buchan, D., Funkuin, Y., Prins, W., Bouckaert, L. and Sleutel, S. (2013). Short-term CO_2 and N_2O emissions and microbial properties of biochar amended sandy loam soils. *Soil Biology and Biochemistry* 57: 401–410.

Antal, J. and Gronli, M. (2003). The art and technology of charcoal production. *Industrial Engineering and Chemistry Research* 43: 1619–1640.

Aubinet, M., Moureaux, C., Bodson, B., Dufranne, D., Heinesch, B., Suleau, M., Vancutsem, F. and Vilret, A. (2009). Carbon sequestration by a crop over a 4-year sugar beet/winter wheat/seed potato/winter wheat rotation cycle. *Agricultural and Forest Meteorology* 149: 407–418.

Baldock, J. and Smernik, R. (2002). Chemical composition and bioavailability of thermally, altered *Pinus resinosa* (red pine) wood. *Organic Geochemistry* 33: 1093–1109.

Brady, N. and Weil, R. (2014). *The Nature and Properties of Soils*, 14th edn. Noida: Pearson.

Brodowski, S., John, B., Flessa, H. and Amelung, W. (2006). Aggregate-occluded black carbon in soil. *European Journal of Soil Science* 57: 539–546.

Bruun, S. and El-Zehery, T. (2012). Biochar effect on the mineralization of soil organic matter. *Pesquisa Agropecuária Brasileira* 47(5): 665–671.

Bruun, S., El-Zehery, T. and Jensen, L. (2009). Carbon sequestration with biochar – stability and effect on decomposition of soil organic matter. *Earth and Environmental Science* 6 (doi:10.1088/1755-1307/6/4/242010).

Bruun, E.W., Hauggaard-Nielsen, H., Ibrahim, N., Egsgaard, H., Ambus, P., Jensen, P.A. and Dam-Johansen, K. (2011a). Influence of fast pyrolysis temperature on biochar labile fraction and short-term carbon loss in a loamy soil. *Biomass and Bioenergy* 35: 1182–1189.

Bruun, E.W., Müller-Stöver, D., Ambus, P. and Hauggaard-Nielsen, H. (2011b). Application of biochar to soil and N_2O emissions: potential effects of blending fast-pyrolysis biochar with anaerobically digested slurry. *European Journal of Soil Science* 62: 581–589.

Bruun, E.W., Ambus, P., Egsgaard, H. and Hauggaard-Nielsen, H. (2012). Effects of slow and fast pyrolysis biochar on soil C and N turnover dynamics. *Soil Biology AND Biochemistry* 46: 73–79.

Case, S.D.C., McNamara, N.P., Reay, D.S. and Whitaker, J. (2012). The effect of biochar addition on N_2O and CO_2 emissions from a sandy loam soil – The role of soil aeration. *Soil Biology AND Biochemistry* 51: 125–134.

Case, S.D.C., McNamara, N.P., Reay, D.S. and Whitaker, J. (2013). Can biochar reduce soil greenhouse gas emissions from a *Miscanthus* bioenergy crop? *GCB Bioenergy* (doi:10.1111/gcbb.12052).

Castaldi, S., Riondino, M. Baronti, S., Esposito, F.R., Marzaioli, R., Rutigliano, F.A., Vaccari, F.P. and Miglietta, F. (2011). Impact of biochar application to a Mediterranean wheat crop on soil microbial activity and greenhouse gas fluxes. *Chemosphere* 85: 1464–1471.

Cayuela, M.L., van Zwieten, L., Singh, B. P., Jeffery, S., Roig, A. and Sánchez-Monedero, M.A. (2014). Biochar's role in mitigating soil nitrous oxide emissions: A review and meta-analysis. *Agriculture, Ecosystems AND Environment* 191: 5–16.

Church, M.J., Dugmore, A.J., Mairs, K.A., Millard, A.R., Cook, G.T., Sveinbjarnardttir, G., Cough, P.A. and Roucoux, K.H. (2007). Charcoal production during the Norse and early medieval periods in Eyjafjallahreppur, southern Iceland. *Radiocarbon* 49: 659–672.

Cross, A. and Sohi, S.P. (2011). The priming potential of biochar products in relation to labile carbon contents and soil organic matter status. *Soil Biology and Biochemistry* 43: 2127–2134.

Cross, A. and Sohi, S.P. (2013). A method for screening the relative long-term stability of biochar. *Global Change Biology Bioenergy* 5: 215–220.

Dalal, R.C., Wang, W., Robertson, G.P. and Parton, W.J. (2003). Nitrous oxide emission from Australian agricultural lands and mitigation options: a review. *Australian Journal of Soil Research* 41: 165–195.

Del Grosso, S.J., Mosier, A.R., Parton, W.J. and Ojima, D.S. (2005). DAYCENT model analysis of past and contemporary soil N2O and net greenhouse gas flux for major crops in the USA. *Soil and Tillage Research* 83: 9–24.

DeLuca, T.H., MacKenzie, M.D. and Gundale, M.J. (2009). Biochar effects on soil nutrient transformations. In J. Lehmann and S. Joseph (eds), *Biochar for Environmental Management*, 251–270. London: Earthscan.

Denman, K.L., Brasseur, G., Chidthaisong, A., Ciais, P., Cox, P.M., Dickinson, R.E. *et al.* (2007). Couplings between changes in the climate system and biogeochemistry. In S. Solomon, D. Qin, M. Manning, Z. Chen, M. Marquis, K.B. Averyt *et al.* (eds), *Climate Change 2007: The Physical Science Basis. Contribution of Working Group I to the Fourth Assessment Report of the Intergovernmental Panel on Climate Change*, 499–587. Cambridge: Cambridge University Press.

Feng Y., Xu Y., Yu Y., Xie Z., Lin X. (2012). Mechanisms of biochar decreasing methane emission from Chinese paddy soils. *Soil Biology and Biochemistry* 46: 80–88.

Forster, P., Ramaswamy, V., Artaxo, P., Berntsen, T., Betts, R., Fahey, D.W., Haywood J., Lean, J., Lowe, D.C., Myhre, G., Nganga, J., Prinn, R., Raga, G., Schulz, M. and Van Dorland, R. (2007). Changes in Atmospheric Constituents and in Radiative Forcing. In S. Solomon, D. Qin, M. Manning, Z. Chen, M. Marquis, K.B. Averyt, M. Tignor and H.L. Miller (eds), *Climate Change 2007: The Physical Science Basis: Contribution of Working Group I to the Fourth Assessment Report of the Intergovernmental Panel on Climate Change*, 129–234. Cambridge: Cambridge University Press.

Haefele, S., Konboon, Y., Wongboon, W., Amarante, S., Maarifat, A., Pfeiffer, E. and Knoblauch, C. (2011). Effects and fate of biochar from rice residues in rice-based systems. *Field Crops Research* 121: 430–441.

Hamer, U., Marschner, B., Brodowski, S. and Amelung, W. (2004). Interactive priming of black carbon and glucose mineralization. *Organic Geochemistry* 35: 823–830

Hofrichter, M., Ziegenhagen, D., Sorge, S., Ullrich, R., Bublitz, F. and Fritsche, W. (1999). Degradation of lignite (low-rank coal) by ligninolytic basidiomycetes and their manganese peroxidase system. *Applied Microbiology and Biotechnology* 52: 78–84.

Hutchinson, G.L. and Davidson, E.A. (1993). Processes for production and consumption of gaseous nitrogen oxides in soil. In L.A. Harper *et al.* (eds), *Agricultural Ecosystem Effects on Trace Gases and Global Climate Change*, 79–93. Madison, WI: Agronomy Society of America,

Inthapanya, S., Preston, T.R. and Leng, R.A. (2012). Biochar increases biogas production in a batch digester charged with cattle manure. *Livestock Research for Rural Development* 24: article 212.

Jones, D.L., Murphy, D.V., Khalid, M., Ahmad, W., Edwards-Jones G. and DeLuca, T.H. (2011). Short-term biochar-induced increase in soil CO_2 release is both biotically and abiotically mediated. *Soil Biology and Biochemistry* 43: 1723–1731.

Kammann, C., Ratering, S., Eckhard, C. and Müller, C. (2012). Biochar and hydrochar effects on greenhouse gas (carbon dioxide, nitrous oxide, and methane) fluxes from soils. *Journal of Environmental Quality* 41: 1052–1066.

Kana, J.R., Teguia, A., Mungfu, B.M. and Tchoumboue, J. (2011). Growth performance and carcass characteristics of broiler chickens fed diets supplemented with graded levels of charcoal from maize cob or seed of *Canarium schweinfurthii* Engl. *Tropical Animal Health and Production* 43: 51–56.

Karhu, K., Mattila, T., Bergström, I. and Regina, K. (2011). Biochar addition to agricultural soil increased CH4 uptake and water holding capacity: results from a short-term pilot field study. *Agriculture, Ecosystems and Environment* 140: 309–313.

Keith, A., Singh, B. and Singh, B.P. (2011). Interactive priming of biochar and labile organic matter mineralization in a smectite-rich Soil. *Environmental Science and Technology* 45: 9611–9618.

Knoblauch, C., Maarifat, A., Pfeiffer, E. and Haefele, S.M. (2010). Degradability of black carbon and its impact on trace gas fluxes and carbon turnover in paddy soils. *Soil Biology and Biochemistry* 43: 1768–1778.

Kolb, S., Fermanich, K. and Dornbush, M. (2009). Effect of charcoal quantity on microbial biomass and activity in temperate soils. *Soil Science Society of America Journal* 73: 1173–1181.

Kumar, S., Jain, M.C. and Chhonkar, P.K. (1987). A note on stimulation of biogas production from cattle dung by addition of charcoal. *Biological Wastes* 20: 209–215.

Kuzyakov, Y., Friedel, J.K. and Stahr, K. (2000). Review of mechanisms and quantification of priming effects. *Soil Biology and Biochemistry* 32: 11–12.

Kuzyakov, Y., Subbotina, I., Chen, H., Bogomolova, I. and Xu, X. (2009). Black carbon decomposition and incorporation into soil microbial biomass estimated by C-14 labeling. *Soil Biology & Biochemistry* 41: 210–219.

Lehmann, J. (2007). A handful of carbon. *Nature* 447: 143–144.

Lehmann, J., Gaunt, J. and Rondon, M. (2006). Bio-char sequestration in terrestrial ecosystems: review. *Mitigation and Adaptation Strategies for Global Change* 11: 403–427.

Lehmann, J., Czimnik, C., Laird, D. and Sohi, S. (2009). Stability of biochar in the soil. In J. Lehmann and S. Joseph (eds), *Biochar for Environmental Management*, 169–182. London: Earthscan.

Lehmann, J., Rillig M. C., Thies, J., Masiello, C.A. et al. (2011). Biochar effects on soil biota: a review. *Soil Biology and Biochemistry* 43: 1812–1836.

Leng R.A., Preston T.R. and Inthapanya S. (2012a). Biochar reduces enteric methane and improves growth and feed conversion in local 'Yellow' cattle fed cassava root chips and fresh cassava foliage. *Livestock Research for Rural Development* 24(11).

Leng R.A., Inthapanya S. and Preston T.R. (2012b). Biochar lowers net methane production from rumen fluid in vitro. *Livestock Research for Rural Development* 24(6).

Leng R.A., Inthapanya S. and Preston T.R. (2013). All biochars are not equal in lowering methane production in in-vitro rumen incubations. *Livestock Research for Rural Development* 25(6).

Liang, B., Lehmann, J., Solomon, D., Sohi, S. et al. (2008). Stability of biomass-derived black carbon in soils. *Geochimica et Cosmochimica Acta* 72: 6069–6078.

Löhnis, F. (1926). Nitrogen availability of green manures. *Soil Science* 22: 253–290.

Masék, O., Brownsort, P., Cross, A. and Sohi, S. (2013). Influence of production conditions on the yield and environmental stability of biochar. *Fuel* 103: 151–155.

Nelissen, V., Ruysschaert, G., Müller-Stöver, D., Bodé, S., Cook, J., Ronsse, F., Shackley, S., Boeckx, P. and Hauggaard-Nielsen, H. (2014a). Short-term effect of feedstock and pyrolysis temperature on biochar characteristics, soil and crop response in temperate soils. *Agronomy* 4: 52–73.

Nelissen, V., Kumar Saha, B., Ruysschaert, G. and Boeckx, P. (2014b). Effect of different biochar and fertilizer types on N_2O and NO emissions from a silt loam soil. *Soil Biology and Biochemistry* 70: 244–255.

Nelissen, V., Rütting, T., Huygens, D., Ruysschaert, G. and Boeckx, P. (2015). Temporal evolution of biochar's impact on soil nitrogen processes – a ^{15}N tracing study. *GCB Bioenergy* 7(4): 635–645.

Neves, E.G., Petersen, J.B., Bartone, R.N. and Silva, C.A.D. (2003). Historical and socio-cultural origins of Amazonian Dark Earths. In J. Lehmann *et al.* (ed.), *Amazonian Dark Earths: Origin, Properties, Management*, 29–50. Dordrecht: Kluwer Academic Publishers.

Novak, J.M., Busscher, W.J., Watts, D.W., Laird, D.A., Ahmedna, M.A. and Niandou, M.A.S. (2010). Short-term CO_2 mineralization after additions of biochar and switchgrass to a typic kandiudult. *Geoderma* 154: 281–288.

Pereira, C.R., Kaal, J, Arbestain C.M., Pardo L.R., Aitkenhead, W., Hedley, M., Macías, F., Hindmarsh, J. and Macià-Agulló, J.A. (2011). Contribution to characterization of biochar to estimate the labile fraction of carbon. *Organic Geochemistry* 42: 1331–1342.

Preston, C.M. and Schmidt, M.W.I. (2006). Black (pyrogenic) carbon in boreal forests: a synthesis of current knowledge and uncertainties. *Biogeosciences* 3: 211–271.

Simek, M. and Cooper, J.E. (2002). The influence of soil pH on denitrification: progress towards the understanding of this interaction over the last 50 years. *European Journal of Soil Science* 53: 345–354.

Singh, B.P., Hatton, B.J., Singh, B., Cowie, A.L. and Kathuria, A. (2010). Influence of biochars on nitrous oxide emission and nitrogen leaching from two contrasting soils. *Journal of Environmental Quality* 39: 1224–1235.

Soinne, H., Hovi, J., Tammeorg, P. and Turtola, E. (2014). Effect of biochar on phosphorus sorption and clay soil aggregate stability. *Geoderma* 219–220: 162–167.

Spokas, K.A. and Reicosky, D.C. (2009). Impacts of sixteen different biochars on soil greenhouse gas production. *Annals of Environmental Science* 3: 179–193.

Spokas, K.A., Baker, J.M. and Reicosky, D.C. (2010). Ethylene: potential key for biochar amendment impacts. *Plant and Soil* 333: 443–452.

Steiner, C., Das, K., Garcia, M., Forster, B. and Zech, W. (2008). Charcoal and smoke extract stimulate the soil microbial community in a highly weathered xanthic ferralsol. *Pedobiologia* 51: 359–366.

Stewart, C.E., Zheng, J., Botte, J. and Cotrufo, M.F. (2013). Co-generated fast pyrolysis biochar mitigates greenhouse gas emissions and increases carbon sequestration in temperate soils. *GCB Bioenergy*, 5: 153–164.

Taghizadeh-Toosi, A., Clough, T.J., Condron, L.M., Sherlock, R.R., Anderson, C.R. and Craigie, R.A. (2011). Biochar incorporation into pasture soil suppresses in situ nitrous oxide emissions from ruminant urine patches. *Journal of Environmental Quality* 40: 468–476.

Topoliantz, S. and Ponge, J.F. (2005). Charcoal consumption and casting activity by *Pontoscolex corethrurus* (Glossoscolecidae). *Applied Soil Ecology* 28: 217–224.

Van Zwieten, L., Singh, B., Joseph, S., Kimber, S., Cowie, A. and Chan, K.Y. (2009). Biochar and Emissions of Non-CO_2 Greenhouse Gases from Soil. In J. Lehmann and S. Joseph (eds), *Biochar for Environmental Management*, 227–249. London: Earthscan.

Van Zwieten, L., Kimber, S., Morris, S., Downie, A., Berger, E., Rust, J. and Scheer, C. (2010). Influence of biochars on flux of N_2O and CO_2 from Ferrosol. *Australian Journal of Soil Research* 48: 555–568.

Verheijen, F.G.A., Jeffery, S., Bastos, A.C., van der Velde, M. and Diafas, I. (2010). *Biochar Application to Soils: A Critical Scientific Review of Effects on Soil Properties, Processes and Functions*. EUR 24099 EN. Luxembourg: Office for Official Publications of the European Communities.

Wardle D.A., Nilsson M.C. and Zackrisson O. (2008). Fire-derived charcoal causes loss of forest humus. *Science* 320(5876): 629.

White, R.E. (2006). *Principles and Practice of Soil Science: The Soil as a Natural Resource*, 4th ed. Oxford: Blackwell Publishing.

Woolf, D. and Lehmann, J. (2012). Modelling the long-term response to positive and negative priming of soil organic carbon by black carbon. *Biogeochemistry* 111: 83–95.

Yanai, Y., Toyota, K. and Okazaki, M. (2007). Effects of charcoal on N_2O emissions from soil resulting from rewetting air-dried soil in short-term laboratory experiments. *Soil Science and Plant Nutrition* 53: 181–188.

Zavalloni C., Alberti, G., Biasiol, S., Vedove, G.D., *et al.* (2011). Microbial mineralization of biochar and wheat straw mixture in soil: a short-term study. *Applied Soil Ecology* 50: 45–51.

Zhang, A., Cui, L., Pan, G., Li, L., Hussain, Q., Zhang, X., Zheng, J. and Crowley, D. (2010). Effect of biochar amendment on yield and methane and nitrous oxide emissions from a rice paddy from Tai Lake plain, China. *Agriculture, Ecosystems and Environment* 139: 469–475.

Zhang A, Rongjun B, Hussain Q, Li L, Pan G, Zheng J, Zhang X, Zheng J. (2013). Change in net global warming potential of a rice–wheat cropping system with biochar soil amendment in a rice paddy from China. *Agriculture, Ecosystems and Environment* 173: 37–45.

Zimmerman, A.R. (2010). Abiotic and microbial oxidation of laboratory-produced black carbon (biochar). *Environmental Science and Technology* 44: 1295–1301.

Zimmerman A.R., Gao, B. and Ahn, M.-Y. (2011). Positive and negative carbon mineralization priming effects among a variety of biochar-amended soils. *Soil Biology and Biochemistry* 43: 1169–1179.

Zimmerman A.R., Bird C., Wurster C., Saiz G., Gooddrick I., Barta J., Capek P., Santruckova P. and Smernik R. (2012). Rapid degradation of pyrogenic carbon. *Global Change Biology* 18: 3306–3316.

Chapter 8

Life cycle assessment
Biochar as a greenhouse gas sink?

Jan-Markus Rödger, Jim Hammond, Peter Brownsort, Dane Dickinson and Achim Loewen

Introduction

To address the real contribution of thermally converted biomass for greenhouse gas mitigation, it is necessary to assess the whole life cycle from the production of the biomass feedstock to the actual distribution and utilisation of the biochar produced in a regional context.

Therefore, this chapter will introduce the idea of life cycle assessment, the methodology of bioenergy assessment and its application to the topic of biochar. We will cover the assessment of the utilisation of six thermally converted types of biomass and conduct a sensitivity analysis for the North Sea region.

Life cycle assessment – a tool for today and the future

There have been numerous efforts to standardise the practice of life cycle assessment (LCA). The goal is to make it possible to compare the outputs of LCAs – the results – to one another in a meaningful way. But as with any accounting system, the answer you get depends on what you count.

Life cycle assessment history

Different international groups have tried to harmonise the assessment of greenhouse gas emissions of products, systems or services. The Greenhouse Gas (GHG) Protocol, the Publicly Available Specification (PAS) 2050 and the International Organization for Standardization (ISO) are well known. LCA covers the environmental performance of products, systems or services during their utilisation. But the manufacturing and the disposal phase also make a significant impact on the environment and have to be taken into consideration as well. As a result, the 'cradle-to-grave' approach, in which no components are removed from the system, was first developed by the Society for Environmental Toxicology and Chemistry (SETAC), which hosted two conferences in 1990. Since then, the LCA approach has developed rapidly and is an important element in any good innovation (see Figure 8.1). But the usual approach is to assess a product from the acquisition of the raw materials, to the transport stage, to the production facility. The utilisation and the end-of-life stage (disposal or recycling) are also taken into consideration, but some material and waste discharge does take place along the life cycle.

ISO standard

The ISO produced the ISO 14040 series to provide a loose framework of guidance on how to conduct an LCA. This series is an approach to quantify the environmental impact of product systems and can assist in:

- identifying opportunities to improve the environmental performance of products at various points in their life cycle;
- informing decision makers in industry, government or non-government organisations (e.g. for the purpose of strategic planning, priority setting, product or process design or redesign);
- the selection of relevant indicators of environmental performance, including measurement techniques; and
- marketing (e.g. implementing an eco-labelling scheme).

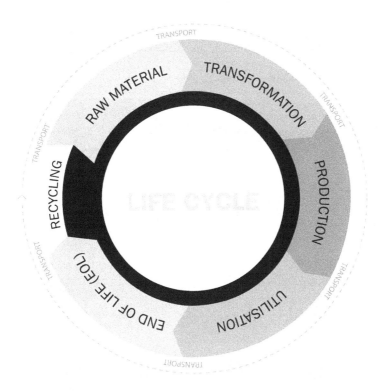

Figure 8.1 Life-cycle approach of products, systems or services.

For practitioners, the ISO 14044 supplies additional information on how an LCA is undertaken. The whole LCA is based on four consecutive phases, although sometimes it's more likely to be an iterative process owing to the emergence of new issues and findings (see Figure 8.2).

The crux of this guidance is that transparency is paramount. From the outset, it should be clearly stated *what* is being assessed and *why*. The system boundary can then be defined on the basis of that statement. The so-called 'goal and scope' stage is the description of the intended application, why, for example, this product is being assessed and to whom the results of the study are to be communicated. The scope defines the product system, the function and the system boundaries. The system boundary is the imaginary line that defines what is included in the LCA

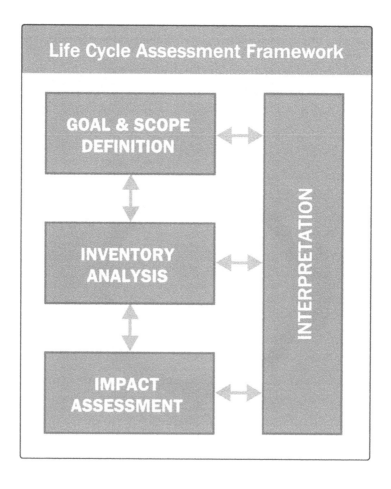

Figure 8.2 Phases of a life cycle assessment according to the ISO 14040.

and what is not (i.e. what is counted and what is ignored). One of the most important parts is the functional unit, which is a measure of the function of the assessed system that is of interest, and defines the performance characteristics to provide a reference to which the inputs and outputs are related (see examples in Table 8.1). The clear selection of functional units greatly aids comparability and interpretation, as well as transparency, which is the most important reporting aspect, according to ISO 14040.

Table 8.1 Examples of functional units

Product/system/process	Functional unit
Car A vs. Car B	Operating life time of 150,000 km over 10 years
Renewable Electricity	Provision of 1 kWh electricity for the grid by various technologies (e.g. onshore, offshore, biogas, solar)
Glass (0.75 l) v. Plastic Bottle (1 l)	Provision of 4 l of water (5 glass bottles and 4 plastic bottles)

Once the system boundary has been set, data must be collected for every step in the process – this is the inventory analysis or life cycle inventory (LCI). The LCI describes the data collection of all relevant inputs and outputs of the product system investigated (e.g. GHG emissions, the energy flow, the flow of pollutants and nutrients and the costs). If, at some point, it becomes necessary to make changes in the boundaries, the analysis has to start from the beginning again. In an attempt to standardise the data used, and therefore enhance the comparability, there are a large number of databases available (e.g. the European Life Cycle Database: EcoInvent), which can used while populating an LCI. Of course, such databases contain only established data; when conducting research into new technologies, data gaps emerge and must be filled by experimental or other cutting-edge information.

Once all the inventory data has been gathered, it can be transferred into an appropriate framework. There are several computer programs that support the development of LCA models (e.g. Umberto, SimaPro, GaBi) and the net environmental impacts of all the emissions (such as gaseous, liquid and solid) can be easily worked out. This is called the life cycle impact assessment phase and several environmental categories can be used to conduct the impact of the system studied. Table 8.2 indicates examples of potential environmental impacts, notably global warming potential (which adds up the combined effect of a number of GHGs, such as carbon dioxide, methane and nitrous oxide) and eutrophication (a measure of the addition of too many nutrients to nutrient-constrained ecosystems). Standardised impact factors are based on yearly evaluations by the Institute of Environmental Sciences in Leiden, the Netherlands.

Table 8.2 Example of the characterisation of inventory data

Environmental impacts

Inventory	Global warming (GWP)		Ozone layer depletion (ODP)		Eutrophication (EP)	
	Element	Factor (kg CO_{2eq}/kg)	Element	Factor (kg RII_{eq}/kg)	Element	Factor (kg PO_{4-eq}/kg)
	CO_2	I	CH_3Br	0,38	P	3,06
	CH_4	25	$C_2Cl_2F_4$	0,94	PO_4-	I
	N_2O	298	CCl_3F	I	H_3PO_4	0,97
	CF_3SF_5	17.700	$CBrClF_2$	6	N	0,42
	SF_6	22.800	$CBrF_3$	12	NH_4+	0,33
	CO_{3-eq}		RII_{eq}		PO_{4-EQ}	

The final interpretation phase is more or less a summary of the results and should be consistent with the scope and goal definition. It provides an understandable, complete and consistent report of the study. Weaknesses, such as missing data, should be mentioned as well as positive aspects, such as data reliability. An additional stage is the process of normalising the results using a weighting system. Each impact category is assigned a weight (such as the average impact per country or inhabitant) reflecting the relative importance of the assessed product. In principle, using these average weighted factors, all the impact categories considered are comparable.

This brief overview of an LCA process as described in the ISO 14040 series applies to all LCAs. Biomass, bioenergy, and even more specifically, biochar LCAs bring up methodological questions, which warrant more detailed attention. How the life cycle approach can be applied to the topic of biochar is described below, along with a consideration of which specific circumstances have to be considered to produce a reliable assessment. The LCA is developed for the North Sea region, accounting for technology, biomass potential, offsetting of energy, transport and other regionally-specific aspects. We also examine how the results differ between several countries in the North Sea region and what overall potential can be expected due to the utilisation of biochar in the agricultural and/or energy sectors in the region.

Life cycle methodology for biochar systems

Conducting an LCA for biochar systems raises problems, some of which are common to the use of LCAs, while others are specific to biochar systems. The major challenges, their potential impacts and possible solutions are described below.

Uncertain data

The greatest problem is likely to be the uncertainty of the data involved. As biochar systems are relatively novel and have not been implemented widely on standardised equipment, there is a definite lack of reliable data. When LCA studies have been conducted, they are, at best, based on a limited set of data, perhaps from one technology provider and from one field experiment. It is impossible to generalise the results on the basis of such a narrow dataset. Missing values in the dataset are often assumed or taken from other sources in the literature.

Uncertainty in data can be accounted for by conducting the LCA with a range of values for any one step in the chain. As a result, the LCA outputs are formatted as a range. This can be a highly time-consuming process and the range of values can be so large as to make the results appear meaningless. Another approach is sensitivity analysis, whereby one factor is changed at a time, and the effect upon the overall result is assessed. This approach can show which uncertainties have a greater influence on the overall result and which do not.

The uncertainties relating to biochar production, material properties and co-products are decreasing as production technology improves. A greater challenge is the uncertainty of the magnitude and duration of biochar effects in the field on crop growth (over a number of growing periods) and on soil nitrous oxide emissions. The duration is even more important than the magnitude: if the effect on crop yield and GHG emissions continues for 10, 20 or even more years, then the impact upon CO_2 (equivalent) emissions is that much greater than if the effect is only evident for one year or a few years.

Methodological problems

Methodological problems are sometimes more confusing than the lack of reliable data. Methodological problems all essentially pertain to deciding what to count and what to exclude from the LCA analysis and from the system boundary. These problems are generally common to other biomass and bioenergy LCA studies and there have been a number of useful articles and guidance documents written (e.g. Cherubini *et al.* 2009; Cherubini 2010; PAS 2050). However, before getting into detail, it is worth noting that each study usually makes methodological decisions, not all of which may be explicitly stated and which limit the comparability to other LCA studies. This situation must be minimised where at all possible.

Clarity of purpose is essential when making decisions on methodology. The purpose of the LCA often defines what decisions should be made. For example, if comparing pyrolysis technologies is the goal of an LCA, the effect of diverting wood chips to pyrolysis away from co-fired power stations does not need to be considered. On the other hand, when investigating the effect of encouraging greater use of pyrolysis where a range of bioenergy technologies are being compared, then the reduction in available fuel for co-firing should be considered. The major methodological issues are described below.

Attributional or consequential approach

Attributional LCA means the assessment of one complete production-use chain, attributing emissions to each step of the process. For example, this could be the growing of biomass, conversion using a particular named technology to biochar and thermal energy and the use of those two products in defined cases. Attributional LCA is useful in benchmarking and comparing different technologies or products.

Consequential LCA means the assessment of a wider system than a single production-use chain; it means assessing the consequences of selecting one use of a material over another. For example, when diverting biomass to biochar production, this may reduce the amount of biomass available for other industries, such as construction, wood panel board or co-firing. More wood may therefore be imported to meet demands from construction, or more coal burned to meet energy demands, when biomass is diverted to pyrolysis and biochar production. Consequential LCA is more useful for decision making at sector-wide or policy levels.

Counterfactuals

Counterfactuals are the 'baseline' situation against which an LCA is compared. Counterfactuals can be included implicitly or explicitly. An example could be the GHG footprint of a wood-fired bio-electricity power station. The GHG emission might be 200kg of CO_2eq/MWh when taking into consideration the production of fuel, transport, building and site overheads, etc. A comparison could be made to a gas-fired power station emitting 580kg of CO_2eq/MWh. The counterfactual is made explicitly. Alternatively, the counterfactual could be made implicitly based on the assumption that for every MWh produced by the wood power station, one MWh is not produced by the gas power station. In that case, the net GHG emission reported would be −380kg of CO_2eq/MWh for the wood-fired power station; i.e. the emission that was avoided is subtracted from the actual emission.

Although this may seem odd, implicit counterfactuals are common practice. In many cases, an LCA may be attributional in all parts except for the assumed 'energy offset'. Other counterfactuals are common too, such as emissions avoided from decaying wastes or alternative uses of feedstocks. Technically, if an LCA is to be attributional, counterfactuals should be included explicitly and not implicitly.

Definition and allocation of products

A system may have multiple products. These can be defined in a material sense: e.g. biochar, syngas and liquids. Or the products may be defined according to their function: GHG reduction, electricity and heat generation, phosphorous equivalent fertilisation, liming value, etc.

Allocation is the process of sharing the burdens of production and benefits from use between each product. Allocation is commonly done by economic value, by energy content, by mass or by substitution. For example, if the GHG impact of growing wheat on one hectare of land is 2t CO_2eq, the burden on production can be divided between the various products of the wheat field: grain, straw, chaff. If the economic value of the grain accounts for 95 per cent of the total economic value of the products (4 per cent to the straw and 1 per cent to the chaff), 95 per cent of the GHG burden of production would be assigned to the grain (with 4 per cent to the straw and 1 per cent to the chaff). Energy content or mass allocation works in a similar way, according to different ratios.

Allocation by substitution works differently and pertains more to consequential LCA. If a co-product substitutes for another material, the impact avoided for that material is taken into account. For example, if the straw generated from the wheat field is combusted in a power station offsetting coal combustion, then an avoided emission would be credited back to the wheat field (Aylott *et al.* 2012).

Timescale issues

There are a number of temporal issues to consider. The duration of the biochar effect in soil is a key uncertainty, as already discussed. The time period over which biochar remains in the soil is another. The regrowth of biomass converted to biochar should also be considered as the standard assumption that biomass is replaced instantly is obviously spurious. This has been the topic of recent debate (Aylott 2012; Cherubini *et al.* 2012).

As scientific consensus converges on the view that climate change is an increasingly urgent problem; actions taken within the next 20 years may prove to be decisive. If that is the case, then actions that increase atmospheric GHGs during the next 20 years but decrease them over the next 100 years would not be sensible. Harvesting timber, converting it to biochar and then regrowing it is one example of such a case. To take the example of a soft wood tree, the tree takes a period of approximately 30 years to grow to full height. During growth, the crop has taken-up (or sequestered) CO_2 from the atmosphere and converted it into solid carbon in the form of wood. When harvested and pyrolysed, approximately half of that carbon is turned into biochar and the other half converted to gases and vapours, probably combusted, and ultimately released back into the atmosphere as CO_2. At the point when the tree has been harvested and converted by pyrolysis, half the CO_2 is back in the atmosphere and half of it is sequestered in soil in the form of biochar. More CO_2 has been emitted into the atmosphere than if the tree had been left standing. However, assuming that the biomass is regrown (converting it into more wood) and begins to take up CO_2 again after a period of years, perhaps fifteen, the carbon balance will be neutral. By the time the next harvest is ready (30 years later), the carbon balance would be negative; i.e. more carbon would have been sequestered than emitted.

The convention in carbon accounting under the United Nations Framework Convention on Climate Change (UNFCCC) is that bioenergy is 'carbon neutral', which effectively assumes that the regrowth of biomass stocks occurs instantly, but this is obviously untrue. This convention creates risks including discounting indirect carbon emissions (e.g. demand for bioenergy, increasing demand for land, pushing up food prices, encouraging conversion of land somewhere else from a non-farmed to a farmed state and releasing potentially very large quantities of carbon stored in vegetation and soils). There is also a risk arising from not responding quickly enough to the threat of climate change.

If responses to climate change are required within the next 10 to 20 years, it may not be appropriate to take actions that emit more in the short term but have a net effect of removing (sequestering) CO_2 from the atmosphere in the longer term. Thus, the focus on the choice of biomass feedstocks could be changed to examine residues and waste streams rather than on longer term carbon initiatives using trees.

Life cycle study of six biochars in the North Sea region

The ultimate goal of the biochar concept is to store carbon in the soil over 100 years and preferably much longer. Converting biomass thermally into stable carbon compounds that would otherwise decompose to form GHGs, such as CO_2 or methane, can affect the net mitigation of GHGs.

In this assessment, six types of biomass (which reflects a common range within the North Sea Region), were chosen for biochar production and the mitigation potential was assessed. It was assumed that some of these potential feedstocks would be cultivated just for biochar production, while others would be organic waste materials arising from a biomass production system.

Each feedstock is transported and needs pre-treatment prior to pyrolysis. There are different pyrolysis technology approaches under development around the world, whereby e.g. different energy inputs and outputs need to be considered. The utilisation of biochar as a soil amendment feeding additive in the animal husbandry industry or as a constituent of animal bedding depends on the effectiveness of the biochar, the status of the market and the availability of alternative products to serve the same market needs. Carbonised biomass can also be used as a fuel substitute (e.g. for coal); however, according to the European Biochar Certificate, in that case it is not called biochar. The use of char as a fuel would depend upon its calorific value and other properties as well as upon energy market dynamics. Figure 8.3 illustrates the minimum number of different steps that must be taken into account to assess the environmental footprint of the production and utilisation of biochar in a consistent and comparable way with respect to other studies.

In this project, all the scientifically proven effects from the cultivation to the actual utilisation as a soil amendment and the energy generation were taken into consideration and applied to the conditions of the North Sea region. This conservative approach was adopted in the selection of values used, but those effects that are still under scientific discussion will be assessed within the potential analysis to identify the additional effects that are still to be examined.

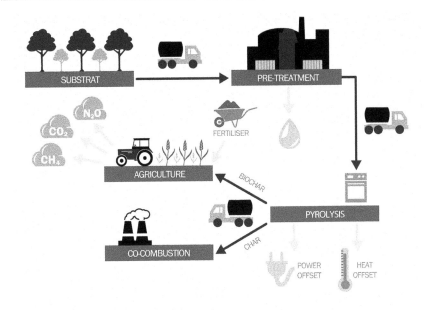

Figure 8.3 Life cycle of biochar production and utilisation.

Biomass feedstock provision

Most of the biomass feedstocks considered for biochar production have a low calorific value due to their high water content. Many lingo-cellulosic materials such as wood and straw have a high moisture content right after harvest (e.g. as high as 50 per cent). To avoid transporting large amounts of water, the feedstocks either have to be dried close to their production site or transport distances to pyrolysis facilities have to be minimised.

In this assessment, a standard European lorry was chosen for transportation with an average utilisation rate. The distance, the road type used and the utilisation rate might differ within the North Sea region and are implemented in this assessment as well (see Table 8.3).

Pyrolysis technologies vary and some require a feedstock with a low moisture content (e.g. 10 to 20 per cent), while others can cope with moisture contents of 50 per cent or even higher (Libra *et al.* 2011). Air drying is the easiest and cheapest approach, but according to Tuch (2006), it takes several years to achieve a water content of 10–20 per cent. Outside storage of biomass from 6 months to one year is generally sufficient to dry biomass for energy purposes under European conditions.

The optimal water content of biomass feedstock for continuously running slow pyrolysis is at about 15–20 per cent – otherwise too much of the product gas has to be used for preheating purposes (Sehn and Gerber 2007; Roberts *et al.* 2010). Two drying systems are considered in this analysis. For relatively dry substrates (e.g.

Table 8.3 Inventory data of the substrate supply (from the field to the next treatment stage)

Substrate		Commercial softwood pellets	Short rotation coppice pellets	Anaerobic digestate	Mixed wood chip	Green waste	Oil seed rape straw pellets
Transport	[–]						
Distance	[km]	80	80	40	80	80	80
Urban	[%]	30	30	30	20	60	30
Rural	[%]	50	50	50	42	30	50
Highway	[%]	20	20	20	38	10	20
Utlilization	[%]	80	60	90	85	60	60
Payload	[t]	17.3	17.3	17.3	17.3	17.3	17.3
Losses	[%/kgFM]	2	2	2	2	2	2
Water content	[%/kgFM]	50	50	94	50	52	20

wood), the most suitable technology is thermal drying. It is assumed that 1.4MJ thermal energy is required to dry 1kg of mixed wood chip. Owing to the high water content of the anaerobic digestate (see Table 8.4), mechanical drying was used prior to the thermal treatment. A large part of the water can be separated mechanically, but 2.7MJ of thermal energy was still used to achieve a water content of around 14 per cent. The low water content (less than 15 per cent) minimises the possibility of decomposition during long term storage, thus avoiding GHG emissions (methane and nitrous oxide; Trippe *et al.* 2010; UBA 2010). Another pre-treatment is pelleting to achieve a homogeneous material for pyrolysis. Table 8.4 indicates all the system combinations considered.

Most pyrolysis units require a feedstock with a relatively high heating value if more energy is to be produced from the process than is added to power the operation.

According to different pyrolysis manufacturers, a heating value of 10–15MJ kg^{-1} is desirable (Bridle 2011); otherwise the process might not run under stable conditions because of the low thermal energy resulting from the combustion of the product gases. Otherwise, additional fossil fuel is needed to maintain the process. Besides the heating value, the particle size of the feedstock (too many large particles might cause blockages) and the surface structure are also very important to achieve a reliable process (Ortwein *et al.* 2010). Some of the state-of-the-art pyrolysis units often have screw feeders for the transport of the feedstock into or even within the kilns to carry the biomass. Using mixed wood chips and green waste, which have heterogeneous properties, may result in a less complete conversion of carbonaceous feedstocks into char or biochar (Trippe *et al.* 2010). Pellets, which have uniform surface structure and size, are perfectly suitable for the pyrolysis process but

Table 8.4 Pre-treatment of the substrates using state-of-the-art technology

Pre-treatment		Commercial softwood pellets	Short rotation coppice pellets	Anaerobic digestate	Mixed wood chip	Green waste	Oil seed rape straw pellets
Watercontent input	[% kg⁻¹]	40	40	94	40	42	16
Mechanical drying	[y/n]	n	n	y	n	n	n
Thermal drying	[y/n]	y	y	y	y	y	y
Pelletising	[y/n]	y	y	n	n	n	y
Watercontent output	[% kg⁻¹]	12.1	12.9	13.9	12.2	11.9	14.4
High heating value	[MJ kg⁻¹%]	19.7	19.2	17.6	20.2	11.4	19.2

pelleting requires energy (e.g. an additional 1.9MJ of power per kilogram output is required for their production; Hagberg et al. 2009).

Pyrolysis

At present, there are few operational slow pyrolysis units throughout Europe. Different technological approaches exist, including batch, continuous or microwave systems, all of which run under limited oxygen supply. The analysis undertaken here considered a continuous process capable of utilising a diverse range of feedstock at the rate of 100 to 150kg of dry matter per hour and an average operating time of approximately 6500h per year. The feedstock is fed into the kiln by a rotary wheel sluice and is conveyed upwards and through by a screw auger. The biomass releases flammable gases and vapours, which are drawn into the combustion chamber. The temperature of the vapours at combustion is usually above 1250°C, and these hot exhaust gases are fed back into a chamber to heat up the kiln (in which the temperature is maintained at approximately 500 to 600°C). A heat exchanger can be installed at the back end of the unit to extract remaining heat from the combustion gases. The pyrolysis unit needs some starting energy (liquefied or natural gas) to heat up the process chamber to the desired process temperature (around 500 to 600°C). Additionally, some power is used for the drive and control technology, which is provided from the electricity grid. A char yield of between 24 and 60 per cent based on dry matter has been modelled during an average residence time of about 20 min in the hot zone and a total residence of up to 45 min in the unit (see Table 8.5). In other studies, the range is smaller and usually a char yield between 20 to 41 per cent is expected (Antal and Gronli 2003; Bridle 2011; Quicker and Schulten 2012). Commonly, 50 per cent of the biomass carbon is converted into char (Lehmann 2007; Gaunt and Lehmann 2008).

Table 8.5 Yields and high heating values of the six different thermal converted biomasses

Pyrolysis		Commercial softwood pellets	Short rotation coppice pellets	Anaerobic digestate	Mixed wood chip	Green waste	Oil seed rape straw pellets
Char yield	[% kg$_{DM}$$^{-1}$]	23.9	24.9	44.0	25.2	60.6	30.4
High heating value	[MJ kg^{-1}]	33.6	31.6	16.9	32.2	8.0	25.7
Carbon	[% kg kg$_{Char}$$^{-1}$]	0.89	0.86	0.52	0.87	0.18	0.69
Liquid yield	[% kg$_{DM}$$^{-1}$]	37.0	33.9	24.3	33.9	12.9	26.2
High heating value	[MJ kg^{-1}]	12.8	11.6	10.9	13.0	13.8	16.0
Syngas yield	[% kg$_{DM}$$^{-1}$]	39.1	41.2	31.7	40.8	26.5	43.5
High heating value	[MJ kg^{-1}]	15.3	13.2	11.3	13.3	11.5	11.9

Source: Interreg IVB North Sea Region Programme, 2007–2013.

The relevant factors for the direct effects on the mitigation potential are shown in Table 8.5. For the utilisation of the biochar as a form of carbon storage in soils, its carbon content is of the greatest interest. This varies between 18 and 89 per cent for the six feedstocks. Usually 20–30 per cent water is added to the hot char to prevent self-ignition and dust emissions. The process itself requires about 5 to 10 per cent of the energy based on the heating value of the different feedstocks; additionally, 10 per cent losses in the process and during the transportation are taken into consideration (Hammond et al. 2011). Bridle (2011) considered a higher loss of roughly 20 per cent, but this does not reflect the real values of the modelled continuous pyrolysis unit. But there is still some useable thermal energy left (about 120 to 150 kW) from the kiln. The offset of GHG emissions is based on a European grid mix average for district thermal heating, which was produced by natural gas combustion in 2010.

Utilisation

The mitigation potential of char application to soil is influenced by various direct and indirect effects. Some of these effects are already scientifically proven and some are still the subject of research or only applicable to some types of soil (Chapters 4 and 7). The amount and the stability of the biochar carbon, as well as the possible enhancement of soil organic carbon, are direct effects. The carbon in the char consists of several fractions – stable, labile and highly labile. The highly labile fraction is described as the part of the biochar with no long-term sequestration effect and which is degraded after a few hours or years (Lehmann et al. 2009). The

Table 8.6 Direct and indirect factors for the chars considered

Agriculture		Commercial softwood pellets	Short rotation coppice pellets	Anaerobic digestate	Mixed wood chip	Green waste	Oil seed rape straw pellets
Direct	Stable carbon [kg C kg_{char}^{-1}]	0.89	0.86	0.52	0.87	0.18	0.69
	Labile carbon [kg C kg_{char}^{-1}]	1.2×10^{-5}	1.8×10^{-3}	2.7×10^{-3}	1.9×10^{-3}	2.2×10^{-3}	5.5×10^{-4}
	Stability factor [kg C_{stable} $100a^{-1}$]	0.89	0.92	0.98	0.87	0.98	1.01
Indirect	Priming effect [kg CO_2 kg_{char}^{-1}]*	1.6×10^{-4}	-1.5×10^{-4}	-1.4×10^{-4}	2.2×10^{-3}	1.4×10^{-3}	-1.4×10^{-4}
	Liming [kg $CaCO_3$ kg_{char}^{-1}]§	0.06	0.57	0.09	0.05	0.10	0.05

Note: * Each year; § every fifth year and equivalent calculation is based on Field et al. (2013).

stable fraction was assessed over a time frame of 100 years in this analysis and varied between 86 and 98 per cent (see Table 8.6) within the different chars based on an accelerated ageing method (Cross and Sohi 2013).

Due to the conservative approach of this assessment, only some indirect effects are taken into consideration. Mineral fertiliser savings have been considered, but in the field trials in the North Sea region, no significant impact was drawn and therefore a likely reduction of 1 per cent N_2O emissions per saved kilogram of the applied nitrogen cannot be counted – according to the Intergovernmental Panel on Climate Change guideline for agricultural emissions (De Klein *et al.* 2006). For both savings, a possible carbon equivalent offset was considered in the model under European conditions, but owing to the lack of evidence, no impact could be assessed. Additional indirect effects such as N_2O suppression, CH_4 flux change or crop yield are assessed here in addition to consideration of different country conditions on the carbon offsetting or rather the utilisation of the carbonised biomass in the energy sector.

Figure 8.4 illustrates the results of the LCA of six carbonised feedstocks applied to the soil. They are directly comparable because they are based on the same system boundaries and the considered pyrolysis processes. Four different bars indicate the contribution to the overall mitigation potential. The production chain of the char covers cultivation, harvest, transport, pre-treatment and the pyrolysis process itself. Feedstocks with high water content and additional conditioning (e.g. pelleting) generate more GHG emissions in supplying one kilogram of feedstock ready for pyrolysis than the same feedstock when it has a lower moisture content or is not processed further.

The highest feedstock supply emissions are associated with anaerobic digestate (0.52t CO_{2eq} per $kg_{Feedstock}$) and wood pellets (e.g. short rotation coppice with 0.41t CO_{2eq} and GWP with 0.36t CO_{2eq} per $kg_{Feedstock}$). One exception is the oilseed rape straw pellets, which is due to the small amount of thermal energy required for their pelleting. Green waste and mixed wood chip feedstock emit less GHG per kg of delivered fuel owing to the relatively low need for pre-treatment.

The mitigation potential of the carbonised feedstock is mainly influenced by the char yield of the different feedstock and the stability factor of the char, compared to the assumed degradation of the feedstocks over a specific time period. The green waste shows the highest char yield, but it consists primarily of ashes rather than carbon. 283g CO_{2eq} per $kg_{Feedstock}$ could be directly mitigated. The char from anaerobic digestate of mixed organic waste shows the highest net mitigation effects, despite the high ash content. Owing to the very high char yield and stability factor, this leads to a possible abatement of 624g CO_{2eq} per $kg_{Feedstock}$. All other chars show a similar effect and it ranges between 536 and 590g CO_{2eq} per $kg_{Feedstock}$.

The soil effects (i.e. owing to saved mineral fertiliser) are negligible over the considered time horizon of five years. But the thermal energy offset is an important factor and reflects the different heating values of the gas and liquid phases. The more char produced, the less thermal offset is achieved. In this first assessment, no electricity generation was considered because of the low technical readiness level

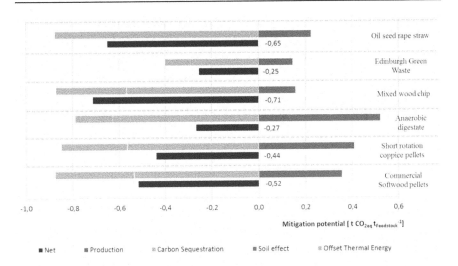

Figure 8.4 Mitigation potential of GHGs from six different biochars produced in state-of-the-art pyrolysis units in EU-25 and their utilisation in the agricultural sector.

of suitable applications. Commercial soft wood has the highest potential with 336g per $kg_{Feedstock}$. The anaerobic digestate and the green waste case can gain significantly less thermal energy offset due to the high solid fraction after the pyrolysis process. As a result, there is less total amount available for the production of energy.

All these assessments clearly indicate that each feedstock should be assessed separately. Some biomasses and their pre-treatment cause more emissions but show a higher mitigation potential at another stage during the life cycle. In this comparative study of the life cycle of the different feedstocks evaluated, mixed wood chips demonstrated the greatest mitigation potential at -0.71kg CO_{2eq} per $kg_{Feedstock}$. The pellet chars have similar amounts from -0.44 to -0.65kg CO_{2eq} per $kg_{Feedstock}$. The green waste and the anaerobic digestate show a relatively low potential of -0.25 to -0.27kg CO_{2eq} per $kg_{Feedstock}$ in total (see Table 8.7). Com-

Table 8.7 Carbon mitigation values for materials along the process chain

Global warming potential		Commercial softwood pellets	Short rotation coppice pellets	Anaerobic digestate	Mixed wood chip	Green waste	Oil seed rape straw pellets
Substrate	[t CO_{2eq} $t_{Substrate}^{-1}$]	-0.21	-0.18	-0.08	-0.30	-0.11	-0.59
Feedstock	[t CO_{2eq} $t_{Feedstock}^{-1}$]	-0.52	-0.44	-0.27	-0.71	-0.25	-0.65
Char	[t CO_{2eq} t_{Char}^{-1}]	-2.57	-2.11	-0.73	-3.35	-0.49	-2.59

Note: Substrate = raw material; feedstock = input in pyrolysis unit; char = at field.

pared to previously conducted LCAs, similar results were obtained to some extent. Roberts *et al.* (2010) identified a mitigation potential of 0.7 to 1.3t of CO_{2eq} per tonne of carbonised biomass (switchgrass and rapeseed straw) under North American conditions.

Additional potential of slow pyrolysis and carbonised biomass

Uncertain direct and indirect effects due to carbonisation of biomass and its actual utilisation so far may have significant impact on the overall mitigation potential. The mitigation potential is sensitive to change in the key variables relating to energy generation, the application of biochar to soil and the co-combustion of carbonised biomass. The offsetting of thermal energy generation is assessed in the context of the North Sea region. Albeit that the power production by hot air gas turbine is in an early technological stage of readiness, the second offsetting concerns the additional benefit of a likely power generation as well (see Figure 8.5). The following assumptions were made to assess the additional impact of further technology enhancement on the mitigation potential due to the use of the pyrolysis unit to produce biochar.

Thermal

The thermal energy mix to be offset is based on natural gas combustion in the seven countries of the North Sea region plus one generic mix, which is valid for the EU-15 countries (55 per cent heating oil, 45 per cent natural gas and 10 per

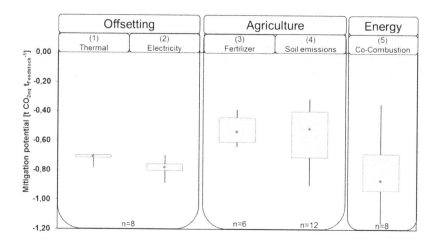

Figure 8.5 Analysis of the carbon mitigation potential of different factors along the production and utilisation of carbonised biomass in six countries in northern Europe.

cent hard coal). The provision of thermal energy differs in each country due to the transport system, availability and the technology used. The additional potential is based on the properties of the carbonisation of mixed wood chips.

Electricity

Some thermal energy can be used via a hot air gas turbine to produce electricity. The technological readiness level of these turbines is low and still under investigation. The amount produced will be credited based on seven different power grid mixes plus one direct substitution for a hard coal power plant (assumed that a pyrolysis unit and a power plant have the same hours of operation or base load). When using the thermal energy in different countries, only minor additional mitigation potential was identified (−3.12 and −0.29 per cent) as long as natural gas was substituted. A comparison to the assumed fossil mix would have an additional mitigation potential of −9.4 per cent. If some electricity is generated during the carbonisation of 1kg of mixed wood chips, an average additional potential of −8.8 per cent (−0.061kg CO_{2eq}) has been identified. Owing to the different shares of fossil fuels in the power grid mixes around northern Europe, different amounts of CO_{2eq} can be mitigated by supplying 1kWh of electricity to the grid. A deviation of +10 per cent and −13 per cent within the North Sea region is possible. Using fast rather than slow pyrolysis, implying a lower char yield, has an effect on the carbon abatement as well, but this has already been discussed by Hammond et al. (2011).

The most important variables concerning carbon abatement are carbon content and biochar carbon stability, but reduced fertiliser use, a change in soil emissions, the effect over a longer horizon and the change in non-biochar soil organic carbon have to be considered as well. Many of these effects are still under discussion, however, and are highly dependent on the actual region and soil conditions in European conditions (Libra et al. 2011). The water holding capacity might be influenced by biochar (see Chapter 3), especially for sandy soils, which leads to less irrigation and therefore less energy consumption from water extraction and pumping. Meyer et al. (2012) reported that the overall climate impact of biochar systems can be reduced between 13 and 22 per cent due to the albedo effect. However, the effect depends strongly on the background soil, the type of biochar and the cultivated crop. Very dense and year-round vegetation as well as soil that's already darker might decrease the negative effect. The likely positive change in biomass yield has an impact as well, but is not analysed owing to its dependency on the actual crop, soil, climate and other conditions.

Fertiliser

The duration of the effects is extended to 100 years. Carbon equivalents offsetting once for saved fertiliser per hectare (N −30kg, P −10kg, K −5kg) and fewer N_2O emissions from the soil (1 per cent of applied N according to IPCC Guidelines) for all biochars − no offsetting of the energy.

Soil emissions

The change of methane and nitrous oxide emissions in comparison to the reference system [CH_4 (−10% − + 5%) and N_2O (−20% + 10%)] might have an impact as well. The range chosen in this part is a best/worst case approach and the study was conducted over a period of 100 years.

In considering less fertiliser use and fewer soil emissions over a time horizon of 100 years, an additional mean benefit of 43 per cent (0.17kg CO_{2eq} per one-off application) has been indicated. An average deviation of ±18 per cent of carbon dioxide equivalent emissions by all different chars are analysed. Due to the likely impact of biochar on the nitrous oxide and methane emissions, an almost negligible increase of 5 per cent (0.03kg CO_{2eq}) was identified. This impact is highly dependent, e.g. on the soil and chars, which is indicated by the identified wide range of about +21 per cent to −11 per cent over a time period of 100 years. In terms of hard coal substitution, the most important factor is the heating value and the milling quality of the char (Quicker and Schulten 2012). In addition to the anaerobic digestate and the green waste char, all others are comparable to fossil coal and might easily be incorporated in energy production.

Co-combustion

Carbonised biomass used in energy production. The offset is either based on the energy grid mix of the North Sea region countries (different specific CO_{2eq} emissions per kWh of electricity due to the diverse shares of power generation technologies) or as one direct substitution of hard coal in a combined heat and power plant. This means in particular that owing to the substitution of fossil fuel, the carbon, which is bound in the char, is released to the short carbon cycle while the carbon from the fossil fuels is not emitted. Owing to the properties of most chars, it is likely to substitute hard coal to some extent in the power plants and gains the highest abatement in the short-term carbon cycle.

Owing to the wide range of power generation sources within the different countries in northern Europe, a wide range of carbon mitigation potential has been indicated. A mean average of −0.86kg CO_{2eq} per $kg_{Feedstock}$ was analysed. The minimum of −0.34kg CO_{2eq} per $kg_{Feedstock}$ can be gained in Norway due to the grid mix having a very low carbon intensity (because of the high use of energy generated from renewable sources). The highest carbon abatement yield is reached by the direct substitution of hard coal (−1.15kg CO_{2eq} per $kg_{Feedstock}$). But co-combustion does not actually save any emissions in the long term, it more or less merely postpones the problem. This is due to the global trading of fossil fuels. One country or company can substitute their fuels with carbonised biomass on the one hand, but the demand is high enough worldwide that the avoided emissions will be produced anyway somewhere in the world. Therefore, as long the electricity or thermal energy is still based on coal firing, the use of carbonised biomass can have a regional impact but has no real impact on the worldwide need to abate GHG emissions.

Bibliography

Antal, M. and Gronli, M. (2003). The art, science, and technology of charcoal production. *Industrial Engineering Chemical Research* 8: 1619–1640.

Aylott, M. (2012). *Bioenergy: A Sustainable Solution for Electricity Generation in the UK?* Heslington: NNFCC.

Aylott, M., Higson, A., Evans, G. and Mortimer, N. (2012). *Measuring the Energy and Greenhouse Gas Balances of Biofuels and Bio-based Chemicals using LCA.* Heslington: NNFCC.

Bridle, T. (2011). *Pilot Testing Pyrolysis Systems and Reviews of Solid Waste Use on Boilers.* North Sydney: Meat and Livestock Australia.

Cherubini, F. (2010). GHG balances of bioenergy systems: overview of key steps in the production chain and methodological concerns. *Renewable Energy* 35: 1565–1573.

Cherubini, F., Bird, N., Cowie, A., Jungmeier, G., Schlamadinger, B. and Woess-Gallasch, S. (2009). Energy- and greenhouse gas-based LCA of biofuel and bioenergy systems: key issues, ranges and recommendations. *Resources, Conservation and Recycling* 53: 434–447.

Cherubini, F., Guest, G. and Strømann, A. (2012). Application of probability distributions to the modeling of biogenic CO_2 fluxes in life cycle assessment. *GCB Bioenergy* 4(6): 784–798.

Cross, A. and Sohi, S. (2013). A method for screening the relative long-term stability of biochar. *Global Change Biology Bioenergy* 5(2): 215–220.

De Klein, C., Novoa, R., Ogle, S., Smith, K., Rochette, P., Wirth, T., McConkey, B., Mosier, A., Rypdal, K., Walsh, M. and Williams, S. (2006). N_2O emissions from managed soils, and CO_2 emissions from lime and urea application. In J Penman, M. Gytarksy, T. Hiarishi, W. Irving and T. Krug (eds), *2006 IPCC Guidelines for National Greenhouse Gas Inventories,* vol. 4, 11.1–11.54. Geneva: IPCC.

Field, J. D., Keske, C., Birch, G., DeFoort, M. and Cotrufo, M. (2013). Distributed biochar and bioenergy coproduction: a regionally specific case study of environmental benefits and economic impacts. *GBC Bioenergy* 5: 177–191.

Gaunt, J. and Lehmann, J. (2008). Energy balance and emissions associated with biochar sequestration and pyrolysis bioenergy production. *Environmental Science and Technology* 42: 4152–4158.

Hagberg, L. Särnholm, E., Gode, J., Ekvall, T. and Rydberg, T. (2009). *LCA Calculations of Swedish Wood Pellet Production Chains.* Stockholm: IVL Swedish Environmental Research Institute.

Hammond, J., Shackley, S., Sohi, S. and Brownsort, P. (2011). Prospective life cycle carbon abatement for pyrolysis biochar systems in the UK. *Energy Policy* 39: 2646–2655.

Hansen, J., Sato, M., Kharecha, P., Beerling, D., Berner, R., Masson-Delmotte, V., Pagani, M., Raymo, M., Royer, D. and Zachos, J. (2008). Target atmospheric CO_2: where should humanity aim? *Open Atmospheric Science Journal* 2: 217–231.

IPCC (2011). *Summary for Policymakers: IPCC Special Report on Renewable Energy Sources and Climate Change Mitigation.* Geneva: IPCC.

Jeffrey, S., Verheijen, F., van der Velde, M. and Bastos, A. (2011). A quantitative review of the effects of biochar application to soils on crop productivity using meta-analysis. *Agriculture, Ecosystems and Environment* 144(1): 175–187.

Lehmann, J. (2007). A handful of carbon. *Nature* 447: 143–144.

Lehmann, J., Czimczik, C., Laird, D. and Sohi, S. (2009). Stability of biochar in soil. In J. Lehmann and S. Joseph (eds), *Biochar for Environmental Management: Science and Technology,* 183–205. London: Earthscan.

Libra, J. Ro, K., Kammann, C., Funke, A., Berge, N., Neubauer, Y. Titirici, M-M., Fühner, C., Bens, O., Kern, J. and Emmerich, K-H. (2011). Hydrothermal carbonization of biomass residuals: a comparative review of chemistry, processes and applications of wet and dry pyrolysis. *Biofuels* 2(1).

Meyer, S., Bright, R., Fischer, D., Schulz, H. and Glaser, B. (2012). Albedo impact on the suitability of biochar systems to mitigate global warming. *Environmental Science and Technology* 46: 12726–12734.

Oguntunde, P., Abiodun, B. and Ajayi, A. v. d. G. N. (2008). Effects of charcoal production on soil physical properties in Ghana. *Journal of Plant Nutrient Soil Science* 4: 629–634.

Ortwein, A., Klemm, M. and Kaltschmidt, M. (2010). Innovative Verfahrensoptionen zur Biomasseverwertung. Conference: Pyrolyse, HTC, Biochar & Co. Klimaschutz mit Biokohle – eine Chance für die Umwelt und Landwirtschaft?, 17 March, Höchst im Odenwald.

Quicker, P. and Schulten, M. (2012). Biokohle: Erzeugung und technische Einsatz-möglichkeiten. *Müll und Abfall* 9: 464–475.

Roberts, K., Gloy, B., Joseph, S., Scott, N. and Lehmann, J. (2010). Life cycle assessment of Biochar-Systems: Estimating the energetic, economic and climate change potential. *Environmental Science Technology* 44: 827–833.

Sehn, W. and Gerber, H. (2007). Pyrolyse mit flammenloser Oxidation kombinieren. *BWK – Das Energie-Fachmagazin* 10: 22–26.

Sohi, S., Krull, E., Lopez-Capel, E. and Bol, R. (2010). A review of biochar and its use and function in soil. *Advances in Agronomy* 105: 47–82.

Trippe, F., Fröhling, M., Schultmann, F., Stahl, R. and Henrich, E. (2010). Techno-economic analysis of fast pyrolysis as a process step within biomass-to-liquid fuel production. *Waste and Biomass Valorization* 1(4): 415–430.

Tuch, U. (2006). *Biomassehöfe: Mittler zwischen Waldbesitzer und Verbraucher.* Hamburg: Institut für Ökonomie der Bundesforschungsanstalt für Forst- und Holzwirtschaft.

UBA (2010). *Klimaschutzpotenziale der Abfallwirtschaft: Am Beispiel von Siedlungsabfällen und Altholz.* Darmstadt: Bundesministerium für Umwelt, Naturschutz und Reaktorsicher-heit.

Woolf, D., Amonette, J. E., Street-Perrott, F. L. J. and Joseph, S. (2010). Sustainable biochar to mitigate global climate change. *Nature Communications* 1: 56.

Chapter 9

The economic viability and prospects for biochar in Europe

Shifting paradigms in uncertain times

Simon Shackley

The dismal science?

Biochar is one of the most intriguing ideas to have emerged in the maelstrom of thinking and dialogue surrounding climate change and carbon reduction in the past two decades. It also happens to be one of the most complex – a heady and often confusing brew of soil science, bio-resources, agronomy, carbon markets, bio-energy technology, regulations and climate change negotiations. Yet, the harsh reality is that producing and applying biochar in European agricultural soils at application rates of several tonnes per hectare across large hectarage from virgin or non-amended organic feedstocks is not economic now or in the foreseeable future. The reasons for this seemingly bold statement are the following:

1   The feedstocks are expensive, with wood pellets trading in European markets at €120 per tonne, straw selling for €50 to €100 per tonne; forestry residues are available but expensive to collect up from the forest site where they are generated, etc. Since, as an approximation, three tonnes of feedstock are required for each tonne of biochar produced, the feedstock costs alone are from €150 to €500 per tonne of biochar. By comparison, synthetic fertilisers cost around €250 to €350 per tonne. Once we factor in the capital, operational and maintenance costs entailed in biochar production, not to mention the expected rate of return on investment, the break-even cost of the biochar is likely to be at least double the feedstock cost.

2   The technology for producing biochar is not at an advanced or mature stage of development. Unit costs for new technologies are always much higher than subsequent production costs, as dominant designs become prevalent and as cycles of learning and economies of scale kick in. Current technologies are focused on biochar production alone, but the biochar is only one-third by weight of the biomass input, meaning that the production of biochar alone is inherently inefficient. Biomass combustion for heat production, by contrast, uses 90–95 per cent of the energy value of the feedstock. Straw burning to generate steam to drive a steam turbine is roughly 35–40 per cent efficient, comparable to the efficiency of burning biomass in place of coal in a

pulverised fuel power plant. Slow or fast pyrolysis technologies which are capable of recovering a much larger proportion of the energy value of the feedstock than the biochar-alone technologies are required; for example, through use of syngas for electricity and pyrolytic liquids as chemical feedstocks. These designs are not yet available and significantly more R&D will be required to design and manufacture efficient, affordable equipment which can generate energy by-products.

3   Inefficient use of biomass where there is a demand for energy brings no reduction to net carbon emissions since some other energy carrier has to be used instead. Consider, for example, where an existing use of biomass for heat production is reduced because some of the biomass is diverted to biochar production instead. Suppose that coal is used as the substitute fuel but because of the high carbon emissions associated with coal combustion, the overall carbon emissions will have increased compared to the situation where biochar is not produced, even taking into account the storage of carbon within the biochar incorporated into soil.

4   The inherent value of the 'pure' biochar is limited. Plain biochar is more like a bulk commodity than a speciality chemical. If the cost per tonne is about €300 to €1000, the evidence from the majority of the field trials undertaken in Europe to date is that no farmer would be anywhere close to realising a good return from investing in pure biochar as a soil amendment. Let us make a highly optimistic assumption that biochar increases yield by 5 per cent per tonne of biochar applied per hectare. This would imply an investment that is similar to, or greater in, size to the cost of synthetic chemical (NPK) fertilisers. For an 8 tonne per ha barley farmer, the value of the 5 per cent uplift would be approximately €85, which is much less than the input costs. Only if the 5 per cent uplift were sustained for several seasons – five or more perhaps – and with no additional inputs or costs, would the plain biochar investment begin to stack up.

5   Based on the agronomic research record to date, the 5 per cent grain yield increases are more typical of 10 to 20t/ha biochar application rates, implying an agronomic value of only €5 to €10 per tonne biochar. The up-front investment might be €3000 to €10,000, which is a huge investment for a farmer to make per hectare. For the barley example above, the gross margin (value of crop minus total variable costs) would go from positive to negative and the crop yield would need to more-or-less triple to maintain the existing gross margin over one year. Alternatively, the investment in biochar addition would break even or be profitable if the improved crop yield is maintained for many years after soil amendment without need for further biochar additions. Agronomic evidence to date is far from conclusive on whether there is a persistent effect of biochar season to season in temperate soils. If there is a year-on-year effect, the value of the biochar would increase commensurately with the longevity of the impact.

6   Assuming the farmer would not be willing to pay more than, say, €200 per tonne biochar, why would the biochar producer choose to sell the material as

biochar when there is a European market for charcoal that is worth from €300 to €1000 per tonne? Why not simply take the extra profit by selling it as charcoal?

7   Carbon financing *could* make a difference but it would have to be set at a much higher level than is currently envisaged. If we assumed a value of €40 per tonne of CO_2, which is widely regarded as a reasonable level up to 2020, and assuming one tonne of biochar abates on average one tonne of CO_2e (equivalent), then carbon financing adds a further €40 per tonne on top of the agronomic value, which, let us assume, ranges from €5 to €85. The total value of the biochar would then range from €45 to €125, but this is still well below the cost of the biochar. Carbon financing would need to be more like €200 per tonne CO_2e before it could begin to drive forward biochar projects based upon the above agronomic findings.

The more detailed calculations underpinning the above assertions can be found in a series of peer-reviewed publications, including McCarl *et al.* (2009), Granatstein *et al.* (2009), Shackley *et al.* (2011), Galinato *et al.* (2011), Field *et al.* (2013), Spokas *et al.* (2013), Hammond *et al.* (2013), Shackley *et al.* (2015), among others.

Economics has been accused of being the 'dismal science' and one conclusion from the above analysis is that pyrolysis–biochar systems (PBS), while a nice concept scientifically, will never 'stack up' financially in the European context. Whether it will stack up in other regions of the world remains to be seen, but one recent study in four Chinese provinces found that even in China it was only likely to be viable in one province, a major reason being rural labour shortages due to higher earnings opportunities in the cities (Clare *et al.* 2014a, 2014b). An analysis of an existing pilot biochar production facility in Malaysia using the extensive bio-residues from oil palm plantations, where there is pressure to utilise such residues more sustainably, indicated that the biochar would need to be worth over $500 per tonne to make a reasonable return on capital investment (Harsono *et al.* 2013), again suggesting that the above analysis might well apply also to some developing and industrialising countries where labour constraints and need for rapid return on capital expenditure are evident.

However, the dismal edict of conventional project investment economics is not an inevitable consequence of the above analysis. There are a few possible routes by which pyrolysis-biochar options could become financially viable now and, possibly more so, in the future. These possibilities are discussed in the following sections.

Towards viability: route one – pyrolysis as a means of treating organic waste materials

European societies and economies generate large quantities of biodegradable waste including food waste, sewage sludge (SSW), manures and slurries (e.g. more than 100 million tonnes each year in just the UK, most of which is currently recycled by spreading it onto land). The formal policy for reducing the quantities of such wastes – called the 'waste hierarchy' – has been (i) reduce, (ii) re-use and (iii) recycle, in that order.

Considerable success has been achieved in source separation of organic from other waste types and in reducing and recycling organic waste arisings from households and businesses through a range of mechanical, biological and chemical treatment processes. The major organic waste treatment options are incineration, AD (both with energy recovery) and composting. This organic waste processing has created large supplies of composts, compost-like outputs, liquids containing bio-solids such as SSW and the products of biological digestion by fermentation and anaerobic decomposition (generating methane as an energy carrier).

The European Waste Framework and Landfill Directives have driven organic waste management policy forwards – in part through requiring the setting of national targets for reduction in organic waste arisings to landfill. EU municipal waste generated in 2012 was 487kg per capita (compared to 523kg in 2007); in 2012 just over 41 per cent of such waste was recycled and 29 per cent was landfilled. There has been a decline in waste that has gone to landfill over the past decade and a commensurate increase in alternative treatment options. Significant reductions in waste sent to landfill have been achieved in Germany, France, Sweden and Spain, where incineration of waste for energy recovery is relatively high. However, even in countries with similar waste generation and lifestyle (e.g. Germany and the UK), the use of waste is very different. Germany has one of the highest recycling and re-use rates in Europe (Mühle *et al.* 2010), a result of stringent regulation and heavy investment in incineration for waste to energy, thus only 1 per cent of waste is sent to landfill. In comparison, the UK has historically been reliant upon landfill as the main route for waste disposal, but recycling is increasing (Mühle *et al.* 2010) and in recent years, major progress has been achieved in many local authorities.[i] Germany has a unified system of waste collection and separation, whereas the UK has more variation.

The UK uses a landfill tax-based system whereby companies generating waste have to pay a set price for each tonne of waste disposed of. At the time of writing, these costs are approximately: €60/t of wood waste, €25/t of garden and green waste, €40/t of food waste and €50/t of SSW. If wood waste can be recycled then its gate fee (the price the waste producer has to pay the operator of the waste processing facility to take the waste) is reduced to approximately €20/t. The attraction of re-use and recycling will increase as landfill taxes continue to rise year-by-year, pushing up the gate fee.

Pyrolysis equipment could be operated as a cost-effective means of treating waste and converting it into a useful product, if so-called end-of-waste criteria can be achieved. The gate-fee income for wastes that are expensive to manage and process can turn around the economics of PBS such that the investment in the technology and biochar application are driven primarily by the receipt of the gate fee.

A possible example is a waste water treatment plant where SSW is the waste disposal problem. The company already has an anaerobic digestion (AD) unit in which SSW is processed to yield an AD digestate (ADD). The company currently disposes of the SSW by application to land, but this is becoming increasingly regulated and costly, hence the firm's interest in exploring whether thermal conversion through gasification or pyrolysis could be a more viable option in the longer term, in particular if the processed SSW and SSW ADDs could become biochar products with an agronomic value through satisfying end-of-waste criteria. A detailed techno-economic study was undertaken to scope out the viability of a PBS investment in this context. The key assumptions and findings of the study are shown in Box 9.1.

Box 9.1 Feasibility analysis for the production of bioenergy and deployment of biochar using sewage sludge waste/digestate: a case study of a water company

As part of a vision for carbon neutrality and environmental sustainability, the company sought to scope the potential for economic gain, improved carbon balance and electricity generation from integration of pyrolysis and biochar into the sewage treatment processes. An assessment was undertaken of the feasibility of using dried SSW and dry sludge digestate as alternative pyrolysis feedstocks, processed at two temperatures (550°C and 700°C) using technologies configured/integrated in five different ways at three possible scales.

Scales of operation considered were, in tonnes of dry solids processed annually (tds/yr), 5000 ('small'), 16,000 ('medium') and 32,000 ('large'). These were based on the experience of technology providers and the current capacity of the selected water works (35,000 tds/year SSW and 24,000 tds/yr digested sludge). Five different slow pyrolysis technology configurations were selected for comparison:

- *Scenario one:* Pyrolytic vapour used as a fuel in a boiler for the production of heat for drying, with no electricity generation.
- *Scenario two:* Pyrolytic vapour used for electricity generation in a Stirling engine and to provide extra heat for drying purposes.
- *Scenario three:* The syngas component of the pyrolytic vapour is combusted and the hot gases recycled to drive the pyrolysis process, while the bio-oil is combusted and the hot gases used to drive a micro-turbine for electricity generation; excess heat is used for drying purposes.
- *Scenario four:* The bio-oil component of the pyrolytic vapour is combusted and the hot gases recycled to drive the pyrolysis process while the syngas is combusted and the hot gases used to drive a micro-turbine for electricity generation; excess heat is used for drying purposes.
- *Scenario five:* The bio-oil component is recycled to the pyrolysis unit and cracked into additional syngas, the syngas then being used in a gas engine for electricity generation and extra heat for drying purposes.

Of these options, scenario one is by far the most realistic given existing technologies. Stirling engines (scenario two) have not found ready commercial applications in pyrolytic or gasification type thermal power plants. The problem in both scenarios one and two would be keeping the pyrolytic vapour at sufficient pressure and temperature that it could be burnt while avoiding condensation of tars. Scenarios three and four assume use of micro-turbine technology but, at the present time, such turbines are not efficient and not widely used for electricity generation. The other challenge with scenarios three and four is that the pyrolytic bio-liquids are a complex mixture of two different tar fractions and an aqueous fraction. At the present time, the best way of separating the three fractions is by leaving the liquids to stand for six months, requiring considerable storage space. The aqueous fraction might have some beneficial applications but could also constitute a waste product if not properly processed. The same liquid separation problem is encountered in scenario five, with the further challenge of how to crack the bio-oil into a syngas. In summary, while scenario one is at a reasonable level of technological readiness, the other four scenarios are all in the R&D stage or, at best, early demonstration stages and by no means mature. It is likely that considerable additional investment in RD&D would be required before scenarios two to five could be made credible from an engineering and economic viability perspective.

Two reference scenarios were used, with different assumptions of feedstock drying prior to pyrolysis. The first assumes that the feedstock (SSW or digested sludge) is not dried. No other fuel, such as natural gas, is required to provide heat for a drying process. In this case, the extra heat from the pyrolysis process does not displace energy provided by natural gas. The second reference scenario assumes that natural gas is used for drying purposes, with excess heat from pyrolysis displacing the energy provided by natural gas, with associated cost and carbon savings.

Key findings from experimental pyrolysis

On a dry mass basis with respect to feedstock, digested sludge yielded more char (54 per cent) than SSW did (40 per cent); however, this mainly reflected the higher ash content of digested sludge. Mass balance calculations showed that dried SSW yielded more gas (35 per cent) than digested sludge did (26 per cent); however, this reflected differences in organic content. Gas mass yields were higher when pyrolysis was undertaken at higher temperature, irrespective of feedstock (for digested sludge, 49 per cent at 700°C against 26 per cent at 550°C). Biochar from both dry digestate and dried sludge created biochar with a high stability; a test tentatively calibrated for equivalence to 100-year degradation in soil removed only between 3–12 per cent of carbon. Biochar created at 550°C sequestered 5–10 per cent less feedstock carbon into stable biochar than biochar made at 700°C. Assuming an application rate of 1 tonne per hectare, both biochar samples readily provided 3kg/ha phosphorus, 6kg/ha Mg and 34–40kg/ha Ca. Metals and metalloids are generally conserved within biochar during pyrolysis; with cadmium being the main exception.

Potential scale of biochar production and use

The maximum amount of biochar that could be produced under the selected scenarios would be 17,000 t/yr (digested sludge feedstock pyrolysed at 550°C at large scale). Grass and arable land within 50km of the potential site that is suitable for biochar deployment (excluding stony and peaty soils, natural protected areas, woodlands and cultural heritage sites) is approximately 70,000 ha; land availability offers no fundamental constraint for biochar deployment at lower (5 t/ha) or higher (20t/ha) application rates.

Assessment of energy balance based on experimental finding

Scenario five yielded the most electricity output (for either feedstock) due to re-circulation of pyrolysis liquids (oil) into the pyrolysis unit and further cracking into additional syngas; with capacities of 350–500kW for small scale, 1,400–2,000kW for medium scale and 2,800–4,000kW for large scale. Electricity outputs were highest when dry sludge digestate or dried SSW were processed at 700°C (3,100kWh and 4,000kWh respectively, for large-scale configurations), offering approximately 25 GWh/yr and 32 GWh/yr; the figure for dried SSW is similar to the current 30 GWh/yr output from the facility's existing AD process.

Scenario one yielded the most heat since the entire pyrolytic vapour stream was used to raise steam for heat with no electricity generation; heat outputs for this configuration ranged from 850–1,300kWh$_{thermal}$ for small scale, 3,400–5,200kWh$_{thermal}$ medium and 6,800–10,300kWh$_{thermal}$ for large scale. In terms of drying requirements, heat output from pyrolysis could provide between 30–80 per cent of the total requirement at the waste water facility, depending on technology configuration, feedstock and pyrolysis temperature. Where natural gas is used for drying purposes, the heat export from the pyrolysis unit would displace 40–100 per cent of the natural gas consumption.

Life cycle analysis

Net carbon abatement (i.e. t CO_2e/t feedstock arising from recalcitrant carbon in the biochar plus other greenhouse gas emission reductions) was calculated for all config-urations and feedstock options (a high carbon abatement implies removal of CO_2 from atmosphere as well as reduction of CO_2 entering atmosphere; hence, a positive carbon abatement value is equivalent to a reduction in climate forcing which leads to human-induced climate change; if expressed in terms of carbon mitigation, the value would be negative). The highest carbon abatement efficiencies were achieved under scenario five (both feedstock types), which also generated the highest electricity output. The highest carbon abatement efficiency under scenario five (1.66 t CO_2e/t feedstock) was achieved when digested sludge was pyrolysed at 550°C; larger indirect effects on the soil were also assumed in this scenario.

Annual carbon abatement ranges achieved under scenario five were 6,200–6,600t CO_2e for small scale, 24,700–26,500t CO_2e for medium scale and 49,400–53,000t CO_2e for large scale, depending on feedstock and pyrolysis temperature. Under scenario five, the impacts of biochar on the soil provided the greatest contribution to total carbon abatement, followed by the electricity offset (grid electricity displaced by electricity generated using pyrolysis by-products) and, thirdly, by carbon stored in biochar in soil. The agricultural impact of this biochar is currently unknown.

Techno-economic assessment

Assuming no value is assigned to the biochar product, the highest revenue would result from scenario five owing to the sale of electricity, with associated economic value of the Renewable Obligations Certificates (ROCs): €40–€95/t feedstock for small scale, €64–€123/t feedstock for medium scale and €76–€135/t feedstock for large scale (depending on the feedstock and pyrolysis temperature). Revenues are higher using dried SSW than dry digested sludge, due to higher gas yield and associated electricity generation. The impact on revenues is approximately doubled if it is assumed that pyrolysis heat displaces natural gas for drying purposes. It is important to note that in no circumstances did scenario one (the only technologically mature design in the study) return a profit, but always ran at a loss. Scenarios two, three and four sometimes ran at a loss, generally at the smaller scale and when natural gas was not being displaced by pyrolysis-derived heat. Hence, profitability is reliant on advances in technological innovation that is inherently uncertain and, to some extent, dependent upon the scale of technology.

An evaluation of the regulatory position

Neither the UK's Safe Sludge Matrix nor the Publicly Available Specification (PAS) 110 are relevant to land spreading of sludge or digested sludge once they have been pyrolysed. Because the biochar has been created from waste, it would be considered a waste unless end-of-waste criteria could be fulfilled. The production and land spreading of char would fall under Tier 3 of the Operational Risk Assessment (OPRA) scheme in the UK, meaning that a bespoke permit would be required. According to the Environment Agency, it is unlikely that the pyrolysis of dried SSW or dry sludge digestate would be exempt from the requirements of the Waste Incineration Directive (WID); consequently, air emissions monitoring would be required – according to a protocol that considers the appropriate level of gas cleaning (abatement) before burning for energy recovery.

The costs of permitting under OPRA were determined and found to be a reasonably modest proportion of total assessed costs (approx. 5–10 per cent); the costs of WID compliance could not be determined but could be (depending on monitoring deemed necessary) considerable. One fast pyrolysis demonstration plant in the UK was mothballed as a consequence of the costs of complying with environmental regulations (Cordner Peacocke, personal communication, 26 June 2013).

Is the project feasible?

If *optimistic* assumptions are made regarding the development of currently immature technology to higher levels of readiness, and assuming the capability and skills for designing, constructing, operating and managing a pyrolysis plant cost effectively and without excessive regulatory costs, a pilot pyrolysis project yielding biochar and bioenergy is feasible, on the basis of this study. There are benefits in terms of carbon storage in soil, though again the higher end values are based on optimistic assumptions about the indirect impacts of biochar in soil, which are currently unvalidated. Revenues are temperature and scale dependent, with a premium on treating SSW at larger scale and higher temperatures. Attractive revenues can also potentially be achieved if pyrolysis is undertaken utilising the digestate of AD only. Given the technological and regulatory uncertainties, however, a potential developer would need to be risk taking and set aside contingency funds such that the company could invest in technological improvements beyond the original budgetary allocation. It is perhaps telling that the company which commissioned the feasibility study decided against proceeding with a pyrolysis plant investment at the current time.

Based on a report for a water treatment company by Ibarrola *et al.* (2012).

Under certain optimistic assumptions, such as the availability of excess heat nearby for drying of the bio-liquids to an appropriate moisture content for pyrolysis, and assuming that electricity can be generated from the pyrolytic vapour generated with reasonable efficiency, plus responding to the growing problem and cost of bio-solids waste disposal, a PBS investment can make economic sense with a pay-back time of 5 to 10 years, may be less in the most favourable situation and certainly more in the less favourable cases. This internal rate of return (IRR) may be insufficient for some investors; however, in the case of regulated utilities, such as water companies, which are used to lower IRRs, or as a public-sector investment with public good benefits, it could be sufficient.

Towards viability: route two – valorisation of biochar

Charcoal retails in Europe for between €0.5 and €1 per kilogram. The few retailers of biochar in the EU sell across a very wide range of €1.1 (Sweden, UK) to €17 per kg (see Table 9.1), though no market data on sales volumes is presently available, so whether the higher (or, for that matter, lower) cost biochar is a realistic market proposition is not currently known. To justify this high cost compared with the cost of charcoal, we need to shift from thinking of biochar as a bulk commodity to considering it instead as a high-value speciality chemical.

There are numerous suggestions on how to valorise biochar and these mostly fall into the following three categories:

- creation of synergies between biochar and other organic or synthetic fertilisers and amendments such that the 'whole is greater than the sum of the parts';
- adding value to biochar through the 'value cascade approach' (Schmidt 2012) whereby the same material gives value to product or process more than one time, hence the economic added gain accumulates across the value chain; and
- identifying new applications for biochar that are inherently higher-value per unit weight, such as: a building material; a replacement for expensive activated carbon substances used as industrial filters or adsorbents; and an animal feed additive.

Table 9.1 Examples of biochar suppliers in Europe and indicative costs of products and biochar

Country	Company	Products	Price of product) as sold (€ or £)
Germany	Terra Preta	Terra Preta	€23 per 1 litre
Germany	Palaterra	Growth medium including biochar	Not known
UK	Carbon Gold	GroChar Biochar Complex (mix)	£9.95 for 1.4kg
		All purpose biochar compost	£8.99 for 20 litre
		Seed compost	£4.99 for 8 litre
UK	Oxford Biochar	Oxford Biochar	
UK	Yorkshire Charcoal	Charcoal	£1 per kg
Sweden	Skogans Kol	Charcoal	€1.1 per kg
UK	Nutrichar	Nutrichar (mix)	£16.6 per kg
Switzerland	Verora GmbH	Biochar compost (10% biochar by volume)	€105 per m³
		Biochar only	€300 per m³
Switzerland	Swiss Biochar	Swiss Biochar	€300 per m³
Austria	Sonnenerde	Black soil (Terra Preta) (70% GW; 20% biochar, 10% stone powder)	€320 per m³

Source: assembled from various presentations and company websites.

The strategy of combining biochar with other organic fertilisers and amendments to create a new family of bio-fertilisers is covered in Chapter 6, while the use of biochar as an animal feed supplement is discussed in Chapter 11 though not the full concept of the value cascade. The evidence presented in Chapter 6 provides strong support for the notion that synergistic benefits arise from combining biochar with nutrient-rich organic matter such as composts, bio-slurries and NPK fertilisers. For instance, Joseph *et al.* (2013) reported rice crop yields of +39 per cent in a trial of biochar-NPK pellets compared to the control (NPK pellets) for an application of around 120kg of biochar per hectare, meaning the biochar is worth roughly €100 per ha, or over €740 per tonne, moving biochar into the right value-territory.

In developing and newly industrialising countries, the economic value of mixtures of biochar and other organic amendments – in terms of increased net profitability to the farmer – is (at least in some circumstances) higher than would be the case in Europe. This is due to low levels of synthetic NPK use (sometimes no NPK is used) and the more ready availability of agri-residues and animal

---

Box 9.2 Biochar production, sale and valorisation: Sonnenerde, Riedlingsdorf, Austria

Sonnenerde ('sun-soil' in German) is a small composting and soil producing company with 14 employers. 10,000 tonnes of SSW and 8000 tonnes of green wastes are used to produce a high quality compost. The compost is used for producing 20 different types of soils such as plant pot-soils, soils for cultivating flowers, soils for use on green roofs, for gardens and so on. There is a single owner and manager (Gerard Dunst) who is responsible for new product development. He also works on a voluntary basis as part of the Ecoregion Kaindorf, the main goal of which is to become carbon neutral by the year 2020. Increasing the carbon content of soils has been the most important part of this project.

In January 2012, Sonnenerde purchased a Pyreg pyrolysis plant for the processing of paper fibre sludge and grain husk, and the biochar produced has been certified by the European Biochar Certificate as being premium quality biochar. All of the surplus energy (150 kWh$_{therm}$) is used to heat the office and to dry the feedstock. The paper fibre sludge is delivered with 80 per cent water content and the first step is to dry it to 25 per cent water content. For this purpose, a very simple floor drying system was installed, where the heated air is blown through channels through the feedstock. After drying, the feedstock is sieved into particles with a size smaller than 15 mm and placed in the 20 m³ feeder using a loader. The material flows into the pyrolysis unit where it is heated up to 600°C. All the produced pyrolytic vapours are burned at 1000°C and the exhaust heat is circulated to heat up the material that enters the pyrolysis unit. After that, the hot flue gas passes through a water heat exchanger, which provides the heat for the drying of the fresh feedstock. The biochar coming out of the unit is moistened to 30 per cent water content.

In the first year (2012) the plant worked nearly 4000 hours and 160 tonnes of biochar were produced. In 2013, 5000 operating hours were recorded and 200 tonnes of biochar produced. Most of this biochar was sold directly to farmers at €450 per tonne. The farmers used it to add to the manure in animal houses or mixed it with other organic matter to form compost piles. In spring 2013, the first biochar-based product was sold (*Riedlingsdorfer Schwarzerde* – black soil from Riedlingsdorf). To produce this soil, 20 per cent (by weight) of fresh biochar is mixed into the fresh composting piles together with 10 per cent stone powder. After four weeks of composting in very small piles (2.5m wide and 1.3m high), 10 per cent crushed bricks and 10 per cent clay are added and composted for another four weeks. The biochar compost is then mixed with a further 10 per cent sand and 10 per cent lava. Before sieving (particle size ≤15mm), 2 per cent soil-activator is added. This soil, containing 20 per cent biochar by weight, is sold for €150 per tonne and has been mostly purchased by individual consumers who have used it in garden vegetable production and in raised plant beds.

The company's next product will aim to simulate the benefits of *terra preta* soils. It will be a biochar-based material that is enriched with nitrogen and the most important trace elements (e.g. Mo, Se, Co). The ingredients are mixed into a matured compost pile (4–6 weeks old) to get the humus building microorganisms into the nutrient enriched biochar. After two weeks, the biochar is sieved out from the compost and is ready to use.

Sonnenerde has invested €1 million in this project and aims to turn it into a profit-making business. In the first year, the production cost per tonne of biochar was too high (€813 per tonne). The main costs were the capital recovery of investment (calculated assuming eight years for the machines and 20 years for the building), the personnel costs (0.5 person was found to be necessary, partly because the machine is a prototype and many adjustments were required) and the feedstock costs (while a gate fee is received for the paper sludge, the grain husks have to be paid for). The balance for the operation is shown in Figure 9.1, indicating that, while there is some way to go before it becomes profitable, there are healthy revenue streams from the sales of biochar and the gate fee. Clearly, the business model depends on the company being able to sell the biochar for €500–600, though valorisation through bio-fertiliser products means that sales of pure biochar are not as likely.

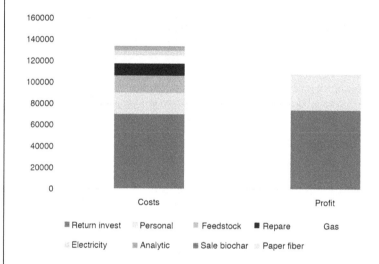

Figure 9.1 Balance Sheet for Sonnenerde (as of late 2013).

Source: Gerard Dunst.

At the time of writing (late 2013), the production costs have been reduced to around €500 per tonne and it is hoped that this production cost can be further reduced to less than €400 per tonne.

Dunst does not believe that it will ever be economic to produce biochar from a high value biomass feedstock like wood chips. There are enough clean wastes on the market which could be used for the pyrolysis process. Dunst also believes that it will be necessary to create different high value products containing biochar. This is because it is very difficult to sell pure biochar for a price which makes its production anywhere near profitable in the European context.

Prepared by Gerald Dunst and edited by Simon Shackley.

manures. Applying large quantities of biochar mixed with organic soil amendments on a yearly basis would not fit too well into many types of conventional agriculture. While organic amendments, such as SSW, spent grain and ADDs, are used, in conventional farming this tends to be once every three-to-five years. Nitrogen fixing cover crops are also cultivated as part of crop rotations, which are then ploughed back into the field to provide soil organic matter and nutrients.

Biochar, together with organic amendments, clearly fits better with organic farming systems in the European context since much larger quantities of organic matter need to be incorporated into soils farmed under organic cultivation on a yearly basis. Hence, mixing biochar with composts, manures and other agri-residues enhances the value of organic amendments. Several of the most prominent producers and users of biochar in Europe are indeed associated with the organic agriculture and food movement.

At this juncture, however, there is insufficient reliable and reproducible evidence to quantify the benefits of biochar in economic terms with confidence; likewise for the use of biochar in a value cascade, in which case the functionality of the biochar has to be matched with the particular needs at each stage of the cascade and not be 'fully used up' or materially modified by prior stages such that the intended functionality no longer remains intact.

It is also premature to be in a position to state much about the economics of using biochar in place of activated carbon filters, as an animal feed, as a replacement for peat in growing media and as a high-quality building constituent. An intriguing proposed application of biochar is as a means of cleaning up contaminated waters produced during fracking for shale gas production. If biochar finds specialised markets in potentially high-value added, and rapidly expanding, industries, valorisation of biochar will be rapid and will likely drive the technology development and maturation as well as the provision of the necessary feedstocks.

Towards viability: route three – the re-emergence of carbon markets

'Carbon markets' refers to a range of mandatory and voluntary markets which allow trade in carbon emission units. The basic concept is to cap the total carbon emissions of a given country and then to allocate the right to produce those emissions between all producers, such as industries, the public and domestic sector. The cap is set by government and is – in principle – a means of ensuring that all carbon emissions remain below a given 'safe' level. A market is then established in which carbon emission units can be traded between those who wish to sell all or part of their allowance and those who wish to purchase allowances. A company may have 100 carbon emission allowances, say, but finds a way to become more efficient in its utilisation of energy, thereby saving 25 of those units that it has been allocated, which it is then free to sell to a buyer who has, perhaps, expanded their operations to the extent that it requires additional emission allowances. A carbon price arises as a consequence of the interplay between buyers and sellers. The caps are intended to get more stringent over time, hence fewer allocations will be handed out by government and the carbon price should go up as allowances becomes scarcer. The carbon price then becomes an incentive for firms and organisations to invest resources in finding ways to reduce carbon emissions. In principle, a market-based approach is a more cost-effective way of reaching a given emission reduction target than a carbon tax, though some economists have argued that a carbon taxation system is better because of information asymmetries (Tol 2014).

At present there are several mandatory carbon markets, of which the largest is the EU Emissions Trading Scheme (EU ETS), established in 2005, and which applies to companies and organisations which combust fossil fuels in installations with a total rated thermal input exceeding 20MW (i.e. it only applies to larger entities and not to small businesses or to the domestic sector). Approximately 50 per cent of the EU28 countries' emissions are covered by the 'traded sector', while the remaining 50 per cent arise from the non-traded sector. The EU ETS is by far the largest statutory carbon market in the world; while there has been considerable debate regarding establishment of a carbon market in the USA, Australia, China, and so on, as yet the EU ETS is the major such market, though the Californian carbon market is now operational.

The Kyoto Protocol (KP) of the UN Framework Convention on Climate Change (UNFCCC) was adopted by Parties in 1997 and entered into force in 2005. The KP includes so-called 'flexibility mechanisms', which are ways in which the carbon emission reductions in one country can contribute to achieving carbon reduction in another country; the Clean Development Mechanism (CDM) is the most important of these. Under the CDM, a low-carbon development project is agreed between a 'donor' country (Annex I countries under the UNFCCC, most of which are OECD countries which have modest emission reduction targets under the KP) and a 'host' country, a non-Annex I country (developing or rapidly emerging, etc.). The project is financed predominantly or entirely by the donor country and the carbon emission savings are shared between the donor and host country. In the case of a CDM project, a projected baseline has to be established into the future in order for emission reductions to be calculated – that is, the projected carbon emissions arising from the project are subtracted from the project baseline emissions without the project ('what would have happened in the absence of the project') in order to calculate the emission savings. These emission savings under the CDM are known as Certified Emission Reductions (CERs). There are no overall greenhouse gas (GHG) emission caps in non-Annex I countries, and so there is no guarantee that there is a net overall reduction in emissions in the host country, but the whole purpose of the scheme is to promote cleaner development.

Under the EU ETS, it has been possible to trade in some of the CERs arising from a CDM project; hence a government or a company based in the EU could meet a certain percentage of its emission reduction targets by purchasing CERs from the global market. The extent of this 'buyout' has been limited to 11–13 per cent of the total allowances of a given operator in order to ensure that the bulk of decarbonisation activities takes place within the EU itself. If or when EU emission reduction targets are firmed up in the future, it is possible that there will be an increasing role for buyout through purchase of international CERs, as this will reduce the overall cost of meeting the target.

Where a CDM project is supporting the financing of a wind farm as opposed to, say, building a gas-fired power plant, there is a tangible reduction in carbon emissions that can be expressed in the difference in the Carbon Emission Factor (CEF), a measure of the CO_2 emissions per unit of electricity (or heat) generated. The CEF for gas-fired power plants is approximately $500kgCO_2$ per MWh (megawatt hours – a measure of electricity generated) while the CEF for wind power is approximately $20kgCO_2$ per MWh, so the direct saving of CO_2 emissions (assuming no change in demand arising from the technology selection) can be reasonably confidently calculated.

In addition to the government-mandated or controlled-carbon markets, there are also several voluntary carbon markets, including the Voluntary Carbon Standard and the Gold Standard. These voluntary schemes allow companies and other organisations to purchase carbon savings arising from projects in any part of the world. Many of the most successful projects in the voluntary market have been North American projects, such as forestry projects, in part because the USA and Canada are not part of a statutory carbon market (excepting California recently) and partly because of the difficulties in authenticating carbon emission reductions in developing countries. The size of the voluntary carbon market is about one tenth that of the EU ETS and the carbon price has tended to be lower on the voluntary market, but it's currently trading higher than the mandatory market price due to the collapse of the carbon price on the EU ETS circa 2010.

In 2008, the EU ETS was modified to allow the inclusion of CO_2 capture and geological storage (CCS) projects. The idea is that CO_2 from power plant flue gases can be scrubbed out, compressed, transported and then injected into a suitable storage site in a rock formation about 1km underground, where it is securely held for thousands of years. This situation is different from renewable energy in the sense that the CO_2 has *already* been released from fossil fuels but is then captured and stored. Hence, it is not an avoided emission but rather captured emissions or, technically, carbon abatement. Under the 2008 revision of the EU ETS, a company which is operating a CCS system can use verified CO_2 storage as an offset to its other emissions. Let's suppose that a large petro-chemical firm captures one million tonnes of CO_2 per year through a CCS project. The same firm has allowances to emit 8 million tonnes (MT) of CO_2 annually. In year X, suppose the firm emits 8 MT. It is able to offset these emissions with the 1 MT from the CCS project (i.e. a net emission of 7 MT CO_2) and can therefore sell 1 MT CO_2 allowances on the market. In year Y, the firm emits 10 MT and can again offset the CCS project emissions against these (i.e. a net emission of 9 MT) and is therefore required to go into the marketplace to purchase 1 MT CO_2. The next logical step would be for a firm that specialised in CCS to capture and store CO_2, obtain the equivalent EU allowances, and trade them into the EU ETS.

The biochar concept is close to CCS in that the carbon stored in the biochar is not avoided emissions but rather carbon removal from the atmosphere. Just as offsetting works in principle for CCS under the EU ETS, it is possible that biochar can be used to offset emissions arising from operations within a firm or directly into the EU ETS. However, at present the only participants in the EU ETS are larger firms, and most biochar projects are nowhere near the scale required for entry. The most credible prospect would be if a large firm that is already within the EU ETS developed a biochar project and then looked to include the captured carbon as an offset against its other emissions. Such an entity could also look to include biochar projects in developing countries in this fashion, through the linkage between the CDM and the EU ETS. A further option is for organisations to develop biochar projects in developing or newly industrialising countries and to sell the carbon credits on the voluntary market.

The period between 2009 and 2015 has been one of disappointment, disillusion and greatly lowered expectations regarding the future of carbon markets in the EU and globally. The reasons are largely threefold: the economic crises of 2008 onwards, which greatly dampened down appetite for investment in low-carbon energy technologies and options; the failure of UNFCCC climate negotiations starting in Copenhagen at COP-15 with all eyes on COP-21 in Paris, December 2015; and the failure of the EU ETS, which was poorly designed and prone to failure from an early stage (e.g. allocating emission permits rather than auctioning them off; allowing some member states too much leeway in estimating required emissions quotas; allowing energy utilities to make windfall profits from the EU ETS while not reducing one kilogram of CO_2; the unfortunate catalogue of failures goes on and on). The reason why all this matters for biochar is that the founders of the concept – Peter Read, Johannes Lehmann, Bruno Glaser and Stephen Joseph – always had in mind that biochar neatly brought together priorities associated with carbon storage and CO_2 emission reduction (carbon abatement) with concerns over soil 'health' and fertility, food production and security. It was this bringing together of two of the world's most serious problems which gave biochar such an exciting aura in around 2005–2010.

The subsequent de-prioritisation of carbon and climate change policies in the face of severe economic problems in many EU member states rather 'cut the legs' from underneath the biochar story. Yet, the key scientific messages in the most recent synthesis report of the Intergovernmental Panel on Climate Change (IPCC), issued in November 2013, have become more ardent, with strong promotion of CCS and, by implication, of other safe and sustainable carbon storage technologies such as biochar.

What has changed, rather, is politicians, industry and wider societies' readiness to take note of what the IPCC is claiming. If we assume that the majority of what the IPCC claims is correct, then it is almost inevitable that at some stage, carbon mitigation technologies will need to be developed and implemented on a much larger scale than at present. It is also highly likely that techniques for removing CO_2 from the atmosphere will be necessary, since atmospheric concentrations will have grown too high. Biochar is one of very few technologies which are capable of removing CO_2 and, potentially, being a carbon negative energy generation technology.

As climate change impacts become more serious, biochar may become one of a handful of 'emergency technologies' which are deployed rapidly and widely, either through direct funding or through introduction of high carbon taxes. There are a whole host of technical issues which need to be resolved before biochar can be included in offsetting carbon markets such as a (revamped) EU ETS and voluntary markets; in particular, a robust method for calculating carbon stability on a climate relevant time scale and a cheap but reliable method for calculating how much carbon has been abated at the project site (e.g. taking due account of the movement of biochar offsite but still in a stabilised form, for instance in sediments in waterways). However, solutions to these technical puzzles would be forthcoming were sufficient time and resource devoted to the problem. The real issue is not the detail of the methodologies, but whether carbon markets can be made to work in the way intended and supported by carbon emission caps at the global scale. Biochar would fit well into Myles Allen and colleagues' (2009) 'carbon abatement requirement' concept, whereby a fossil fuel producer would need to demonstrate that they had abated a given percentage of the CO_2 emitted from combustion of the fuels that it supplies. (This percentage would gradually increase and become 100 per cent as the 'one trillion' tonnes of carbon that can be safely emitted into the atmosphere is reached). Through using pyrolysis to produce biochar and bioenergy from organic matter, fossil fuel firms could earn a carbon abatement credit which could be traded for the equivalent net CO_2 emissions from fossil fuel combustion in calculating the total abatement requirement.

Biochar policy analysis in an uncertain world

Biochar emerged at the interface of soil science, bio-energy technology, climate change and biodegradable waste management. The founders of the field came from each of these domains and biochar was the creative outcome of the exchange of ideas in the mid-2000s. Advocates began referring to biochar as a 'win–win–win–win' option (sustainable carbon abatement, economic development, energy production, sustainable waste management, etc.). The balance between the four areas has shifted since then, owing to a serious lack of investor appetite for development of risky environmental and low-carbon energy technologies since 2008 and the stalemate on international climate policy development. Bio-energy technology development in general has been hit by the take-off of other renewable energy technologies such as wind and solar power which were 'further up the queue'; their development was more advanced, with well-organised industry and sectoral lobby organisations involved in their promotion. Renewable energy subsidy schemes in European countries worked out more effectively for wind and solar power owing to their greater technological maturity and more flexibility with respect to scale up (i.e. relatively low unit capital costs but scalable through multiple units). Public opposition to new and innovative bioenergy facilities has also slowed down developments in Europe. Mature bio-energy technologies, such as methane recovery from waste water treatment plants and from landfill sites, and some bio-waste to combined heat and power schemes, have also benefitted from the availability of renewable energy subsidies.

Planned-for large bioenergy combustion plants in Europe, which replaced coal with chipped or pelletised biomass, have also emerged in the past five or so years, with some large facilities in the Netherlands and UK (approx. 300 to 750 MW capacity). The driver has been the EU's Renewable Energy Directive (RED), which has perceived a major role for biomass in meeting the EU's ambitious renewable energy targets (58 per cent of total RED by 2020, constituting 12 per cent of total gross energy demand within the EU). Such scales of use require millions of tonnes of biomass per year – volumes that are not obtainable at an economically sustainable price from within north west Europe. In the UK, for example, the current consumption of biomass for energy generation stands at 1.685 million tonnes per annum. If all biomass energy plants in planning and proposed were to go ahead, demand would increase by over 20 times to 35.5 million tonnes per annum. This is approximately three times the total annual production of round and sawn wood in the UK; it is about 40 per cent of the total annual production of round and sawn wood in Sweden, Europe's largest timber producer.

Hence, major European power utilities have begun to develop biomass supply chains from other parts of the world with large land areas for forestry, in particular North and South America, Russia, Africa and the Asia-Pacific region. These initiatives have been heavily criticised by environmental and nature conservation organisations because of the net impacts on carbon emissions and biodiversity arising from the extraction of millions of tonnes of wood from primary or secondary forests. A highly complex and technical debate has emerged within the wider forestry, wood products, bio-energy sector and climate change and carbon communities regarding the life cycle impacts of wood extraction for bioenergy (Manomet 2010; Zanchi et al. 2011). An easy resolution of this complex debate is very unlikely and is tied up with politically-charged climate negotiations.

The EU's RED of 2009 set a 10 per cent target for liquid biofuels, driving tropical deforestation in Malaysia and Indonesia for palm oil plantations, directly and indirectly creating huge 'carbon debt' as highly carbon rich landscapes (especially tropical forests on peat soils) were decarbonised (e.g. Fargione et al. 2008; Searchinger et al. 2008). The increased demand for biofuels in Europe and the USA also helped in pushing up food prices with adverse impacts on the poor, who have to spend a larger proportion of their income on food purchases. The adverse effects of the biofuels target in the EU's RED led to much criticism and scrutiny from environmental NGOs and national government agencies, and as a consequence of which, much more demanding sustainability criteria were put into operation, including a moratorium on biodiesel from palm oil and the requirement of a net 35 per cent reduction in life cycle GHG emissions for the biofuel relative to the liquid fossil fuel that it substitutes for. More recent revisions of the EU RED have: scaled back the target for liquid biofuels to 5 per cent; imposed a requirement for a 60 per cent reduction in net GHG emissions relative to the fossil fuel; included indirect land-use change factors in calculating the life cycle GHG emissions; and proposed extending the requirements to the use of solid and gaseous forms of biomass.

In the build up to UNFCCC COP-15 in Copenhagen (December 2009) the International Biochar Initiative (IBI) engaged in lobbying to try and get biochar included in the negotiations and as a legitimate option under the 'flexibility mechanisms' such as the CDM. Statements were circulating in the public domain to the effect that biochar had already been endorsed as a credible and important option by the IPCC. It turns out that the IPCC had not endorsed biochar; rather, one IPCC panel meeting had had a discussion about biochar but had not come to any formal evaluation or position on the topic.

Probably in a direct response to this rather strong 'push' by the IBI and other biochar advocates, Biofuel Watch, an NGO, launched a remarkably effective campaign in 2009 against the idea of biochar, raising many questions about its sustainability and desirability as a climate mitigation option. An article by pioneering biochar scientists published in 2006 (Lehmann et al. 2006) had proposed that biochar could abate between 5.5 and 9.5 billion tonnes of carbon per year by 2100; this value is not dissimilar from the total human-caused carbon emissions of about 8 billion tonnes per year. Such a large value is only credible if large-scale land-use change took place such that extensive biomass plantations were planned explicitly as part of PBS development. In the light of the debacle over liquid biofuels and the adverse effects of direct and indirect land-use change, very high estimates of carbon abatement from biochar were seized upon by Biofuel Watch as evidence that leading biochar scientists advocated massive monocultural plantations of fast growing trees or energy crops.

Approximately 150 NGOs from all around the world signed a petition coordinated by Biofuel Watch to oppose the inclusion of biochar in the climate negotiations and in the CDM. This wave of criticism culminated in the UK with a withering critique of the biochar concept in an article by George Monbiot in 2009, which appeared in the *Guardian* newspaper, entitled 'Woodchips with everything'. The article began: 'It's the Atkins plan of the low-carbon world: The latest miracle mass fuel cure, biochar, does not stand up; yet many who should know better have been suckered into it.'

A few years on, both sides in this debate have gone through a learning process and the expectations of biochar promoters have been lowered and made more realistic. For example, no longer do biochar advocates talk about using energy crops or plantation biomass for producing biochar. The feedstocks now under discussion are waste wood and other biodegradable organic waste materials, as well as agri-residues such as rice husks and arable straws that would otherwise be wasted or used inefficiently. The total credible and sustainable annual carbon abatement by 2050 has been reduced from 5.5–9.5 billion tonnes carbon per year to the far more modest 1 to 1.6 billion tonnes per year (Woolf *et al.* 2010): this new calculation assumes that no additional land-use change for producing biochar occurs.

The benefits to the small farmer and to the local community of biochar have also come to the foreground. The agronomic benefits of biochar in producing more sustainable food from 'healthier' and climate-friendly soils have now become centre stage – more so than the much vaunted recalcitrant carbon storage benefits of biochar of the earlier 2006–2010 era. This has also led to a newly invigorated role for scientists, since there is a plethora of fascinating and important research questions which need to be addressed to make biochar work in agronomic and agricultural settings (also taking into account environmental impacts) as illuminated in Chapters 5–7. Alongside the applied and consultancy research, entrepreneurs are busy engaging in a two-way transfer of knowledge and experience from users and researchers/scientists. This approach is not confined to Europe and other industrialised regions, but is also very much found in emerging and developing economies, where agricultural R&D on biochar is becoming a sort of marketplace between different kinds of experts – scientists and agronomists, technologists, local firms and entrepreneurs and the farmers, resource owners and managers.

It is important that this dynamic and innovative marketplace where ideas and concepts can be shared and exchanged is cultivated. The role of big business as biochar RD&D and opportunities expand, and attempts by individuals and firms to take intellectual property (IP) positions, will need to be constantly scrutinised. Large investors are likely to be necessary in order to make progress on the development of technology by bringing sufficient financial and human capital to the problem and to provide capital for scaling up successful systems. On the other hand, expansive IP positions and consolidation of biochar companies could end up inhibiting innovation and providing 'solutions' which are not sufficiently flexible and adaptable to the myriad of complex contexts within which biochar can and should made a positive difference to people's livelihoods and to the environment.

Bibliography

Allen, M., Frame, D. and Mason, C. (2009). The case for mandatory sequestration. *Nature Geoscience* 2: 813–814.

Clare, A., Barnes, A., McDonagh, J. and Shackley, S. (2014a). From rhetoric to reality: Farmer perspectives on biochar economics in China. *International Journal of Agricultural Sustainability*, 12(4): 440–458.

Clare, A., Shackley, S., Joseph, S., Hammond, J., Pan, G. and Bloom, A. (2014b). Competing uses for China's straw: the economic and carbon abatement potential of biochar. *GCB Bioenergy* online first (doi: 10.1111/gcbb.12220).

Fargione, J., Hill, J., Tilman, D., Polasky, S. and Hawthorne, P. (2008), Land clearing and the biofuel carbon debt. *Science* 319: 1235–1237.

Field, J.L., Keske, C.M.H., Birch, G.L., DeFoort, M.W. and Cotrufo, M.F. (2013). Distributed biochar and bioenergy co-production: a regionally specific case study of environmental benefits and economic impacts. *GCB Bioenergy* 5: 177–191.

Galinato, S.P., Yoder, J.K. and Granatstein, D. (2011). The economic value of biochar in crop production and carbon sequestration. *Energy Policy* 39: 6344–6350.

Granatstein, D., Kruger, C., Collins, H., Garcia-Perez, M. and Yoder, J. (2009). *Use of Biochar from the Pyrolysis of Waste Organic Material as a Soil Amendment*. Final project report. Wenatchee, WA: Center for Sustaining Agriculture and Natural Resources, Washington State University.

Hammond, J., Shackley, S., Prendergast-Miller, M., Cook, J., Buckingham S. and Pappa, V. (2013), Biochar field testing in the UK: outcomes and implications for use. *Carbon Management* 4(2): 159–170.

Harsono, S., Grundman, P., Lau, L., Hansen, A., Salleh, M., Meyer-Aurich, A., Idris, A. and Ghazi, T. (2013). Energy balances, greenhouse gas emissions and economics of biochar production from palm oil empty fruit branches. *Resources, Conservation and Recycling* 77: 108–115.

Ibarrola, R., Sohi, S., Brownsort, P., Ahmed, S., Cross, A., Peters, C., Masek, O., Hammond, J. and Shackley, S. (2012). *Feasibility Analysis for the Production of Bioenergy and Deployment of Biochar using Sewage Sludge/Digestate*. Edinburgh: UK Biochar Research Centre, University of Edinburgh.

Joseph, S., Graber, E., Chia, C., Munroe, P., Donne, S., Thomas, T. and Hook, J. (2013). Shifting paradigms: development of high-efficiency biochar fertilizers based on nanostructures and soluble components. *Carbon Management* 4(3): 323–343.

Lehmann J., Gaunt, J. and Rondon, M. (2006). Bio-char sequestration in terrestrial eco-systems: a review. *Mitigation and Adaptation Strategies for Global Change* 11: 403.

Manomet (2010). *Biomass Sustainability and Carbon Policy Study*. NCI-2010-03. Boston, MA: Manomet Centre for Conservation Sciences.

McCarl, B.A., Peacocke, C., Chrisman, R., Kung, C. and Sands, R.A. (2009). Economics of biochar production, utilization and greenhouse gas offsets. In J. Lehmann and S. Joseph (eds), *Biochar for Environmental Management: Science and Technology*, 341–358. London: Earthscan.

Monbiot, G. (2009). Woodchips with everything. *The Guardian*, 24 March.

Mühle, S., Balsam, I. and Cheeseman, C. (2010). Comparison of carbon emissions associated with municipal solid waste management in Germany and the UK. *Resources, Conservation and Recycling* 54: 793–801.

Schmidt, H.-P. (2012). 55 uses of biochar. *Ithaka Journal for Ecology, Winegrowing and Climate Farming*. Available at www.ithaka-journal.net/55-anwendungen-von-pflanzenkohle?lang=en.

Searchinger, T., Heimlich, R., Houghton, R., Dong, F., Elobeid, A., Fabiosa, J., Tokgoz, S., Hayes, D. and Yu, T-H. (2008). Use of US croplands for biofuels increases greenhouse gases through emissions from land-use change. *Science* 319: 1238–1240.

Shackley, S., Hammond, J., Gaunt, J. and Ibarrola, R. (2011). The feasibility and costs of biochar deployment in the UK. *Carbon Management* 3(2): 335–356.

Shackley, S., Clare, A., Joseph, S., McCarl, B. and Schmidt, H.-P. (2015). Economic evaluation of biochar systems – current evidence and challenges. In J. Lehmann and S. Joseph (eds), *Biochar Science and Technology for Environmental Management*, 2nd edn, 813–851. Abingdon: Routledge.

Spokas, K., Cantrell, K., Novak, J., Archer, D., Ippolito, J., Collins, H., Boateng, A., Lima, I., Lamb, M., McAloon, A., Lentz, R. and Nichols, K. (2013). Biochar: a synthesis of its agronomic impact beyond carbon sequestration. *Journal of Environmental Quality* 41: 973–989.

Tol, R. (2014), *Climate Economics*. Cheltenham: Edward Elgar.

Woolf, D., Amonette, J., Street-Perrott, A., Lehmann, J. and Joseph, S. (2010). Sustainable biochar to mitigate global climate change. *Nature Communications* 1: article 56.

Zanchi, G., Pena, N. and Bird, N. (2011). *The Upfront Carbon Debt of Bioenergy*. Graz: Joanneum Research.

Note

1   The amount of biodegradable municipal waste sent to landfill in England has decreased from 11.3 million tonnes in 2007/08, to 10.3 million tonnes in 2012. The Landfill Allowance and Trading Scheme (LATS) sets a limit on the amount of biodegradable municipal waste that each unitary and waste disposal authority in England can send to landfill. However, currently only 40 per cent of councils have separate food waste collections.

Chapter 10

The legality of biochar use

Regulatory requirements and risk assessment

Jim Hammond, Hans-Peter Schmidt, Laura van Scholl, Greet Ruysschaert, Victoria Nelissen, Rodrigo Ibarrola, Adam O'Toole, Simon Shackley and Tania van Laer

Introduction

One of the major motivations for the use of biochar is environmental improvement, both to improve soil quality and to reduce atmospheric greenhouse gas (GHG) concentrations. However, if biochar is produced or used inappropriately it could cause more harm to ecosystems, human health and GHG concentrations than benefit them – a situation to be avoided. Regulation and legislation, as well as voluntary standards, aim to ensure that biochar use is not environmentally counterproductive.

This chapter covers two main topics: first, the currently known potential risks – or undesirable outcomes – from biochar production and use are identified and explained. The authors offer no advice on the likelihood of a risk occurring, nor the magnitude of the impact, should it occur. Each risk should be evaluated on a case-by-case basis by producers and users of biochar and by regulatory authorities.

Second, the regulations and legislation applicable to biochar in some European countries is outlined. While every care has been taken to ensure that the guidance offered in this chapter is correct, comprehensive and up to date, it is not a substitute for professional legal advice. This is also due to the fact that legislation is evolving rapidly.

Finally, it should be stressed that biochars can be very different to one another. Depending on the substance the biochar is made from and the process of pyrolysis it has undergone, biochar's material properties and material content can be very different. The environment to which the biochar is added should also be considered and is, of course, highly variable.

The potential risks from biochar production and use

Providing that biochar is produced and used in a responsible way, the foreseeable risks can be controlled by appropriate action. Unforeseen risks, such as a long-term negative influence on soil nutrient cycling, could be more difficult to mitigate. In order to minimise and control the risks related to the production and use of biochar, it is important to understand the risks. The major risks to health and the environment are illustrated in Figure 10.1. These risks can be divided into three categories:

1   Potential risks to human health: ingestion of soil (including biochar) particles; contaminants (toxins) in biochar entering the food chain and being consumed by humans; inhalation of fine particles of char entering the lungs of humans; dermal uptake of contaminants from biochar dust particles; improperly stored char catching fire; or airborne toxic or particulate emissions from poorly controlled pyrolysis.

2   Potential risks to agriculture and soil function: contaminants in biochar entering the food chain to have an adverse effect on crop quality, crop production, quality of animal products or the health of livestock (for an overview of risks from contaminants in agriculture see De Vries *et al.* 2007a, 2007b, 2008); unforeseen effects of biochar on soil nutrient cycling, soil biota, soil contaminant bio-availability or some other soil function. If some adverse impact on soil functioning were to occur, the impact could be long term and it would be difficult and expensive to remove the biochar once it had been mixed into the soil.

3   Potential risks to the wider environment: biochar may increase atmospheric GHG levels if it is not sufficiently stable, increase soils GHG emissions, or result in the release of CO_2 from native soil organic matter; biochar made from unsustainably or unethically sourced feedstocks could increase atmospheric GHG levels and cause further environmental and social problems; contaminants in biochar or unforeseen effects on soil functioning may increase leaching and run-off of nitrogen, phosphorous or other contaminants; poorly controlled airborne emissions from pyrolysis.

Contaminants in biochar: what they are and where they come from

Contaminants in biochar are either in the form of heavy metals (otherwise known as potentially toxic elements or PTEs) or in the form of organic compounds which can cause toxic effects (Shackley and Sohi 2010; van den Bergh 2009).

Heavy metals in biochar come directly from the feedstock, and in most cases almost all the metal in the feedstock becomes concentrated in the biochar (with the exception of mercury which has a very low boiling point).

Potentially toxic organic compounds that are generally formed during the pyrolysis reaction include polyaromatic hydrocarbons (PAHs), polychlorinated biphenyls (PCBs) and polychlorinated dibenzo-p-dioxins (PCDDs) and polychlorinated dibenzofurans (PCDFs). The latter three compounds require the presence of chlorine atoms during pyrolysis to form, whereas PAHs do not. In addition, there may be a great number of other volatile organic compounds of lower toxicity risk, which are as yet poorly characterised (Spokas *et al.* 2011). There is some evidence that these volatiles can stimulate plant growth (Graber *et al.* 2010; Joseph

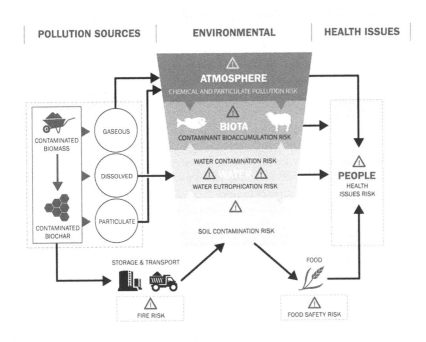

Figure 10.1 The risks associated with biochar production and use.

et al. 2013) although if present in high concentrations they could have deleterious effects.

The risk of producing contaminated biochar can be reduced first by using feedstock that is low in metals and chlorine, and second by using a well-controlled and reliable pyrolysis process. Providing that the biochar has been made from relatively uncontaminated feedstocks, contamination is generally not a cause for concern (Freddo *et al.* 2012; Hale *et al.* 2012).

This type of 'direct' contamination of soils through contaminated biochar is probably the most obvious risk to human and ecosystem health, and may be of the most interest to environmental regulators. Common methods to assess contaminants in biochar have been suggested (International Biochar Initiative 2013; Schmidt *et al.* 2012), and research papers have been published which characterise biochars and include methodologies for characterisation (Singh *et al.* 2010; Freddo *et al.* 2012; Hale *et al.* 2012). Not all of the commonly accepted analysis methods of contaminants in soils work well for biochar because biochar and soil have differing material properties. The impact can be assessed using the well-developed methodology of ecotoxicological risk assessment, which assesses the quantity of contaminants, the relative toxicity and the exposure pathways to plants, animals and humans. Such assessments are commonly performed for other soil amendments such as composts, slurry or other wastes which are applied to land.

A further issue to consider is the bioavailability of contaminants (i.e. how likely they are to be released from the biochar and available for uptake or transported to other soil compartments in the short and long term; Reid *et al.* 2000; Hale *et al.* 2012). Most legislation uses the most cautious approach, which is to measure the total amount of contaminants in a soil amendment, assuming that in the worst-case scenario all would become bioavailable. In practice this is unlikely to occur with biochar as the stable carbon matrix prevents many contaminants from becoming bioavailable. Indeed, many types of biochar have been shown to absorb and immobilise additional organic and inorganic contaminants from soils and are under consideration as a tool in soil remediation (Beesley *et al.* 2011; Uchimiya et al 2010; Khan *et al.* 2013). However, scientific consensus has not been reached as these approaches do not remove the toxin from the soil and it could, potentially, become bioavailable at some point in the future (Quilliam *et al.* 2013; Jakob *et al.* 2012).

Possible effects on soil function

Charcoal and black carbon can be found in fertile, well-functioning soils around the world (Skjemstad *et al.* 2002). Current interest in biochar was sparked by the discovery of exceptionally fertile soils in the upper Amazon region which contained very high levels of black carbon (see Chapter 6; Glaser *et al.* 2001; Glaser and Birk 2012). Although this implies that black carbon, including biochar, *can* be beneficial to soils, it does not mean that *all* biochars will be beneficial to soils. Furthermore, the quantity and rate of biochar application may disrupt the equilibrium of soil processes in the short term, even if in the long term beneficial effects are achieved. This type of risk is more difficult to assess and is not covered by the standard ecotoxicological risk assessment. Until such time as our scientific understanding develops well enough to be able to understand the effects of a particular biochar in a particular soil, it may be prudent to trial a biochar before it is deployed more widely.

Biochars have been shown to affect nutrient cycling in both positive and negative ways. Biochar-nutrient interactions between ammonium, nitrate and phosphorous have been most thoroughly investigated in terms of direct nutrient release from the biochar (Yao *et al.* 2012; Mukherjee and Zimmerman 2013); physical absorption/desorption of nutrients in the biochar matrix (Taghizadeh-Toosi *et al.* 2011a; Hale *et al.* 2013); and by altering the microbial populations which modulate the nitrogen cycle (Ducey *et al.* 2013; Lehmann *et al.* 2011). Confusingly, the magnitude and direction of these effects are modulated by the material content of the biochar (i.e. the quantity of a particular nutrient contained within the biochar), the production conditions (in particular peak temperature) and the age and weathering of the biochar. To generalise on the incomplete evidence referenced in this paragraph, it may be said that most of the ash-based nutrients in a biochar are released relatively quickly, but nitrogen-based nutrients are not. Biochars produced at a higher temperature tend to better absorb nutrients out of the soil solution, and older biochars tend to exchange nutrients more easily.

In the short term, ammonium immobilisation has been observed (see Chapters 5 and 7; Bruun *et al.* 2011, 2012; Novak *et al.* 2010; Yao *et al.* 2012), which can cause crop growth reduction under some conditions. An overall increase in the speed of the nitrogen cycle due to biochar addition has also been reported, but the increased release of ammonium was buffered by the biochar's tendency to absorb bioavailable ammonium (Nelissen *et al.* 2012). However, experiments have also shown that ammonium which is absorbed by biochars can become bioavailable again (Taghizadeh-Toosi *et al.* 2011a), and that older biochars have a higher anion exchange capacity, and therefore are better able to exchange absorbed ammonium (Mukherjee *et al.* 2011; Cheng *et al.* 2008; Mukherjee and Zimmerman 2013).

As noted above, biochar has been proposed as a tool in land remediation to reduce the bioavailability of heavy metal elements, particularly lead and cadmium, and trials have yielded positive results (Namgay *et al.* 2010; Moon *et al.* 2013; Beesley and Marmiroli 2011). However, it should be noted that increased soil pH due to biochar addition can in some circumstances increase the bioavailability of certain contaminants, such as copper (Ahmad *et al.* 2012) and arsenic (Beesley *et al.* 2010); although these effects are inconsistent (Namgay *et al.* 2010).

Some biochars have been shown to absorb pesticides, herbicides and similar compounds from soils. This means that less of the pesticide reaches its intended destination, and the function of the pesticide may therefore be disrupted (Yu *et al.* 2009; Sopeña *et al.* 2012; Kookana 2010).

Finally, it is possible that other effects could occur which we do not yet know about. These would be the unforeseen effects or 'unknown unknowns'. Whether the state of knowledge is advanced enough yet for widespread biochar use is a matter of discussion and there will inevitably be a certain amount of 'learning through doing'. However, caution is urged and smaller scale trials with well-thought-out designs to optimise learning are advised before widespread deployment. It is likely that the marginal profitability of biochar use will prevent run-away deployment, although it is also possible that if a profitable application of biochar were found it could be deployed with rapidity.

Safety concerns

There are a few additional risks of biochar production and use; for example, biochar can break down into a very fine dust which can be harmful if inhaled, and it could even be carcinogenic (Parliamentary Office of Science and Technology 2010). Such impacts are usually associated with prolonged occupational exposure, such as in coal mining, to airborne black carbon. When handling dusty biochar, a facemask should be used. When applying biochar to fields, dry, dusty material should be wetted or combined with some other more moist material to prevent it from becoming airborne. Even if the material is dry, however, the level of exposure to a member of the public arising from biochar particles would be much lower than the levels associated with occupational exposure (e.g. in coal mines or during traditional charcoal making), and is more likely to cause a nuisance rather than a health risk. Of more concern would be long-term exposure by operatives of biochar-producing equipment or by those applying biochar to fields.

Airborne contaminants can also be released through the production of biochar. Most modern pyrolysis technology is required by European and national legislation (European Parliament 2008a) to monitor and control harmful emissions. This can be achieved through scrubbing technologies and appropriate equipment design. Old-style charcoal ring kilns or earthen kilns are generally below the scale which is set by legislation, but they can release particulates and contaminants in quantities that may pose health risks (Kammen and Lew 2005; Sparrevik *et al.* 2013). Old-style charcoal kilns also yield lower proportions of charcoal and release gases with high global warming potentials; for these reasons, old-style charcoal kilns are not recommended as an environmentally appropriate production method for biochar. Simple modifications of relatively basic pyrolysis equipment can render them clean burning, as seen in the adoption of retort devices where the hot pyrolytic gases are fed into a burner which combusts the gas, greatly reducing the prevalence of larger molecules and yielding mostly CO_2 and H_2O.

Fire can be a potential risk when storing biochar in a dry state, and more so if it has a large proportion as fine particles. It could be ignited by an external source or self-ignition is possible, and slow, smouldering combustion may go unnoticed until it becomes uncontrollable (Rein 2009). To combat these risks, char should be appropriately stored, e.g. in piles of suitable size, dimension and moisture content.

The risk of increasing greenhouse gas emissions

One of the major motivations for using biochar is to reduce GHG concentrations and combat global warming (Lehmann *et al.* 2006; Woolf *et al.* 2010). However, there are a number of ways in which biochar could increase GHG emissions, thus posing a threat to the wider environment and undermining one of the major motivations for biochar deployment. The major sources of risk in this regard are from using a biochar which has been incompletely pyrolysed and contains a large proportion of unstable carbon or from the use of unsustainable or inappropriate feedstocks. A lesser risk (in terms of the quantity of potentially emitted GHGs) is from soil priming and subsequent increase of soil GHG emissions (see Chapter 7; Hammond *et al.* 2011; Roberts *et al.* 2010).

If biomass is incompletely pyrolysed, it is stable only on a decadal timescale. Much of the biochar carbon would be decomposed back to CO_2 before it contributed to reducing global warming. The solution to this is to ensure that biochar producers follow guidelines to ensure that the biochar is properly and completely pyrolysed (International Biochar Initiative 2013; Schmidt *et al.* 2012; British Biochar Foundation 2014). They should also ensure that the stability of biochar is tested for, including assessing the ratio of organic carbon to hydrogen (see Chapters 3 and 7) and performing aggressive decompositions (Cross and Sohi 2013; Crombie *et al.* 2013).

Unsustainable harvesting of feedstocks could entail a high carbon debt (for example, if old growth forests or peat lands were disturbed), or long time-periods for biomass regrowth, resulting in unfavourable GHG balances. However, emissions attributed to the harvesting, processing and transportation of feedstocks are generally low (Hammond et al. 2011; Ibarrola et al. 2012). For more detail on this, see Chapter 8.

Finally, increased GHG emissions may come from soil. There are two routes: soil priming (the release of CO_2 from native organic matter in the soil) and biochar effects upon soil N_2O and CH_4 emissions (see Chapter 7). Increased soil priming has been seen in some high carbon soils (Wardle et al. 2008) and it therefore seems prudent to avoid adding biochar to these soils. In agricultural soils, priming is generally short term (Cross and Sohi 2011; Novak et al. 2010; Luo et al. 2011), and negative priming (enhanced levels of native soil carbon) has also been seen in several studies (Bruun et al. 2012; Laird et al. 2008; Lehmann et al. 2006).

Experiments with biochar have shown both reductions in soil-emitted N_2O (Angst et al. 2013; Liu et al. 2012; Taghizadeh-Toosi et al. 2011b; Sarkhot et al. 2008; Castaldi et al. 2011; Felber et al. 2012) and no change or increases of soil-emitted N_2O (Troy et al. 2013; Case et al. 2012; Scheer et al. 2011; Bruun et al. 2011). A recent laboratory incubation experiment showed that nitrous oxide suppression decreased with aged biochar compared with fresh biochar (Spokas 2013). Generally, no change has been seen in soil methane emissions. The long-term effects of biochar on soil N_2O emissions are still uncertain but if suppression could be consistently achieved for a period of 10 to 100 years, it would be a major benefit in reducing GHGs. Conversely, there is a risk that nitrous oxide emissions could be increased in some circumstances.

Unsustainable use of land and of biomass

The risk of unsustainable land use applies to any use of biomass, not just for biochar. There is a risk of using harvested material unsustainably and a separate risk of removing existing biomass and converting land to a different use, usually agriculture. Sustainability entails not just GHG emissions or environmental hazards, but also ecosystem integrity, species conservation, long-term productivity and human considerations such as rights of access, rights of use, continuity of culture, food security and so on.

Various voluntary certification schemes exist to assure consumers of the sustainability of harvested biomass; for example, the Forest Stewardship Council or Programme for the Endorsement of Forest Certification (PEFC). In March 2013, the European Timber Regulation (European Parliament 2010b), which aims to ensure that all timber imported into the European Union is from certified sustainable sources, came into effect.

Compliance with these and similar measures should go a long way towards ensuring that biomass used for biochar is sustainable. Of course, waste materials or residues can also be used to make biochar and may entail less competition than the use of virgin materials or purpose grown crops.

Risk control measures: regulation and legislation

Appropriate legislation is required to minimise risks from producing and using biochar. However, the modern concept of 'biochar' has not yet been specifically legislated for, and at the time of writing, there has been little, if any, case law built up. To fill the 'legislative gap' a number of voluntary standards have been produced, and discussions are under way at national and international level to enact mandatory standards.

Even so, there is plenty of legislation which already applies to biochar and must be complied with if biochar use is to remain lawful. Until such time as dedicated legislation brings greater clarity to the regulations applicable to biochar production and use, it will remain something of a grey area, and each case may well have to be negotiated individually with the responsible (environmental) regulatory authorities.

This section summarises the existing EU legislation most pertinent to biochar production and use; some of the important similarities and differences in national legislation and regulation; guidance on how best to position biochar within existing legislation; and a summary of the voluntary standards developed.

EU legislation

European legislation, conceived as European Regulations and Directives, is proposed by the European Commission and (in general) enacted by the European Parliament and the European Council. After entering into force, it applies to all member states of the EU and possibly to the member states of the European Economic Area (EEA) (i.e. an internal market which unites the member states of the European Union and three member states of the European Free trade Association (EFTA), namely Norway, Iceland and Liechtenstein) in as far as the legislation is considered to be EEA relevant.

Regulations and Directives are essentially different in nature. An EU Regulation is a binding legislative act which applies in all member states without further transposition in national law and is as such binding for its citizens. It thus replaces or overrules national legislation. Directives, on the other hand, are not directly applicable in the member states; they merely set out goals to be achieved. Member states are free to decide on the measures to be taken to achieve these goals and need to transpose the Directives in national legislation within an agreed timeline. Although common goals are set by the Directives, national legislation may thus vary considerably in the different member states.

However, the variety in final application of European legislation is not only due to the transposition of the Directives. The national legislation which aims to trans-pose the Directives, as well as the content of the Regulations, is often put into practice and concretised by regulatory authorities (such as environmental protection agencies, food standard authorities, etc.). These authorities elaborate and attempt to implement the (transposed) European legislation in daily practice. Since these authorities are often sub-national in their scope, multiple regulatory authorities and thus multiple interpretations of legislation may exist in one country, each aiming to enforce the same European legislation.

Finally, the case-law build up by courts and sanctioning administrative authorities may contribute to the variety in interpretation. These institutions decide whether or not an infringement of the legislation has occurred and thus also contribute to the delineation of the scope of the legislation. Insofar as legal gaps occur or legislation is unclear, the point of view of these institutions is decisive. It is through this build-up of case law that legislation becomes better defined and actions are proven to be lawful or unlawful. Until such case law builds up around biochar, the interpretation of legislation as it applies to biochar will remain unproven, especially since biochar is not mentioned by name in European nor national legislation (see below).

EU legislation and biochar

At the time of writing, no EU legislation refers to biochar by name. Nevertheless, certain Regulations and Directives might potentially have a bearing on the production and use of biochar in agriculture, such as the Fertiliser Regulation (European Parliament 2003), the Waste Framework Directive (European Parliament 2008b), the Industrial Emissions Directive (European Parliament 2010a), the Renewable Energy Directive (European Parliament 2009), the Registration, Evaluation, Authorisation and Restriction of Chemicals Regulation (commonly referred to as REACH) (European Parliament 2006). Each Regulation or Directive applies to biochar production or use only under particular circumstances. For example, the waste incineration directive might only apply if the feedstock used is a waste, as defined by the Waste Framework Directive.

The Regulations and Directives mentioned above are at first sight potentially the most relevant ones for the production and use of biochar. Nevertheless, other EU legislation may have a direct or indirect impact on the production and use of biochar. However, an all-encompassing enumeration of these Regulations and Directives is not the aim of this Chapter. Moreover, given the scientific uncertainty surrounding biochar, providing such an overview seems currently unfeasible.

The key points of each Regulation and Directive mentioned above, are summarised in Table 10.1 (for a more information, see van Laer et al. 2015).

National legislation and regulation relevant to biochar

Although all member states of the European Union are bound by the same Regulations and Directives, there is considerable difference in the way they are put into practice in the different member states. As mentioned above, this difference results from the transposition into national (or even regional) binding legislation

Table 10.1 Summary of EU legislation relevant to biochar

EU legislation	Aim of legislation	Potential application to biochar	Potential impact	Notes
Fertiliser Regulation (EC) No 2003/2003	Harmonisation of technical characteristics of fertilisers (e.g. regarding the composition and definition types of fertilisers) so as to facilitate their trade (EC fertilisers).	At present this only applies to inorganic fertilisers and not to biochar. The 2015 review will include organic fertilisers as well as soil improvers. Including biochar by name, categorised as an inorganic soil improver, has been proposed.	At present no impact, but potentially legally binding guidance for trading biochar as an EC fertiliser, if biochar is included in the planned update.	
Waste Framework Directive (2008/98/EC)	Protecting the environment and human health through the prevention of the harmful effects of waste generation and waste management, reducing the use of resources and favours the practical application of the waste hierarchy. This Directive should help move the EU closer to a 'recycling society' seeking to avoid waste generation and to use waste as a resource.	Applies to any biochar which is to be produced from waste (remainders of biomass which one wants to discard (e.g. waste streams from agriculture) and to any biochar produced as by-product in a production process (i.e. not as the primary aim of the production process).	Defines what is a waste and what is not, determines how to get rid of the waste status (end-of-waste product) and the corresponding obligations for waste (e.g. concerning waste transport), determines at a general level how to deal with waste (waste hierarchy, separate collection, permitting of treatment plants, etc.)	
Industrial Emissions Directive (2010/75/EU)	Provide for an integrated approach to prevention, reduction and control of emissions into air, water and soil, to waste management, to energy efficiency and to accident prevention arising from industrial activities.	Applies to biochar production plants in which waste undergoes a thermal treatment, such as pyrolysis and gasification if the substances resulting from the treatment are subsequently incinerated.	Sets stringent operational conditions (e.g. permit requirements) and technical requirements through setting emission limit values.	

Table 10.1 Continued

EU legislation	Aim of legislation	Potential application to biochar	Potential impact	Notes
Renewable Energy Directive (2009/28/EC)	Establishes a common framework for the production, use and promotion of energy from renewable sources in order to limit GHG emissions and to promote cleaner transport.	Applies to electricity, heating and cooling or transport fuels produced from renewable energy sources.	The financial support mechanisms resulting from the Renewable Energy Directive are only applicable if energy is produced from biofuels and bioliquids that comply with certain sustainability criteria.	Discussions on how to account for land use change and indirect land use change are under way.
REACH (EC 1907/2006)	Establishes requirements for companies to register chemical substances, identify and manage the risks associated with chemical substances which they create, import or use.	Manufacturers of a chemical substance (such as biochar) in excess of one tonne per year, must meet the requirements of the legislation.	Substances must be formally identified and registered with the European Chemical Agency,[a] and information regarding risk management must be passed down the supply chain with the product.	Considerable support for the process is available from the European Chemical Agency, and companies are encouraged to work together.

Note: [a] See http://echa.europa.eu.

and further interpretation and application of this national and European legislation by national authorities and courts. Hence one Directive, or even Regulation, may be applied in many different ways.

Moreover, a lot of issues are not (entirely) regulated at European level. Insofar as these issues have not been regulated, member states have considerable discretion to regulate them. Since the production and use of biochar is currently not explicitly mentioned in EU legislation, member states thus rely on their own discretion when producing and using biochar. Nevertheless, when regulating biochar, they do have to take into account existing legislation (e.g. the Directives mentioned above). In some member states, biochar would not fit into existing national legislation and an exemption or special permission would have to be applied for; whereas in other member states, the regulation of biochar might be inserted in the existing national legislation for organic soil amendments or charcoal application to soil, which is already regulated.

Even so, the regulation of biochar in the different member states share some common factors. As shown below, all member states examined in light of this study impose maximum permissible limits (MPLs) for PTEs and compounds in substances which can be applied to soils, even though no member state mentions biochar by name in general legislation. This implies that some negotiation with regulatory authorities will be necessary until such time as the use of biochar is more common and specifically regulated for.

Summaries on how European member states currently deal with biochar are given in the case studies below.

United Kingdom

No regulation or legislation refers to biochar by name, nor is it included in the list of 'other agricultural soil amendments' specified in regulation (Shackley and Sohi 2010). The environmental regulatory authorities consider that there is a 'regulatory gap' where biochar is concerned and they are willing to work with biochar producers, users and researchers to fill that gap, providing sufficient demand is evident. Each country in the UK (England, Scotland, Wales and Northern Ireland) has a different environmental regulator, and so advice and regulation differs between each country.

If biochar is made from non-waste materials then there is no regulatory reason why it should not be applied to agricultural soils, providing that no environmentally negative effects occur. These negative effects might include the exceeding of nutrient application limits, the pollution of soils with unacceptable levels of potentially toxic metal, organic or biological substances, the pollution of waterways or other environmental nuisance such as clouds of airborne dust or bad smells. The guidance for the use of other soil amendments such as manures and slurries (Defra 2010), composts (British Standards Institution 2005) and digestates (British Standards Institution 2010) serve as useful pointers. MPLs are usually defined as a maximum quantity in the soil amendment itself (see below).

Biochar production must abide by the pollution prevention and control legislation (Defra 2009) which sets limits on the resulting emissions. If the biochar is made from a waste feedstock, then regulation is more arduous. The biochar producer can either aim for an exemption under waste regulations – a difficult and potentially expensive process – or attempt to obtain an 'end-of-waste status'[1] for the biochar, which then enables equivalent use to a non-waste material.

Following sufficient number of applications for end-of-waste status, generalised exemptions from waste controls and position statements may be produced by the regulatory authorities. An interim position statement has been released by the Scottish Environmental Protection Agency which facilitates the small scale production and use of biochar from clean woody wastes (SEPA 2012) and a similar draft document has been prepared by the Environment Agency in England. United Kingdom specific guidance on what constitutes a high quality biochar has been published, in collaboration with UK environmental authorities (British Biochar Foundation 2014).

The Netherlands

In the Netherlands, the use of fertilisers and soil improvers is regulated via the Fertiliser Act ('*Meststoffenwet*'). Next to it, Environmental Management Act ('*Wet Milieubeheer*') and the Soil Protection Act ('*Wet Bodembescherming*') are of relevance. In the Fertiliser Act, 'fertilisers' are defined as 'products which are meant for the application to a soil ... and which consist of compounds or a mixture of compounds, which are intended to make soils suitable or more suitable as growing medium for plants'. In the Environmental Management Act, a 'waste material' is defined as following the EU Waste Directive: 'any substance or object which the producer or the person in possession of it discards or intends or is required to discard'.

The Dutch Fertiliser Act makes a distinction between various types of fertilisers and waste materials which are used as fertilisers. If the legal status of the feedstock is a waste, the resulting biochar also gets the status of waste. In that case, it is possible to obtain fertiliser status by sending in a request to the Ministry of Economy of the Dutch National Government. A Committee of Experts of the Fertiliser Act assesses requests. The committee ascertains that no objections from an agricultural or environmental point of view exist, following a defined protocol (van Dijk *et al.* 2009). If an application is successful, the material will be added to Annex A of the Fertiliser Act, and may be traded and used as a fertiliser.

If a biochar is made from feedstock which is not waste, and which is not considered to be a residue from energy production, the biochar may be considered to be a fertiliser, provided that it fulfils all the requirements of fertilisers.

The Fertiliser Act defines different categories of fertiliser according to the main benefit which the product delivers (e.g. nitrogen, phosphorus, potassium or other nutrients, neutralising value, organic matter). Biochar can be considered to be an organic fertiliser as the main material effect of applying biochar is addition of organic matter (even though the organic matter in biochar is in a different state to most organic fertilisers). Maximum thresholds are formulated for the concentration of heavy metals and arsenic and for organic micro contaminants, which are expressed as a fraction of the valuable compounds (in this case, organic matter). For example, the MPL of cadmium is 0.8mg/kg of carbon. In order to calculate the MPL of cadmium in a given biochar, the carbon content of that biochar must be taken into account. If a biochar contained 80 per cent carbon, the MPL of cadmium would be 0.64mg/kg of biochar (0.8 × 80% = 0.64). Other MPLs are shown in Table 10.2, all assuming a biochar of 80% carbon (for illustrative purposes only).

Belgium (Flanders)

Belgium is a country with separate regions – namely, Flanders, Brussels and Wallonia – disposing of a large autonomy and proper competencies in matters such as environmental regulation. Although some issues with regard to the environment are regulated at federal level, environmental legislation is mainly enacted at regional level. Consequently, environmental legislation, including the transposition of European legislation, is applied differently in each Region. The following summary only applies to Flanders. For a more in-depth review see van Laer et al. (2015).

If biochar is deliberately produced from waste the Public Waste Agency of Flanders (OVAM) seems to take into consideration the end-of-waste status of biochar. This means biochar may under certain circumstances be used as a soil-improving product or as a fertiliser.

Table 10.2 Calculated MPLs for some contaminants in biochar in four countries, derived from guidelines for other soil amendments

PTE	Britain (PAS100) (mg/kg compost)	Belgium (Flanders) (mg/kg soil improving amendment)*	Norway (Class 1) (mg/kg amendment)	Netherlands (mg/kg amendment)
As		150		8
Cd	1.5	6	0.8	0.64
Cr	100	250	60	40
Cu	200	375	150	40
Pb	200	300	60	53.6
Hg	1	5	0.6	0.4
Ni	50	50	30	16
Zn	400	900	400	160

Note: *The thresholds in the table are the ones set for soil improving amendments (with a dry matter content of >2 per cent) produced from waste.

In particular, one has to obtain a 'raw material declaration' ('*grondstofverklaring*') or application form provided by the Public Waste Agency. It requires (among other declarations), the personal data of the applicant, the personal data of the biochar producer, identification of the biochar, an overview of the production process, a copy of the environmental permit of the plant producing the biochar, one or more recent (less than one year old) sampling, analyses of the biochar performed by a certified laboratory and a description of the use of the biochar. After receipt of the application, the Agency takes 20 days to declare the application acceptable and complete. If the application is acceptable and complete, the Agency will then decide within 45 days to grant or refuse the 'raw material declaration'. In the case of refusal, one can appeal with the relevant minister within 30 days. The minister will then make a decision within three months after receiving the appeal.

The 'raw material declaration' contains conditions for the use of the biochar and may also impose specific conditions; for example, relating to the origin of the material used or the thresholds for contaminants. As regards contaminants, Flanders has an exceptionally precautionary system in place for soil-improving amendments produced from waste, with MPLs given for a large number of substances (see below) and acceptable methods to be used. Contaminants are measured in terms of absolute quantity in the amendment and in terms of annual dose to soil per year. As noted in Chapter 3, however, biochar cannot be characterised using the same methods as those used to characterise soil samples and most other soil amendments. Officially designated methods for PAHs have already been shown to be unsuitable for such measurements.

If biochar becomes widely used, it is possible that a system similar to the one for compost may be established, thus avoiding the need for every producer to obtain a 'raw material declaration'. As regards compost, no 'raw material declaration' is required, but the compost needs to comply with certain standards and has to be produced in a certified plant that is permitted to produce compost.

In order to trade biochar and use it as a fertiliser or soil-improving product, an exemption should be applied for at the Federal Public Service for Health, Security of the Food Chain and Environment. The application should contain information regarding the composition, nature and origin of the biochar; a description of the production process; a description of the agricultural value of biochar; and a report from a certified laboratory describing the relevant parameters of the biochar. If the application is complete, a decision regarding the exemption is taken within 12 months. This exemption can only be granted if the biochar produced from waste has obtained a 'raw material declaration' as described above.

Norway

Although not specifically mentioned, biochar application to soils in Norway is regulated under the laws regarding the application of organic materials and fertilisers to soil.[2] These regulations are designed to control the use and application of composts, sewage sludge and organic waste materials in order to prevent, for example, the excessive accumulation of heavy metals in soil, the loss of excess nutrients such as phosphorus and nitrogen into waterways and air and the spreading of diseases and pests to plants, animals and humans.

Currently these laws are undergoing review as there has been a need to move away from criteria based on the concentration of heavy metals in a given organic amendment to annual loadings of the heavy metals from the applied organic materials. The regulations divide organic amendments into four quality classes and restrict their end-use application. The four quality classes are:

- *Quality class 0:* Can be used on agricultural land, private gardens and parks. The applied amount should not exceed the plants' nutritional needs.
- *Quality class I:* Can be used on agricultural land, private gardens and parks with up to 40 tonnes of dry matter per hectare per 10-year period.
- *Quality class II:* Can be used on agricultural land, private gardens and parks with up to 20 tonnes of dry matter per hectare per 10-year period.
- *Quality class III:* Can be used for parks, ovals and other areas where food or fodder crops are not grown. The product should be applied to the surface at a maximum of 5cm thickness and mixed into the soil in situ. It can be used as a top cover for landfills at a maximum thickness of 15cm.

The content of heavy metals in produced biochars will be the main criteria which determines which quality class a biochar falls into, and thereby where and how much biochar can be applied (see below).

Most commercially produced biochars that are being used in research in Europe currently fall within the highest quality class in Norway. The highest quality class, in addition to heavy metals, stipulates that one can apply as much of the material to the soil as desired as long as it does not negatively affect plant growth. Therefore, upper limits for biochar application would have to be determined according to different crop and soil types in order to conform to the current regulations within this quality class.

Denmark

There is currently no Danish legislation specifically for the use of biochar for soil improvement. Regulators may grant an exemption if producers provide chemical analysis which follows the guidelines from the bio-ash (BEK 2008) or the sludge notice (BEK 2006). These guidelines provide MPLs for contaminants in soil amendments. In order to use residues from thermal conversion in the soil, standard chemical analysis needs to be done by a certified lab, including C, N, P, K, ash, pH, surface area, heavy metals, PAHs and any dioxins/furans. This information also contributes to an overall assessment of the physical and biological effects that biochar could potentially have on soil properties and the general soil environment.

The possible commercial utilisation of biochar will be regulated as the market grows. Regulation will be between two authorities, namely the Danish Plant Directorate and Danish Environmental Protection Agency. The Danish Plant Directorate is a government institution under the Danish Ministry of Food, Agriculture and Fisheries, whereas the Danish EPA is part of the Danish Ministry of the Environment.

The regulation will take place under law 318 of 31 March 2007 (decree on fertilisers and soil conditioners) and involves the transfer of more general information regarding the product in question to the Plant Directorate. In addition, the content of critical compounds (e.g. PAHs) will be a point of regulation. Highly toxic PAHs may have individual cut-off limits, and it should be noted that naphthalene – a common PAH in biochar – is not a regulated PAH, but it could still be controlled as an 'organic volatile', although with a significantly higher cut-off limit. Other problematic features, e.g. heavy metal elements, may be individually regulated.

Comparing maximum permissible limits

MPLs for potentially dangerous substances in soil amendments are calculated in different ways in different countries. Quantity per unit of soil amendment is the most common (UK, Flanders, Norway); but maximum application to soil over a defined period of years (UK); maximum amount added per year per hectare (Flanders); or maximum amount per unit of carbon in the soil amendment (Netherlands) are also used. Even when the different methods of calculation are taken into account, the thresholds differ between countries. Some of these thresholds are shown in Table 10.2.

The MPLs for organic pollutants (PAHs, PCBs, dioxins, etc.) are even more variable, as there is scientific uncertainty as to what level to set limits at and what methods to use to test for these compounds. For example, Belgium (Flanders) and the Netherlands both use detailed thresholds for 30 or more compounds; the UK provides thresholds for acceptable concentrations of five compounds in soils, and Norway advises that said compounds should not be at levels that cause risk of harm to human health, but it does not state what those levels may be.

For a much more detailed assessment of the different approaches to regulating fertilisers and other soil amendments in northwestern Europe, see Ehlert et al. (2013).

Interpretation of national legislation

Although the modern concept of 'biochar' has not yet been specifically legislated for, there are many similarities in the legislation and regulation relevant for biochar in different countries. The definitions and controls of waste material are similar; there are usually requirements on meeting MPLs of contaminants and producers must meet soil amendment criteria. Production process controls are broadly similar too, although they have not been addressed in detail here. However, there are differences too: either in the detail of regulation (such as the numbers in the MPLs), or procedural issues such as the process by which 'end-of-waste' criteria can be achieved.

Achieving regulatory approval

In order for a biochar producer to be approved by the environmental regulators in their region, it is likely that, at least at first, a detailed and lengthy negotiation will have to be entered into. Hopefully the above summary of the risks relating to biochar and the existing regulatory approaches will help to equip both producers and regulators in their forthcoming negotiations.

There are two more approaches which may be useful in negotiations between biochar producers and regulatory authorities. The first is to use existing regulation, and not to forget that charcoal has been used for a very long time and that biochar is not necessarily that different (but may sometimes be very different). The second approach is to make use of the voluntary standards for biochars. These standards prescribe MPLs of contaminants, and various other safeguards to ensure that any certified biochar is environmentally sound and of a reliable quality.

Placing biochar within existing regulation: stories from the field

As concluded above, there are many similarities in the legislation and regulation between different countries. Hence, it may be possible to place biochar within the existing legal framework rather than creating new legislation specifically for biochar.

Biochar is a new term which does not exist in any regulation. However, other wording exists which refers to the same pyrolysed material. In English, it is 'charcoal' and 'vegetal carbon' or 'carbon black', sometimes 'activated charcoal'. In French, it is '*charbon de bois*' or '*charbon vegetal*'; in German, it is '*Holzkohle*' and '*Pflanzenkohle*'; and in Spanish, '*carbon vegetal*'.

Charcoal, by one name or another, is often used as a soil amendment, compost additive and fertiliser additive. There is no real definition of charcoal with regards to feedstock, production parameters, carbon content, etc., and so it is not clear to what extent it does or does not include biochar. Although the material can be the same, the intended purpose can be different: charcoal is intended for use as a fuel but biochar is intended for use as a soil amendment. Charcoal applied to soil could therefore be defined as biochar. All of this leads to a grey area in legislation as it currently stands. Note that pyrolysis of non-woody feedstocks typically produces a material that is not comparable to charcoal but can be defined as biochar – hence the definition of biochar is wider than that for charcoal.[3]

Voluntary regulation schemes

Proponents of biochar use have produced voluntary sets of standards for its safe production and use. Voluntary standards are a common and well-accepted method by which industries assure consumers and regulators of the quality of their products. These voluntary standards may or may not inform legislation and mandatory regulation as it develops; however, until such time as legislation exists, the voluntary standards are the only comprehensive assessment suites designed especially for biochar, and may be useful to producers of biochar or regulatory authorities faced with biochar related questions.

At the time of writing three such standards have been published. The International Biochar Initiative (IBI) has published a second version of Standardized Product Definition and Product Testing Guidelines for Biochar that is Used in Soil (International Biochar Initiative 2013); and it has announced a certification body which is able to verify which biochars meet that standard. The original document was developed through a series of expert workshops and then a one-year-long public consultation period with a series of webinars that were open to the public.

The European Biochar Certificate (EBC) Guidelines for Biochar Production was first published in January 2012, certified by the European Biochar Foundation and written by a team of six expert authors (Schmidt *et al.* 2012).

The EBC is based in Europe, while the IBI guidelines are based in North America. The content of the documents reflects to some degree the region of origin. For example, the IBI guidelines are highly concerned with liability issues. Another important difference is that the EBC uses a positive list of feedstocks which are permitted for biochar production, while the IBI does not. The EBC offers two standards of biochar certification, where the higher standard demands more stringent limits on contaminants, higher carbon content and stability and more environmentally friendly biochar production conditions than the lower standard. The EBC requires biochar to have a minimum of 50 per cent by mass organic carbon, while the IBI definition has three categories of organic carbon content:

- $\geq 10\%$ to $<30\%$;
- $\geq 30\%$ to $<60\%$; and
- $\geq 60\%$.

Otherwise the IBI and EBC documents are fairly similar: both contain MPLs for contaminants; limits for minimum carbon content and stability; require that nutrient content and pH be reported; require that pyrolysis emissions be minimised; and require that a full set of information be delivered to the consumer.[4]

The third voluntary standard is the Biochar Quality Mandate (BQM) (British Biochar Foundation 2014). The BQM is national in scope and was developed in the UK by experts from research in consultation with the regulatory authorities. In addition to covering the same issues as the IBI and EBC guidelines, the BQM also requires that the sustainability of biomass sources be proven and offers advice tailored to meet the requirements of the UK regulators. The BQM has also adapted a methodology from regulations on sewage sludge disposal to calculate the allowable application of biochar to a given land area over a given time period.

MPLs from each of the voluntary standards are compared in Table 10.3 below.

Conclusions

In the eyes of the law, biochar remains something of a grey area. Biochar made from wood is physically identical to charcoal, which has been used by humans since before legal systems were established, and which has numerous applications within legal structures. Pyrolysed carbon made from other organic matter is also a naturally occurring substance, a product which is legislated for in terms of use as a feed amendment and as a waste material from certain industrial processes.

Biochar is novel only in the function for which the substance is intended to be used. Biochar is not yet specifically mentioned in legislation in any country within the European Community or at EU level. Various legislations can be interpreted to apply to biochar, and none of it is immediately restrictive. The burden of interpretation and justification falls on the producer of the biochar.

Table 10.3 MPLs of contaminants from three voluntary biochar standards

PTE	IBI Guidelines (mg/kg dry matter)	European Biochar Certificate premium biochar (mg/kg dry matter)	European Biochar Certificate basic biochar (mg/kg dry matter)	Biochar Quality Mandate (high grade) (mg/kg dry matter)	Biochar Quality Mandate (standard grade) (mg/kg dry matter)
Arsenic	12–100	n/a	n/a	10	100
Cadmium	1.4–39	1	1.5	3	39
Chromium	64–100	80	90	15	100
Copper	63–1500	100	100	40	1500
Lead	70–500	120	150	60	500
Mercury	1 to 17	1	1	1	17
Manganese	n/a	n/a	n/a	3500	n/a
Molybdenum	5 to 75	n/a	n/a	10	75
Nickel	25	30	50	10	600
Selenium	1–100	n/a	n/a	5	100
Zinc	200–2800	400	400	150	2800
PAHs	6 to 20	4	12	<20	<20
Dioxins/Furans(ng/kg)	9	20	20	<20	<20

In order to support the producers, there are a number of voluntary certification schemes which have generally been developed with opportunity for input from the relevant regulatory authorities, and which biochar producers can sign up to at a minimal cost.

Providing that biochars are produced from materials which are not contaminated with high levels of heavy metals or plastics, they are likely to meet the requirements outlines in voluntary certification schemes, and as such are likely to meet with approval from local regulatory authorities.

Bibliography

Ahmad, M., Moon, D., Lim, K., Shope, C., Lee, S., Usman, A., Kim. K-R., Park, J-H., Hur. S-O., Yang, J. and Sik, Y. (2012). An assessment of the utilization of waste resources for the immobilization of Pb and Cu in the soil from a Korean military shooting range. *Environmental Earth Sciences* 67(4): 1023–1031.

Angst, T., Patterson, C., Reay, D., Anderson, P., Peshkur, T. and Sohi, S. (2013). Biochar diminishes nitrous oxide and nitrate leaching from diverse nutrient sources. *Journal of Environmental Quality* 42(3): 672–682.

Beesley, L. and Marmiroli, M. (2011). The immobilisation and retention of soluble arsenic, cadmium and zinc by biochar. *Environmental Pollution* 159(2): 474–480.

Beesley, L., Moreno-Jiménez, E. and Gomez-Eyles, J.L. (2010). Effects of biochar and greenwaste compost amendments on mobility, bioavailability and toxicity of inorganic and organic contaminants in a multi-element polluted soil. *Environmental Pollution* 158(6): 2282–2287.

Beesley, L., Moreno-Jimenez, E., Gomez-Eyles, J., Harris, E., Robinson, B. and Sizmur, T. (2011). A review of biochars' potential role in the remediation, revegetation and restoration of contaminated soils. *Environmental Pollution* 159(12): 3269–3282.

BEK (2006). *Executive Order No. 1650 of 13/12/2006: Sludge Notice.* Available at www.retsin-formation.dk/forms/R0710.aspx?id=13056.

BEK (2008). *BEK nr 818 af 21/07/2008 Gældende.* Available at www.retsinformation.dk/forms/R0710.aspx?id=116609&exp=1.

Bonten, L.T.C., Römkens, P.F.A.M. and Brus, D.J. (2008). Contribution of heavy metal leaching from agricultural soils to surface water loads. *Environmental Forensics* 9(2): 252–257.

British Biochar Foundation (2014). *Biochar Quality Mandate.* Available at www.british-biocharfoundation.org/wp-content/uploads/BQM-V1.0.pdf.

British Standards Institution (2005). *PAS 100:2005 Specification for Composted Materials.* London: British Standards Institution.

British Standards Institution (2010). *PAS 110:2010 Specification for Whole Digestate, Separated Liquor and Separated Fibre Derived from the Anaerobic Digestion of Source-Segregated Biodegradable Materials.* London: British Standards Institution.

Bruun, E., Müller-Stöver, D., Ambus, P. and Hauggaard-Nielsen, H. (2011). Application of biochar to soil and N_2O emissions: potential effects of blending fast-pyrolysis biochar with anaerobically digested slurry. *European Journal of Soil Science* 62(4): 581–589.

Bruun, E., Ambus, P., Egsgaard, H. and Hauggaard-Nielsen, H. (2012). Effects of slow and fast pyrolysis biochar on soil C and N turnover dynamics. *Soil Biology and Biochemistry* 46: 73–79.

Case, S.D.C., McNamara, N.P., Reay, D.S. and Whitaker, J. (2012). The effect of biochar addition on N₂O and CO₂ emissions from a sandy loam soil: the role of soil aeration. *Soil Biology and Biochemistry* 51(April): 125–134.

Castaldi, S., Riondino, M., Baronti, S., Esposito, F., Marzaioli, R., Rutigliano, F., Vaccari, F. and Miglietta, F. (2011). Impact of biochar application to a Mediterranean wheat crop on soil microbial activity and greenhouse gas fluxes. *Chemosphere* 85(9): 1464–1471.

Cheng, C.-H., Lehmann, J. and Engelhard, M.H. (2008). Natural oxidation of black carbon in soils: Changes in molecular form and surface charge along a climosequence. *Geochimica et Cosmochimica Acta* 72(6): 1598–1610.

Cornelissen, G., Gustafsson, O., Bucheli, T., Jonker, M., Koelmans, A. and van Noort, P. (2005). Extensive sorption of organic compounds to black carbon, coal, and kerogen in sediments and soils: mechanisms and consequences for distribution, bioaccumulation and biodegradation. *Environmental Science and Technology* 39(18): 6881–6895.

Crombie, K., Masek, O., Sohi, S., Brownsort, P. and Cross, A. (2013). The effect of pyrolysis conditions on biochar stability as determined by three methods. *GCB Bioenergy* 5(2): 122–131.

Cross, A. and Sohi, S.P. (2011). The priming potential of biochar products in relation to labile carbon contents and soil organic matter status. *Soil Biology and Biochemistry* 43(10): 2127–2134.

Cross, A. and Sohi, S.P. (2013). A method for screening the relative long-term stability of biochar. *GCB Bioenergy* 5(2): 215–220.

Defra (2009). *Environmental Permitting Guidance the Directive on the Incineration of Waste For the Environmental Permitting (England and Wales) Regulations 2007*. London: Defra.

Defra (2010). *Fertiliser Manual (RB209)*. London: Defra. Available at www.gov.uk/government/publications/fertiliser-manual-rb209.

De Vries, W., Lofts, S., Tipping, E., Meili, M., Groenenberg, J. and Schütze, G. (2007a), Impact of soil properties on critical concentrations of Cadmium, Lead, Copper, Zinc and Mercury in soil and soil solution in view of ecotoxicological effect. *Reviews of Environmental Contamination and Toxicology* 191: 47–89.

De Vries, W., Romkens, P. and Schütze, G. (2007b), Critical soil concentrations of Cadmium, Lead and Mercury in view of health effects on humans and animals. *Reviews of Environmental Contamination and Toxicology* 191: 91–130.

De Vries, W., Romkens, P. and Bonten, L. (2008), Spatially explicit integrated risk assessment of present soil concentrations of cadmium, lead, copper and zinc in the Netherlands. *Water Air Soil Pollut* 191: 199–215.

Ducey, T.F., Ippolito, J., Cantrell, K., Novak, J. and Lentz, R. (2013). Addition of activated switchgrass biochar to an aridic subsoil increases microbial nitrogen cycling gene abundances. *Applied Soil Ecology* 65: 65–72.

Ehlert, P.A.I. *et al.* (2013). Appraising fertilisers: origins of current regulations and standards for contaminants in fertilisers. Available at www.wageningenur.nl/wotnatuurenmilieu.

European Parliament (2003). *Regulation (EC) No 2003/2003 of the European Parliament and of the Council of 13 October 2003 Relating to Fertilisers*. Brussels: European Parliament.

European Parliament (2006). *Regulation (EC) No 1907/2006 of the European Parliament and of the Council of 18 December 2006 Concerning the Registration, Evaluation, Authorisation and Restriction of Chemicals (REACH), Establishing a European Chemicals Agency, Amending Directive 1999/45/EC and Repealing Council Regulation (EEC) No 793/93 and Commission Regulation (EC) No 1488/94 as well as Council Directive 76/769/EEC and Commission Directives 91/155/EEC, 93/67/EEC, 93/105/EC and 2000/21/EC*. Brussels: European Parliament.

European Parliament (2008a). *Directive 2008/1/EC of the European Parliament and of the Council of 15 January 2008 Concerning Integrated Pollution Prevention and Control.* Brussels: European Parliament.

European Parliament (2008b). *Directive 2008/98/EC of the European Parliament and of the Council of 19 November 2008 on Waste and Repealing Certain Directives.* Brussels: European Parliament.

European Parliament (2009). *Directive 2009/28/EC of the European Parliament and of the Council of 23 April 2009 on the Promotion of the Use of Energy from Renewable Sources and Amending and Subsequently Repealing Directives 2001/77/EC and 2003/30/EC.* Brussels: European Parliament.

European Parliament (2010a). *Directive 2010/75/EU of the European Parliament and of the Council of 24 November 2010 on Industrial Emissions (Integrated Pollution Prevention and Control) (Recast).* Brussels: European Parliament.

European Parliament (2010b). *Regulation (EU) No 995/2010 of the European Parliament and of the Council of 20 October 2010 Laying Down the Obligations of Operators who Place Timber and Timber Products on the Market.* Brussels: European Parliament.

European Parliament (2011). *Commission Regulation (EU) No 575/2011 of 16 June 2011 on the Catalogue of Feed Materials.* Brussels: European Parliament.

Felber, R., Huppi, R., Leifeld, J. and Neftel, A. (2012). Nitrous oxide emission reduction in temperate biochar-amended soils. *Biogeosciences Discussions* 9(1): 151–189.

Freddo, A., Cai, C. and Reid, B.J. (2012). Environmental contextualisation of potential toxic elements and polycyclic aromatic hydrocarbons in biochar. *Environmental Pollution* 171: 18–24.

Glaser, B. and Birk, J.J. (2012). State of the scientific knowledge on properties and genesis of Anthropogenic Dark Earths in Central Amazonia (terra preta de Índio). *Geochimica et Cosmochimica Acta* 82: 39–51. Available at http://linkinghub.elsevier.com/retrieve/pii/S001670371100144X (accessed 2 October 2013).

Glaser, B., Haumaier, L., Guggenberger, G. and Zech, W. (2001). The 'Terra Preta' phenomenon: a model for sustainable agriculture in the humid tropics. *Naturwissenschaften* 88(1): 37–41.

Graber, E., Harel, Y., Kolton, M., Cytryn, E., Silber, A., David. D., Tsechansky, L., Borenshtein, M. and Elad, Y. (2010). Biochar impact on development and productivity of pepper and tomato grown in fertigated soilless media. *Plant and Soil* 337(1–2): 481–496.

Hale, S.E., Lehmann, J., Rutherford, D., Zimmerman, A., Bachmann, R., Shitumbanuma, V., O'Toole, A., Sunqvist, K., Arp, H. and Cornelissen, G. (2012). Quantifying the total and bioavailable polycyclic aromatic hydrocarbons and dioxins in biochars. *Environmental Science and Technology* 46(5): 2830–2838.

Hale, S., Alling, V., Martinsen, V., Mulder, J., Breedveld, C. and Cornelissen, G. (2013). The sorption and desorption of phosphate-P, ammonium-N and nitrate-N in cacao shell and corn cob biochars. *Chemosphere* 91(11): 1612–1619.

Hammond, J., Shackley, S., Sohi, S. and Brownsort, P. (2011). Prospective life cycle carbon abatement for pyrolysis biochar systems in the UK. *Energy Policy* 39(5): 2646–2655.

Ibarrola, R., Shackley, S. and Hammond, J. (2012). Pyrolysis biochar systems for recovering biodegradable materials: A life cycle carbon assessment. *Waste Management* 32(5): 859–868.

International Biochar Initiative (2013). *Standardized Product Definition and Product Testing Guidelines for Biochar That Is Used in Soil.* Westerville, OH: International Biochar Initiative. Available at www.biochar-international.org/characterizationstandard.

Jakob, L., Henriksen, H., Elmquist, M., Brandii, R., Hale, S. and Cornelissen, G. (2012). PAH-sequestration capacity of granular and powder activated carbon amendments in soil, and their effects on earthworms and plants. *Chemosphere* 88(6): 699–705.

Joseph, S., Graber, E., Chia, C., Munroe, P., Donne, S., Thomas, T. and Hook, J. (2013). Shifting paradigms: development of high-efficiency biochar fertilizers based on nano-structures and soluble components. *Carbon Management* 4(3): 323–343.

Kammen, D.M. and Lew, D.J. (2005) *Review of Technologies for the Production and Use of Charcoal*. Berkeley, CA: Renewable and Appropriate Energy Laboratory, University of California. Available at http://rael.berkeley.edu/old_drupal/sites/default/files/old-site-files/2005/Kammen-Lew-Charcoal-2005.pdf.

Khan, S., Wang, N., Reid, B., Freddo, A. and Cai, C. (2013). Reduced bioaccumulation of PAHs by Lactuca satuva L. grown in contaminated soil amended with sewage sludge and sewage sludge derived biochar. *Environmental Pollution* 175: 64–68.

Kookana, R.S. (2010). The role of biochar in modifying the environmental fate, bioavailability, and efficacy of pesticides in soils : a review. *Soil Research* 48(7): 627–637.

Laird, D., Chappell, M., Martens, D., Wershaw, R. and Thompson, M. (2008). Distinguishing black carbon from biogenic humic substances in soil clay fractions. *Geoderma* 143(1–2): 115–122.

Lehmann, J., Gaunt, J. and Rondon, M. (2006). Bio-char sequestration in terrestrial ecosystems: a review. *Mitigation and Adaptation Strategies for Global Change* 11(2): 395–419.

Lehmann, J., Rillig, M., Thies, J., Masiello, C., Hockaday, W. and Crowley, D. (2011). Biochar effects on soil biota: a review. *Soil Biology and Biochemistry* 43(9): 1812–1836.

Liu, X., Qu, J-J., Li, L-Q., Zhang, A-F., Jufeng, Z. and Pan, G-X. (2012). Can biochar amendment be an ecological engineering technology to depress N_2O emission in rice paddies? A cross site field experiment from South China. *Ecological Engineering* 42: 168–173.

Luo, Y., Durenkamp, M., DeNobili, M., Lin, Q. and Brookes, P. (2011). Short term soil priming effects and the mineralisation of biochar following its incorporation to soils of different pH. *Soil Biology and Biochemistry* 43(11): 2304–2314.

Moon, D., Park, J., Chang, Y., Ok, Y., Lee, S., Ahmad, M., Koutsospyros, A., Park, J. and Baek, K. (2013). Immobilization of lead in contaminated firing range soil using biochar. *Environmental Science and Pollution Research International* 20(12): 8464–8471.

Mukherjee, A. and Zimmerman, A.R. (2013). Organic carbon and nutrient release from a range of laboratory-produced biochars and biochar–soil mixtures. *Geoderma* 193–194: 122–130.

Mukherjee, A., Zimmerman, A.R. and Harris, W. (2011). Surface chemistry variations among a series of laboratory-produced biochars. *Geoderma* 163(3–4): 247–255.

Namgay, T., Singh, B. and Bhupinder, P.S. (2010). Influence of biochar application to soil on the availability of As, Cd, Cu, Pb, and Zn to maize (*Zea mays* L.). *Australian Journal of Soil Research* 48: 638–647.

Nelissen, V., Rutting. T., Huygens, D., Staelens, J. Ruysschaert, G. and Boeckx, P. (2012). Maize biochars accelerate short-term soil nitrogen dynamics in a loamy sand soil. *Soil Biology and Biochemistry* 55: 20–27.

Novak, J., Busscher, W., Watts, D., Laird, D., Ahmedna, M. and Niandou, M. (2010). Short-term CO_2 mineralization after additions of biochar and switchgrass to a Typic Kandiudult. *Geoderma* 154(3–4): 281–288.

Parliamentary Office of Science and Technology (2010). *Post Note 358: Biochar*. London: Parliamentary Office of Science and Technology.

Quilliam, R., Rangecroft, S., Emmett, B., Deluca, T. and Jones, D. (2013). Is biochar a source or sink for polycyclic aromatic hydrocarbon (PAH) compounds in agricultural soils? *GCB Bioenergy* 5(2): 96–103.

Reid, B.J., Jones, K.C. and Semple, K.T. (2000). Bioavailability of persistent organic pollutants in soils and sediments: a perspective on mechanisms, consequences and assessment. *Environmental Pollution* 108(1): 103–112.

Rein, G., 2009. Smouldering combustion phenomena in science and technology. *International Review of Chemical Engineering* 1: 3–18.

Roberts, K., Gloy, B., Joseph, S., Scott, N. and Lehmann, J. (2010). Life cycle assessment of biochar systems: estimating the energetic, economic, and climate change potential. *Environmental Science and Technology* 44(2): 827–33.

Santore, R., Di Toro, D., Paquin, P., Allen, H. and Meyer, J. (2001). Biotic ligand model of the acute toxicity of metals. 2. Application to acute copper toxicity in freshwater fish and Daphnia. *Environmental Toxicology and Chemistry* 20(10): 2397–2402.

Sarkhot, D.V., Berhe, A.A. and Ghezzehei, T.A. (2008). Impact of biochar enriched with dairy manure effluent on carbon and nitrogen dynamics. *Journal of Environmental Quality* 41(4): 1107–1114.

Scheer, C., Grace, P., Rowlings, D., Kimber, S. and van Zwieten, L. (2011). Effect of biochar amendment on the soil-atmosphere exchange of greenhouse gases from an intensive subtropical pasture in northern New South Wales, Australia. *Plant and Soil* 345(1–2): 47–58.

Schmidt, H-P., Bucheli, T., Kammann, C., Glaser, B., Abiven, S. and Leifeld, J. (2012). The European Biochar Certificate. Available at www.european-biochar.org/en.

SEPA (2012). Position statement: manufacture and use of biochar from waste. Available at www.sepa.org.uk/media/156613/wst-ps-031-manufacture-and-use-of-biochar-from-waste.pdf.

Shackley, S. and Sohi, S. (eds) (2010). *An Assessment of the Benefits and Issues Associated with the Application of Biochar to Soil.* SP0576. Report to DECC and DEFRA. London: UK Government,

Singh, B., Singh, B.P. and Cowie, A.L. (2010). Characterisation and evaluation of biochars for their application as a soil amendment. *Australian Journal of Soil Research* 48(7): 516. Available at www.publish.csiro.au/?paper=SR10058.

Skjemstad, J., Reicosky, D., Wilts, A. and McGowan, J. (2002). Charcoal carbon in US agricultural soils. *Soil Science Society of America Journal* 66(4): 1249–1255.

Sopeña, F., Semple, K., Sohi, S. and Bending, G. (2012). Assessing the chemical and biological accessibility of the herbicide isoproturon in soil amended with biochar. *Chemosphere* 88(1): 77–83.

Sparrevik, M., Field, J., Martinsen, V., Breedveld, G. and Cornelissen, G. (2013). Life cycle assessment to evaluate the environmental impact of biochar implementation in conservation agriculture in Zambia. *Environmental Science and Technology* 47: 1206–1215.

Spokas, K.A. (2013). Impact of biochar field aging on laboratory greenhouse gas production potentials. *GCB Bioenergy* 5(2): 165–176.

Spokas, K., Novak, J., Stewart, C., Cantrell, K., Uchimiya, M., DuSaire, M. and Ro, K. (2011). Qualitative analysis of volatile organic compounds on biochar. *Chemosphere* 85(5): 869–882.

Spurgeon, D. and Hopkin, S. (1996). Effects of variations of the organic matter content and pH of soils on the availability and toxicity of zinc to the earthworm *Eisenia fetida*. *Pedobiology* 40: 80–96.

Taghizadeh-Toosi, A., Clough, T., Sherlock, R. and Condron, L. (2011a). Biochar adsorbed ammonia is bioavailable. *Plant and Soil* 350(1–2): 57–69.

Taghizadeh-Toosi, A., Clough, T., Condron, L., Sherlock, R., Anderson, C. and Craigie, R. (2011b). Biochar incorporation into pasture soil suppresses in situ nitrous oxide emissions from ruminant urine patches. *Journal of Environment Quality* 40(2): 468–476.

Troy, S., Lawlor, P., O'Fynn, C. and Healy, M. (2013). Impact of biochar addition to soil on greenhouse gas emissions following pig manure application. *Soil Biology and Biochemistry* 60: 173–181.

Uchimiya, M., Lima, I., Klasson, K. and Wartelle, L. (2010). Contaminant immobilization and nutrient release by biochar soil amendment: roles of natural organic matter. *Chemosphere* 80(8): 935–940.

Van den Bergh, C. (2009). Biochar and waste law: a comparative analysis. *European Energy and Environmental Law Review* 18(5): 243–253.

Van Dijk, T. *et al.* (2009). Protocol for the evaluation of products for the Fertiliser Law, version 2.1. Available at http://edepot.wur.nl/15311.

Van Laer, T., de Smedt, P., Ronsse, F., Ruysschaert, G., Boeckx, P., Verstraete, W., Buysee, J. and Lavrysen, L. (2015). Legal constraints and opportunities for biochar: a case analysis of EU law. *GCB Bioenergy* 7(1): 14–24.

Wardle, D.A, Nilsson, M.C. and Zackrisson, O. (2008). Fire-derived charcoal causes loss of forest humus. *Science* 320(5876): 629.

Woolf, D., Amonette, J., Street-Perrott, F., Lehmann, J. and Joseph, S. (2010). Sustainable biochar to mitigate global climate change. *Nature Communications* 1(5): 56.

Yao, Y., Gao, B., Zhang, M., Inyang, M. and Zimmerman, A. (2012). Effect of biochar amendment on sorption and leaching of nitrate, ammonium, and phosphate in a sandy soil. *Chemosphere* 89(11): 1467–1471.

Yu, X.-Y., Ying, G.-G. and Kookana, R.S. (2009). Reduced plant uptake of pesticides with biochar additions to soil. *Chemosphere* 76(5): 665–671.

Notes

1   See, for example www.environment-agency.gov.uk/business/sectors/124299.aspx.
2   The Norwegian Gjødsel Forskrift ('Fertiliser law') is available from www.lovdata.no/for/sf/ld/ld-20030704-0951.html.
3   Vegetal carbon is used as a food additive and colouring agent, in the form of E153, or as an additive to animal feed (European Parliament 2011; see also www.food-info.net/uk/e/e153.htm). Vegetal carbon is defined as a product obtained by carbonisation of organic vegetal material.
4   For a point-by-point comparison of the two protocols, see www.european-biochar.org.

Chapter 11

Current and future applications for biochar

Adam O'Toole, David Andersson, Achim Gerlach, Bruno Glaser,
Claudia Kammann, Jürgen Kern, Kirsi Kuoppamäki, Jan Mumme,
Hans-Peter Schmidt, Michael Schulze, Franziska Srocke,
Marianne Stenrød and John Stenström

Introduction

There are multiple applications for biochar for agricultural and environmental management. This chapter describes some of these applications which are currently being studied or are in the early phases of implementation. These include:

1 biochar addition to soil as a part of a voluntary or state-backed carbon credit programme;
2 as a multi-use material to be used on farms, e.g. as an additive to feed, silage, animal bedding and manures;
3 as a growing medium in nursery production and for green roofs;
4 as a sorbent material for removing contaminants and nutrients from water and gas; and
5 as an additive in biogas reactors to improve performance and to help flocculate and separate nutrients and organic matter from the biodigestate.

While not an exhaustive analysis, this chapter points the reader towards possible uses and commercialisation opportunities for biochar in the agricultural and environmental management sector.

Biochar application to soil as part of voluntary or mandatory carbon trading programmes

The principles and key features of carbon markets have been dealt with in Chapter 9. In the following section, further detail is provided on the challenges surrounding inclusion of biochar as a component of carbon markets. Typically, soil carbon storage (also called sequestration) has been absent in emission offset programmes for a number of reasons:

• *No guarantee of permanence* – soil organic carbon is a carbon pool undergoing constant change and is vulnerable to loss from changes in agricultural practices such as tillage methods and crop residue management. Therefore, it has been difficult to argue within a carbon sequestration context that additions of carbon to the soil will represent a permanent avoidance of CO_2 loss from soil to the atmosphere. On the flip side, farmers may expose themselves to economic risk by accepting payments for soil carbon sequestration if they cannot guarantee that the carbon they capture in their soils is not lost again. In other words, will carbon offset programmes require a refund from farmers if accrued carbon levels drop below baseline levels?

- *Additionality* – farmers return organic residues and organic fertilisers to the soil as a normal practice to improve soil structure and fertility and therefore it is uncertain whether payment for soil carbon sequestration would represent an *additional* effort to sequester carbon and avoid soil CO_2 emissions. Yet, the Clean Development Mechanism (CDM) and the voluntary carbon markets only accept projects which would not have happened in the absence of the carbon credits (a requirement known as 'additionality') in order to avoid subsidies for projects that are already financially viable.
- *Verification transaction costs* – typically, soil carbon stocks change slowly over time and require thorough soil sampling to detect changes in the soil carbon pool. The spatial variability of soil carbon also requires taking many soil samples across a field in order to arrive at a representative measurement for soil organic carbon. The sampling costs can add a high cost to projects, especially when the land area is large.
- *Leakage issues* – carbon leakage refers to where 'a carbon sequestration activity on one piece of land inadvertently, directly, or indirectly, triggers an activity which counteracts the carbon effects of the initial activity' (IPCC 2001). Consider the example of a leaking bucket; one cannot increase the storage of water in a bucket by filling it if at the same time water is leaking out from a hole in the bottom.

How does biochar address the permanence issue?

The enhanced biological stability of biochar carbon (Chapter 7) offers a possible opportunity to store carbon in soil for much longer periods than would be possible by adding other organic materials such as straw or compost to soil. While it is true that part of these organic materials can become incorporated in stable humus fractions in soil, this is often dependent on climatic conditions such as temperature, moisture, and also the extent to which mineral soil has aggregated and protected the soil organic carbon from microbial mineralization (Schmidt *et al.* 2011). In other words, soil carbon sequestration via humus accumulation will require specific soil management practices on behalf of the farmer to minimise the conditions by which carbon is degraded to CO_2, and it is also dependent on location and climate. In typical agricultural soils, formation of organic matter that is stable in the long term represents a very small percentage of total organic matter added. The recalcitrance of biochar-carbon, by contrast, means that farmers can begin to store lots more carbon than would be otherwise possible from incorporating degradable organic matter.

However, not all of the carbon in biochar is stable; a smaller labile fraction often decomposes to CO_2 within the first weeks of it being added to soil (Chapter 7; Smith *et al.* 2010). Stable carbon fractions in biochar have been reported to be between 70 and 90 per cent for biochar pyrolysed at temperatures higher than 450°C (Mašek *et al.* 2011). A number of methods are being developed or borrowed for the purpose of accounting for the stable carbon fraction in biochar in soil. One method is an accelerated biochar ageing method developed by Cross and Sohi (2013), which correlates stability to the oxygen-to-carbon ratio content in the biochar. Another protocol, developed by the International Biochar Initiative, estimates stability by looking at the hydrogen-to-carbon (H:C) ratio in the biochar (Budai *et al.* 2013). This is based on scientific studies where H:C ratio is correlated with biochar stability (Figure 11.1).

How does biochar address the additionality and baseline criteria?

The additionality criterion is designed to ensure that investment of funds in projects to reduce emissions or increase sinks actually lead to a real net decrease in atmospheric CO_2 levels. The additionality criterion asks whether a CO_2-

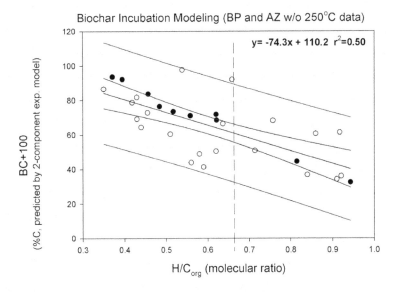

Figure 11.1 Correlation between molecular ratio of hydrogen to carbon and the fraction of biochar which is more stable than 100 years measured in two incubation studies. Black and white dots relate to two different incubations of biochar.

Source: Budai *et al.* (2013); A. Zimmerman and B.P. Singh, used with permission.

reducing activity would have occurred in the absence of a monetary incentive. As biochar additions to soils are not a conventional practice within agriculture, it could be argued that the use of biochar does represent an additional sequestration of carbon in the system. However, in order to confirm this, a baseline scenario would have to be documented to show what would have happened to the biomass feedstock had it not been converted to biochar. It would be prudent in carbon credit schemes involving biochar only to use feedstocks that otherwise would have degraded to CO_2 without any bioenergy capture, e.g. forestry residues which are not currently collected, crop straw that is burnt or ploughed in, or green wastes collected at municipal waste depots (compost or landfill). In the case of using agricultural straws for biochar production, the extra gains in soil carbon from biochar would have to subtract the amount of carbon that would have otherwise been present in stabilised humus fractions after incorporation of straw into the soil.

How does biochar address the high transaction costs of verifying soil carbon changes?

Owing to the high stability of biochar compared to unmodified organic matter, the stable fraction of biochar is expected to remain in the soil for up to 500–2000 years (Kuzyakov *et al.* 2009; Cheng *et al.* 2006; see Chapter 8). Owing to biochar's slow degradation rate, the need to frequently sample and document the quantity of it left in the soil is less than would otherwise be necessary for additions of unmodified organic amendments. This reduces the transaction costs linked to soil sampling and to the overall cost of a carbon credit project.

How does biochar address the leakage issue?

Biochar, as with bioenergy in general, shares the challenge to prove that the supply of biomass feedstock to the bioenergy system does not result in extra emissions outside the project boundaries. For example, if a virgin wood feedstock is used for biochar production, it could mean that other biomass users, for example a furniture maker, would need to cut additional trees in order to fulfil their demand. Another scenario could be if purpose-grown energy crops are grown on prime agricultural land to supply biomass for biochar production, thereby stimulating the need to find additional food growing areas (known as indirect land use change), which may impact existing carbon stocks. To address this possibility, it will be necessary for direct and indirect land use change to be monitored and for methodologies to be in place for measuring carbon emissions arising from land use change. Such methods are being discussed in the context of revisions to the EU Renewable Energy Directive. In other countries, sustainability protocols may be required for biochar carbon offset programmes that stipulate what types of biomass feedstocks can and cannot be used.

Status of biochar in different carbon offset programmes around the world

Among the many voluntary and compliance carbon offset programmes around the world, there are a few where the inclusion of biochar is either under investigation or where biochar may be included in the near future.

The Carbon Farming Initiative (CFI) in Australia is a sub-programme under the National Carbon Offset programme which financially rewards Australian farmers for activities which enhance the sequestration of carbon in Australian farmlands. Activities such as tree planting and reducing methane emissions from manure ponds are some of the things farmers can do to attract carbon payments. Biochar use has been included in the list of possible future activities, and currently a large biochar research programme financed by the Australian government is looking into a reliable methodology to institutionalise the production, application and verification of biochar carbon sequestration in this programme. Australia is in a unique position to lead the way in large scale biochar carbon sequestration because it has an enormous land surface with low carbon soils needing rejuvenation, a wealthy economy that can pay for it and well equipped and educated farmers who can execute the changes needed to restore farmlands. The Australian government was planning to link its carbon farming initiative to the European Union Emissions Trading Scheme (EU ETS) so that emitters in Europe can offset their emissions by sequestering carbon or reducing CO_2 emissions in the Australian agricultural sector (DOE 2012). However, recent political changes in Australia have meant a turn away from climate and carbon policies in the country, at least for the time being.

Other offset programmes which have considered or are considering the inclusion of biochar are The Alberta Offset System (Regional Compliance Scheme) in Canada and the American Carbon Registry (voluntary scheme). The American Carbon Registry is developing a protocol for voluntary biochar carbon offsets and this is currently under scientific peer review.

Case study: Ecoera Biosfair platform

Swedish company Ecoera, started at Chalmers University of Technology in 2007, has been developing a business model for biochar, including voluntary carbon

Table 11.1 Examples of carbon offset schemes and status of biochar inclusion therein

Carbon credit scheme	Cost per tonne CO_2	Status of biochar in scheme
Carbon Farming initiative (AUS)	AU$23 per tonne	Research and methodology development
EU Emission Trading Scheme	€2–5 per tonne	Not included (see Chapter 9 for more on the ETS)
American Carbon Registry	–	Under scientific peer review
Ecoregion Kaindorf, Austria	€45 per tonne	Under consideration; applies to humus at the moment

sequestration offset and sales of biochar. They started in 2009 with a customer wanting to offset their carbon footprint using biochar. The company first used a birch charcoal created from standard charcoal production. The carbon dioxide sequestered was calculated by a simple 1:1 factor ratio for the offset of biochar-carbon to CO_2 equivalent (reduced factor ratio due to a minor labile biochar carbon content). The biochar–carbon was sequestered in a clayey soil on the west coast of Sweden, similar to the site in Figure 11.2. The spreading technique was by using a lime spreader, while more recent applications have used a discer (Figure 11.3).

Figure 11.2 Biochar field site in Ås, Norway.
Source: Adam O'Toole.

Figure 11.3 Biochar application using a discer.
Source: Adam O'Toole.

The char was ploughed down one month later. The second large-scale sequestration was done for the Swedish meteorologist Pär Holmgren and his company. This was a larger sequestration offset and required the spreading of biochar at different sites. To keep track of the offset certificates and the land they are connected to, a web-based database was developed. This system links the carbon offset buyer with the biochar they have paid for and records the GPS coordinates to where the biochar carbon has been sequestered, including images verifying the soil application. By this, an audit trail can be kept to comply with future verification protocols, and it is also possible for customers to visit their carbon sequestration after prior agreement with farmers. In 2010 and 2011, several biochar carbon sequestration voluntary offsets were sold. The programme is still in development but shows that there is potential to create viable local biochar-carbon sequestration projects that are paid for by local businesses that seek to offset their CO_2 emissions.

As we have just outlined, the inclusion of biochar in carbon-offset programmes is still under development and is not a widespread practice. However, this may change as the findings of the Fifth Assessment Report of the Intergovernmental Panel on Climate Change (IPCC 2014) are digested by policy makers in government, industry and civic sectors. Namely, that limiting global temperature increases to 2°C or less will require the widespread implementation of bioenergy systems which can also capture and store carbon – that is, carbon-negative emission technologies, of which bioenergy with CO_2 capture and storage (BECCS) is a strong contender, but so is biochar (IPCC 2014).

The cascading benefits of biochar in animal production systems

While the biochar community waits for the long promised 'carbon credits' for biochar sequestration, a growing number of biochar practitioners and farmers in Europe are investigating a multitude of beneficial uses for biochar. A 'cascaded' use of biochar implies that biochar should ideally provide multiple benefits in a variety of on-farm applications prior to its deposition to soil – an idea first developed by Hans-Peter Schmidt in Switzerland. These include using it as an animal feed additive, a material for adding to animal bedding and composts (Steiner *et al.* 2011), for retaining nutrients in slurry (Hua *et al.* 2009) and even as a building material (Schmidt 2014). In the following section, we provide a description of historical evidence and some research for the use of biochar in livestock farming.

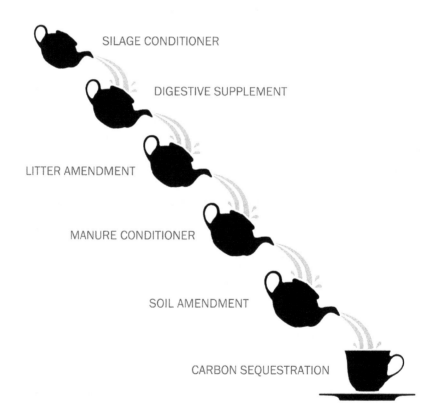

SILAGE CONDITIONER

DIGESTIVE SUPPLEMENT

LITTER AMENDMENT

MANURE CONDITIONER

SOIL AMENDMENT

CARBON SEQUESTRATION

Figure 11.4 Using biochar in a 'cascading' manner shows that biochar can be used for multiple on-farm applications prior to its final utility as a long-term carbon sequestration material.

The use of biochar as a feed additive

Charcoal is one of the oldest homespun remedies against digestive disorders, not only for humans but also for livestock. Cato the Elder (234–149 BC) mentioned it in his classic *On Agriculture*: 'If you have reason to fear sickness, give the oxen before they get sick the following remedy: 3 grains of salt, 3 laurel leaves … 3 pieces of charcoal, and 3 pints of wine' (Cato 1935). Beside the administration of medicinal herbs, oil or clay, charcoal was widely used by traditional farmers all over the world for stomach disorders. While for some animals like chicken or pigs the charcoal was administered pure, for others it was mixed with butter (cows), with eggs (dogs) or with meat (cats).

At first glance, it might seem somewhat unnatural to feed char to animals, but in fact even wild mammals occasionally eat charcoal if it is available to them in nature (e.g. charcoal that can still be found years after a forest or prairie fire). Struhsaker *et al.* (1997) reports observations of deer and elk eating charcoal from charred trees in Yellowstone National Park and others report domestic dogs eating charcoal briquettes. Famous in this respect is the Zanzibar red colobus (*Procolobus kirkii*), a small ape which regularly eats charcoal to help digest young Indian Almond (*Terminalia catappa*) or mango (*Mangifera indica*) leaves with their toxic phenolic compounds (Cooney and Struhsaker 1997; Struhsaker *et al.* 1997). Struhsaker *et al.* (1997) observed that individual colobus monkeys consume about 0.25–2.5g of charcoal per kg body weight daily. Additional adsorption tests performed by Cooney and Struhsaker (1997) indicated that the African kiln charcoals (which the monkeys also ate) were surprisingly good at adsorbing hot-water-extracted organics from the above-mentioned tree leaves, whereas charcoals from wildfires performed much less well in this regard. Thus, the authors concluded that the monkeys' charcoal consumption was likely a self-learned behaviour that increases the digestibility of their typical leaf diet.

The use of activated and non-activated charcoal for feeding purposes was already being researched and recommended by German veterinarians at the beginning of the previous century. Since 1915, research into activated charcoal has revealed its effect in adsorbing pathogenic clostridial toxins such as those produced by *C. tetani* and *C. botulinum* (Luder 1947; Starkenstein 1915). Mangold (1936) presented a comprehensive study on the effects of charcoal in feeding animals, concluding that 'the prophylactic and therapeutic effect of charcoal against diarrhoeal symptoms attributable to infections or to the type of feeding is known. In this sense, adding charcoal to the feed of young animals would seem a good preventive measure.' Volkmann (1935) described an effective reduction in excreted oocysts through adding charcoal to the food of pets with coccidiosis or coccidial infections. Haring (1937) recommends mixing charcoal into cattle feed; while Barth and Zucker (1955) were not able to detect any negative growth effects in poultry when the level of added charcoal was kept at around 1 per cent.

Recent studies report that growth rates of Vietnamese goats improved when their feed included 0.5–1g of bamboo biochar per kg body weight per day (Van *et al.* 2006). Kana *et al.* (2010) have shown that 0.2–0.6 per cent maize cob biochar added to chicken feed resulted in significant weight increases. In Vietnam, Leng *et al.* (2013) demonstrated that live weight gain was increased by 25 per cent when 0.6 per cent biochar by weight was added to the feed. It is not yet clear what biochar function could cause the live weight gain and increased feed efficiency, though Struhsaker *et al.* (2007) state that phenolics adsorb better to charcoal and can thus partly be immobilised by the biochar, while the digestibility of smaller proteins and amino acid molecules might be improved (Cooney and Struhsaker 1997). In the case of ruminants, it can be supposed that biochar immobilised lignin and tannins, while the digestibility of cellulose, smaller proteins and amino acids might be improved, as was shown by (Mohan and Karthikeyan 1997) for activated charcoal. However, it is not clear to what extent biochar also adsorbs essential feed compounds and metabolites of feed compounds. It is supposed that biochar is a selective adsorber for longer polar and apolar molecules, thus blocking surface site vacancies for feed-relevant shorter molecules like amino acids and vitamins. This rather widely accepted explanation has not, however, been systematically investigated yet for animal digestion. As long term trials with activated carbon as feed and as food additive have not shown any nutrient deficiency (Friedman *et al.* 1978; Hayden and Comstock 1975; Yatzidis and Oreopoulos 1976) and since there are no known side effects of activated carbon applications (Greensher *et al.* 1979), it might be supposed that immobilisation of beneficial feed compounds on biochar is not of relevant extent.

Most of the effects of biochar are thought to be based on the following mechanisms: adsorption, co-adsorption, competition, chemisorption, adsorption followed by a chemical reaction and desorption. From a toxicology perspective, classifiable distinctions between these different mechanisms in the digestive tract of animals need to be made. With regard to the specific mechanisms, more detailed research is urgently needed to beneficially implement the use of target-designed biochars in animal feeding.

Of particular importance is the specific colonisation of the biochar with gram-negative bacteria with increased metabolic activity (Schirrmann 1984) as most pathogens in animal digestive systems are gram-negative. This results, on the one hand, in a decrease in endotoxins produced by these bacteria, and on the other hand, in the adsorption of the toxins in the char. One major advantage in the use of biochar is to be found in its 'enteral dialysis' property. Schirrmann (1984) found that lipophilic and hydrophilic toxins already adsorbed by the blood plasma can be removed from the plasma by the ingested char since the char interacts with the permeability properties of the intestine.

No research has (to our knowledge) been conducted yet on the effects of soluble organic carbons and other potential pyrolytic condensates like polycyclic aromatic hydrocarbon (PAHs) from biochar on the biota of the digestive system. Some condensates may even have stimulating effects at low dosages (hormesis effect: Steinberg 2012), but toxic or inhibiting effects cannot be excluded and thus need intensive research before a biochar may be administered on a regular basis. It needs to be stated that certified biochars (EBC 2012) – meaning those investigated using appropriate methods that desorb not only lower-molecular-weight PAHs such as naphthalene, but also higher-molecular PAHs for quantification (Hilber *et al.* 2012) – will have only extremely low bioavailable PAH concentrations and thus act more as an adsorbent than a source for PAHs (Quilliam *et al.* 2012). The existing long practical experience to date with biochar as a feed additive seems to indicate that no immediate harm to livestock is to be feared; however, there is a huge difference between biochar (or activated charcoal) administration as acute medication, and a rather regular feed additive. For this reason, biochar-feed producers advise interrupting the biochar treatment every 10 days for at least five days (Swiss Biochar 2013); however, farmers often do not respect this precautionary advice and no adverse effects have been reported so far. As an alternative approach, farmers can make charcoal freely available to farm animals for intake at their own liking. Most farm animals voluntarily include charcoal in their daily feed when it's made freely available to them (author's own observation on horses, goats, sheep, and chicken over two years).

Although the activity of higher temperature biochars is only about one third compared to the most widely used activated carbon (own repeated analysis with phenazone adsorption tests), the adsorption capacity is still high (Cooney and Struhsaker 1997). Luder (1947), who studied the adsorption capacity of *carbo ligni* and *carbo adsorbens* (medical terms for charcoal) and came up with an efficiency of 1:3–4 for charcoal/biochar compared to activated carbon. To reach the same adsorption capacity as activated carbon, one can adjust the biochar-to-feed ratio accordingly. In the future, optimised medical/pharmaceutical biochars, or biochar-clay complexes, could be designed although they would need research, certification and quality control schemes to ensure safe use.

In Vietnam, Leng *et al.* (2013) demonstrated that methane formation was reduced by 20 per cent when 0.6 per cent biochar was added to the feed and more than 40 per cent when combined with 6 per cent potassium nitrate. This confirms earlier results from in vitro trials by Leng *et al.* (2012) where biochar administered at 0.5 per cent and 1 per cent reduced methane emissions by 10 per cent and 12.7 per cent, respectively. With 50 per cent nitrate as nitrogen and 50 per cent urea nitrogen plus biochar administered at a rate of 1 per cent (wt) each, the reduction in methane was 40.5 per cent. In another in vitro trial, Hansen *et al.* (2012) showed a non-significant decrease of rumen methane emissions between 11 per cent and 17 per cent at 9 per cent biochar (such a dosage is much higher than practised in all other biochar feeding trials, and likely not practicable anyway). If significant reduction in rumen methane emission could be confirmed, the use of biochar as feed additive for cattle and other ruminants could have more profound effects for the reduction of agricultural greenhouse gas (GHG) emissions than the pure carbon sink of biochar incorporated into soils. However, more research is needed to understand the function and the potential of biochar in reducing enteral methane production, how it can be optimised with other feed blends, and if it can be generalised or optimised.

The use of biochar as a feed ingredient is subject to strict food quality rules under Commission Regulation (EC) 178/2002, and to the strict regulations for organic livestock feed under Commission Regulation (EC) 834/2007. In particular, the concentrations of heavy metals, dibenzodioxines and dibenzofuranes play an important role as limiting factors, thus only certified biochars (EBC 2012) should be considered for administration to farm animals.

Practical use and experience of biochar in feeding cattle

Very few scientific studies exist to date on the long-term effect of biochar in animal farming (Hayden and Comstock 1975; Nau *et al.* 1960); however, the following observations by farmers indicate further benefits of its use. They may trigger further scientific studies on the subject as these are certainly overdue.

The veterinarian Achim Gerlach conducted a survey with 21 farm managers in Northern Germany, each with an average herd of 150 cows. The farmers reported their impressions of the initial effects they had observed 1–4 weeks after starting biochar administration (Gerlach and Schmidt 2012). The farmers reported generally improved health and appearance; improved vitality and udder health; decreased cell counts in the milk (interrupting the administration of biochar led to higher cell counts and a drop in performance); minimisation of hoof problems and stabilisation of post-partum health; reduced diarrhoea within 1–2 days, with faeces subsequently generally more solid; an increase in milk protein and/or fat; a marked improvement of slurry viscosity; and the slurries not smelling as bad as they used to.

In an as yet unpublished survey of the Ithaka Institute conducted with 30 farmers using biochar as a feed additive for more than one year in Switzerland, Germany and Austria, the following findings were reported: 77 per cent noted an improvement in diarrheal symptoms; 62 per cent reported that the animals became calmer; 77 per cent noticed less bad odours in the stable. Further observations from Swiss farmers feeding a fermented biochar-bran (CarbonFeed with 20 per cent Swiss Biochar) were obtained by feed consultants of EM Schweiz AG: one year after the start of administration, none of the cows receiving the administration had needed veterinary treatment; none of the pigs needed antibiotic treatment during the six months of administration; in a chicken farm the mortality rate decreased, while at the same time a high and continuous increase in weight of 90–100 g per day was observed. A comprehensive survey of 80 to 100 farmers using biochar in livestock systems is currently being conducted (Ithaka Institute 2013), with statistically relevant data to be expected in 2015.

In conclusion, the use of biochar as a feed additive may provide interesting and economically feasible solutions to the increasingly complex problems of modern-day farming. The adsorptive qualities of biochar permit a wide range of toxic substances to be bound in the gastrointestinal tract. They may also have the potential to help detoxify already resorbed toxins (in particular, lipophilic toxins) in the blood plasma via 'enteral dialysis' (Schirrmann 1984). First studies indicate a reduction in methane emissions for ruminants (Leng *et al.* 2012, 2013).

Box 11.1 Testimonial from Swedish farmers about biochar use in Sweden in the 1920s

In the old Swedish animal husbandry it was common to let the cows and cattle graze in the nearby woods. They eat all sorts of grasses, ferns and fungi to keep their health in an optimum. Sometimes it happened that they nibbled on the old wooden roofs of older barns.

One curious thing observed by farmers in the old days was the cattle browsing on charcoal remains from old char kilns out in the woods.

This observation led the farmers to add charcoal to the cows feed – during wintertime. The charcoal was not only appreciated by the cows, but also by the pigs.

Therefore, a strong, wooden box was nailed to wall of the pig sty for them to easily feed on charcoal whenever they needed. The height was just within reach for halfgrown pigs. The pigs stopped returning to the 'char-box' as they felt better.

Like the old Sami people of Sweden said regarding reindeer: 'A reindeer must move freely, so it can cure itself.' I say: 'If a pig or cow can get charcoal – they can cure themselves.'

The charcoal was retained from the stove in the home. It was mainly from leafy trees, such as Birch. The wood did not undergo complete combustion in the open stove, and pieces of charcoal invariably remained in the ashes.

Source: Personal communication with a Axel Hugo Andersson, a farmer, in 2013.

Animal bedding

The main reason for putrescence and bad odours in animal housing is the intermixture of liquid and solid excreta (Otto *et al.* 2003). When faeces and urine mix together on the floor, exponential microbial proliferation results in massive putrescence of the faeces, multiplication of pathogens and unhealthy gas emissions such as ammonia (NH_3) and excess GHG emissions (nitrous oxide, methane).

When animals graze outside on pastures, urine and faeces liquids soak into the soil while the faeces dry out on the soil surface and are degraded by soil organisms, thus producing much less or no putrescence. Ideally, this should also be the case on the floor of animal housing. The urine should be soaked up by the bedding material and thus be separated from the faeces. Usually this is done via a thick layer of straw. However, the adsorption capacity of straw is too low to avoid extensive contact between the nutrient-rich urine and carbon-rich faeces. But if a thick layer of biochar, bentonite (a clay mineral) and compost is bedded under the straw, this could have an effect comparable to pasture in soaking up the liquids. Experiments at the Ithaka Institute in Switzerland in a chicken barn and in a horse stable have shown that the high biological activity of the biochar-compost, and the sufficiently high oxygen content in the bedding layers, help to decompose and rapidly humify the solid excretions, and thus to avoid bad odours and bacteriological infestations. In an as yet unpublished survey by the Ithaka Institute 25 out of 30 farmers using biochar as bedding agents noted a reduction of smells in the stable, as well as the minimisation of hoof problems.

The biochar-bentonite-compost-straw bedding enriched with animal manure and urine is a very suitable feedstock for composting.

The use of biochar in livestock farming has the potential to lead to improved animal health and production capacity, to stabilise and improve the climate in animal housing, to disinfect the liquid manure, to prevent nutrient losses and GHG emissions, and, ultimately, leads to more efficient nutrient retention and fertilisation of farmland. Using biochar as a feed additive has been traditional practice and the object of various (older) scientific publications in German, French and in the main East Asiatic languages; however, adding biochar to silage and bedding, or the treatment of slurry with biochar (the whole cascade of biochar use in animal farming), is not yet backed up by published scientific research, although it is already in practical use. Here, the large discrepancy between existing farming practice and lack of sound scientific knowledge should be bridged by reinforced research efforts.

Biochar as an addition to grass roofs, nursery soil and commercial composts

Biochar as a growing medium in grass roofs

Green roofs or turf/sod roofs are a traditional roofing material in many countries. Modern green roof technology has evolved since the turn of the twentieth century, especially in Germany, to mitigate the damaging effects of solar radiation on roof structures (Oberndorfer *et al.* 2007). Recently, vegetated roofs have received renewed attention as a way to alleviate many environmental problems in urban areas, such as buildings' energy consumption and the volume of urban water runoff (Oberndorfer *et al.* 2007; Pataki *et al.* 2011; Ahiablame *et al.* 2012). However, green roofs have been shown to leach nutrients from the growing media, especially phosphorus (Berndtsson 2010; Buccola *et al.* 2008), thus causing a new source of surface water pollution. The amendment of biochar into green roof substrate could help reduce this nutrient loading from green roofs and also improve the retention of water. Beck *et al.* (2011) found that the addition of 7 per cent biochar to a green roof growing medium planted with sedum and ryegrass resulted in significant retention of both water and nutrients (nitrate and phosphate present from the growing media) compared to the control without biochar. In a field experiment in southern Finland, one of the authors of this chapter (Kirsi Kuoppamäki) used green roof prototypes assembled on 40 × 50cm trays. Following an addition of 7 per cent biochar (of birch origin) in ready-made Sedum and meadow green roofs, the monitoring of water quantity and quality was carried out for one year during natural precipitation events. Biochar amendment improved the retention of water, total phosphorus and total nitrogen. Only during extremely rainy weather in autumn 2012, when green roofs were saturated by water, was the impact of biochar less effective. This experiment showed that water retention capacity can be increased and the leaching of nutrients reduced, although not eliminated, by adding biochar into green roof media.

Biochar as a peat substitute or additive in growing media

The ability of biochar to increase the water holding capacity and reduce the bulk density of soils (Chapter 4) is of increasing interest to commercial horticultural producers. Generally, high-quality sphagnum peat is still used as base substrate for the production of growing media. Peat substrates are hard to replace by peat-free fibres due to their superb growth-supporting properties such as slightly acidic pH, high cation exchange capacity, good aeration and a very high water holding capacity (e.g. 3.3g H_2O g^{-1} dry peat substrate; Busch et al. 2012). However, drainage of bogs for turf and soil production destroys natural peatlands, releasing stored carbon into the atmosphere both when it is extracted from soil and when the extracted peat eventually degrades at its new location. Alternative materials such as vermiculite have a poor carbon footprint, and prices for this material have risen by 50 per cent since 2004 (Dumroese et al. 2011). For these reasons, some commercial nurseries are becoming more interested in using biochar as a replacement for vermiculite and peat in soil mixtures. One experiment (Dumroese et al. 2011) produced biochar pellets from biochar and wood flour (43 per cent each) with starch and polyacetic acids as binding agents (7 per cent each). They reported that mixtures of 75 per cent peat substrate and 25 per cent biochar pellets gave ideal physicochemical properties in soil growing media. Benefits included improved water availability and reduced shrinkage of the soil media. Recently, Altland and Locke (2012) studied the impact of biochar on nutrient retention and leaching from a soil-free substrate with application rates of 1, 5 or 10 per cent biochar to peat substrate. They found that biochar might be effective in moderating extreme fluctuations of nitrate levels in container substrates over time (Altland and Locke 2012). Tian et al. (2012) used pure peat substrate, pure biochar or a 1:1 mixture of peat and biochar, and observed the largest growth of the ornamental plant *Calathea* with the peat–biochar mixture (>22 per cent compared to peat alone). In addition, they reported reduced degradation of the growth media by biochar addition (Tian et al. 2012). In a pot trial with basil (*Ocimum basilicum*) at University Gießen, Germany, reduced shrinking in peat-biochar mixtures with up to 60 per cent (vol/vol) biochar addition was observed.

Problems such as dustiness of the biochar may easily be avoided by pelletising (Dumroese et al. 2011), by wetting or spraying it with nutrient solutions or pure water when it leaves the pyrolysis unit (H. Gerber, Pyreg GmbH, pers. comm. 2014). Theoretically, the wetting may also be achieved by using the biochar to catch nutrient-rich effluents (Sarkhot et al. 2012), which can at the same time load the biochar with nutrients. Adding large amounts of freshly produced biochar with low amounts of nitrogen may increase the need for fertilisation, as the fresh biochar will compete with plant roots to adsorb the available nitrogen in the soil. In a pot trial in Norway which tested wheat straw biochar and different rates of nitrogen fertiliser on perennial ryegrass growth, it was found that plants in biochar amended soils took up significantly less nitrogen than plants in unamended soil (O'Toole *et al.* 2013).

Another promising use of biochar in plant nursery or horticulture is the addition of small doses to improve plant health (Elad *et al.* 2010; Elmer and Pignatello 2011; Zwart and Kim 2012), e.g. by induction of systemic resistance to fungal pathogens (Elad *et al.* 2010; Harel *et al.* 2012). The latter was achieved with quite small doses of biochar addition of 1–3 per cent to growth media without soil.

In summary, biochar use in plant nurseries and in horticulture is a promising field since accrued benefits may quickly pay for the costs associated with biochar. These benefits may include reduced incidence and severity of plant diseases, reduced need for fungicide, potential to re-use biochar in new batches of soil media and reduced need for watering of potted plants due to the higher water holding capacity in biochar soil mixtures. A side benefit to using biochar as a commercial growth media is that the goal of biochar carbon sequestration will occur without the need for additional financial incentives (e.g. in soils used for tree and bush seedlings and ornamental flowers sold via garden centres; Dumroese *et al.* 2011).

Biochar as a sorption material for environmental applications

The use of activated carbon for filtering drinking water and in air filters has been a commercial activity for over 100 years. Although it does not have the same adsorption capacity as activated carbon, biochar can also be used for adsorbing a variety of nutrients and contaminants from soils, water and gas (Beesley *et al.* 2011). The premise behind using biochar instead of activated carbon is cost effectiveness. Owing to chemical or steam activation, activated carbon is a much more expensive material than biochar. Biochar may therefore be a cost-effective filter material for a variety of applications including wastewater treatment, land reclamation and controlling the transport of pesticides and herbicides from soils to waterways.

Adsorption of pesticides

To illustrate some of the possibilities and challenges in using biochar as a sorption material, we focus on its potential use for the adsorption of pesticides at critical source points. When pesticides are sprayed in the field, a large amount of the active ingredient does not reach its intended target (Ravier *et al.* 2005). These pesticides and their transformation products can subsequently be transported to ground and surface waters (Bach and Frede 2012). Such off-target movement of crop protection compounds is increasingly coming under public and regulatory pressure – as shown for instance by current European policy that focuses on sustainable use of pesticides (Commission regulation 2009/128). Sorption of pesticides to biochars has been studied (e.g. Graber *et al.* 2012; Wang *et al.* 2010) in relation to the effect of biochar feedstocks, pyrolysis temperatures and in a variety of soils. The results demonstrate the phenomenon of sorption–desorption hysteresis (adsorption hysteresis is when the quantity of gas adsorbed is different when gas is being adsorbed than it is when being desorbed). They also indicate the effects of aging of biochars on sorption properties. Aged biochars are shown to desorb contaminants more easily than fresh biochars, but there is dispute as to whether this difference is of practical concern in the field. Many different conventional methods can be used for enhanced degradation of pesticides (e.g. treatment with ozone or hydrogen peroxide and using modified clays or activated carbon to bind pesticides, metals and nutrients; Bacaoui *et al.* 2002). However, these methods are expensive, which makes the development of new methods that rely on cheap and easily available materials interesting. From an environmental perspective, biochar can adsorb pesticides and herbicides and prevent them from leaching to waterways; however, one must balance this against the chemicals that are bound to biochar so that they do not fulfil their original purpose in killing pests and weeds (Graber *et al.* 2012).

Biochar adsorption of phosphorus to improve water quality

The transport of phosphorus (P) to rivers, lakes and marine environments leads to algal blooms and poor water quality. Several reviews show that diffuse losses of P from agricultural soils is one of the largest contributors of P to water bodies (Humborg *et al.* 2007; Sharpley *et al.* 2009) and the adoption of the Water Framework Directive (Commission Regulation 2000/60) has increased the need to deal with such losses. New and innovative mitigation strategies are thus urgently needed to protect water resources from pollution and eutrophication (addition of too many nutrients to water ways that are naturally nutrient-constrained; Albelda *et al.* 2012). Research to date has shown varying results regarding biochar's ability to adsorb P. Because biochar can be made from a wide range of organic materials with varying levels of P content, it is logical to expect that biochar can be both a source and sink for P in soils. For example, biochar made from animal bones is high in P and biochar from hardwoods is relatively low in P. Therefore, the proper choice of feedstocks will be needed in order to make biochars which are suitable for sorption applications. There exist techniques to also upgrade the P adsorption capacity of biochar. For example, one study has shown that mixing biochar made from orange peel with iron particles in solution prior to pyrolysis can enhance the P adsorption capacity of the biochar (Chen *et al.* 2011). The process yields a product comprised of nano-size magnetite particles attached to the biochar surface. Interestingly, the adsorption capacity increases were greater than if one summated the adsorption capacities of the biochar and iron particles alone. An added benefit of magnetic biochars for wastewater treatment is the ability to separate the biochar from the filter material at the end of the process via magnetism (Safarik *et al.* 2012). Another technique for upgrading the P adsorption capacity of biochar is mixing the feedstock with magnesium nanoparticles prior to pyrolysis (Yao *et al.* 2011).

The use of biochar as a sorbent, rather than a soil amendment, will involve concentrating the biochar in one spot rather than spreading it over the soil. Owing to biochar's low density and its tendency to float in water, it will need to be encased in filter materials such as geotextile if it is to serve this purpose. Biochar encased in geotextile materials such as Filtersoxx could be used to intercept nutrients flowing in surface water from vulnerable hotspots (e.g. subsurface drainage systems outlets and sloping agricultural land close to surface water bodies). Biochar could also be used as an alternative to gravel, which is often used as a backfill over subsurface drainage pipes in agricultural fields. The larger surface area and pore volume of biochar compared to gravel would give more opportunity for the development of microbial biofilms that degrade dissolved organic material and pollutants.

Biochar's role in biogas systems

Although biogas production by anaerobic digestion (AD) has been established for some decades, there is still a need for optimisation of the fermentation process in terms of process stability, methane yields and inhibition problems (Y. Chen *et al.* 2008; Ward *et al.* 2008; Yadvika *et al.* 2004). An effective approach to enhance AD is the addition of inorganic particles to the reactor. It was recently shown by Adu-Gyamfi *et al.* (2012) that the availability of a high surface area for attachment of microorganism has a crucial impact on methane production. In their study, they revealed that increasing the amount of molecular sieve zeolites by 50 per cent led to a 47 per cent increase in methane yield. The correlation between biofilm formation and methane yield in anaerobic fixed film reactors has also been reported (Michaud *et al.* 2002, 2005). In general, immobilised microorganisms seem to be more resistant to inhibitors and changing pH values (Hanaki *et al.* 1994). Dissolved, unionised ammonia is very toxic because it can penetrate through cell membranes changing the intracellular pH value (Kadam and Boone 1996; Sprott and Patel 1986; Sung and Liu 2003). For mitigation of ammonia inhibition, zeolites have been added to organic sludge in ADs of piggery waste or cattle manure (Borja *et al.* 1996, 1993; Montalvo *et al.* 2005; Tada *et al.* 2005).

Publications dealing with the use of carbonised materials in anaerobic reactors are scarce. Both charcoal and activated carbon have been tested for increasing the efficiency of AD of cattle dung or swine manure, respectively (Geeta *et al.* 1986; Hansen *et al.* 1999; Kumar *et al.* 1987). Geeta *et al.* (1986) found an elevated biogas yield of up to 15 per cent at charcoal concentration of 20g per litre in batch trials with cow slurry. Kumar *et al.* (1987) reported an increase of the biogas yield of cow dung in batch and semi-continuous digestion up to a respective 17 per cent and 34.7 per cent when a powdered charcoal was applied with best results at a charcoal to dung ratio of 1:20 (dry weight basis). Both articles speculate that the higher yields could be caused by biofilm formation on the particle surface. Proof for this hypothesis, however, was not shown. Positive effects on thermophilic digestion of swine manure by the addition of activated carbon were attributed to sulphide adsorption (Hansen *et al.* 1999).

Recently, Mumme et al. (2014) reported on the use of biochars on AD. They showed that biochars from pyrolysis can mitigate mild forms of ammonia inhibition, while biochars from hydrothermal carbonisation can serve as a carbon source for the anaerobic microflora. Both biochars were found to support growth of archaeal microorganisms, which are responsible for methane production. Thus, biochar seems to be a feasible solution to counteract inhibition of methane production caused by high ammonia concentrations. Like zeolites, biochar can act as a cation exchanger depending on its point of zero charge, and therefore accumulates ammonium ions on its surface, although it has been shown to be less effective than zeolites (Hina et al. 2013). However, as biochar can contain various potentially inhibitory volatile organic compounds such as benzenes, aldehydes and phenols (Y. Chen et al. 2008; Spokas et al. 2011) significant negative effects on biogas formation are also possible. It is conceivable that, after biogas production, the charcoal can be used as soil fertiliser since it can release bioavailable plant nutrients such as NH_3 (Taghizadeh-Toosi et al. 2012) into the soil. A difficulty of using biochar is the high variability of its composition depending on feedstock and production conditions.

Integrated production of biogas and biochar can result in various synergies. The residual digestate of AD can serve as a feedstock for biochar production, while the use of biochars in the biogas process can improve both the process and the biochars for soil use. Such a concept was introduced by Mumme (2015) (Figure 11.5) under the acronym APECS (anaerobic pathways to renewable energies and carbon sinks).

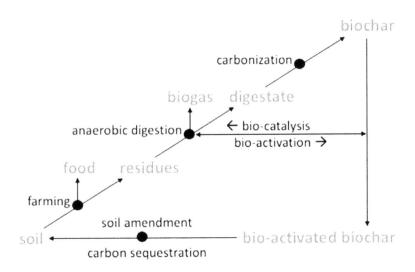

Figure 11.5 Integrated production of biogas and biochar for soil application according to the APECS concept.

Source: based on Mumme (2015), with permission.

Table 11.2 Summary of applications and future considerations

Maturity – time horizon	Estimated market value for biochar (per tonne oven dry)	Barriers/risks to implementation	Market/policy trends which support this application	Biochar application
Biochar accepted as a method in Australian Carbon Farming Initiative, 2014, and will be connected to the European ETS by the end of 2015 and therefore be available for European carbon emitters to purchase	Australia CFI~15 EURO/tonnCO$_2$ = ~25 EURO /tonne oven dry Biochar (assuming 80% carbon content of which 60% is stable carbon content)	Technology development and prohibitive costs. In Australia, the demand for farmers to sign a contract for 100 years adds considerable risk to the farmer	Merging themes of tackling climate change, increasing food production and securing livelihoods for farmers lends support for the payment of soil carbon sequestration	Carbon credits
Currently being used for feed additive although in limited geographic areas	Approximately 250 €/ tonne biochar as part of an animal feed mix at 15% (Swiss biochar)	Regulation which for one reason or another prohibits the use of biochar in animal feed. Lack of scientific data to support practical experience may delay market development	The lack of implementation of biochar so far may encourage the development of niche applications such as animal feed additives and other applications in animal husbandry	Animal husbandry
Established as filter material within five yearsEstablished as an on-site application within 10 years (for the most promising areas)	Need to be a lower cost/price than competing products (activated carbon) i.e. less than € 5/kilo (e.g. Charcoal Green® SOIL DeTOX™)	General: need for activation giving too high priceField: possible negative effects on nutrient availability in short term, reduction in pesticide efficacy. Biochar filters may need post-processing of spent material	EU sustainable use directive for pesticides and water directive – focusing on measures to reduce pollution from pesticides If the cost is low then the application in already established fields will come (waste water treatment, land reclamation/ contaminated soil treatment) when the efficiency is well documented	Sorption material

Table 11.2 Contined

Maturity – time horizon	Estimated market value for biochar (per tonne oven dry)	Barriers/risks to implementation	Market/policy trends which support this application	Biochar application
Already been sold to consumers under a variety of brands as a blended soil compost/biochar	Example: 20 € /L (7 € /kg) of Gro char® biochar (consumer price) Extra 2 € /L for biochar blended 20 L compost bag compared to conventional compost blend (Amazon.co.uk)	Commercialisation may be premature with claims for a product that have not yet been substantiated or scientifically proven	Biochar, being an expensive material to produce, is most likely to be introduced in high value consumer end soil blends as the first application area	Soil and growing media
Very early stage. Further R&D needed	Costs need to be lower than alternative biogas additives such as zeolite, iron salts and stone dust	Biochar needs to be free of inhibitory substances as biogas production is a very delicate process. Competition with a broad range of established biogas additives. Possible legal barriers in countries that have a restricted range of input materials for biogas plants	EU renewable energy policy. Increased production and use of biogas from problematic wastes that require addition of additives	Biogas additive

As digestates are usually very rich in water, a water-based char production process such as hydrothermal carbonisation (HTC; see Chapter 2) could be more feasible than pyrolysis. The use of HTC for converting digestate into HTC-derived char was studied by Mumme *et al.* (2011). Adding the energetic value of the digestate-based biochar to the produced biogas can increase the overall energy efficiency of biomass conversion by a large extent. Corresponding results from lab-scale experiments have been reported for wheat straw, where its fermentation into biogas (net energy efficiency of 1/3) was compared with a cascaded fermentation into biogas followed by HTC of the digestate into biochar (net energy efficiency of 2/3; Funke *et al.* 2013). In addition, AD was shown to recover energy from the HTC wastewater, while cleaning it at the same time (Wirth and Mumme 2013). This indicates that AD might also be applicable for treating the condensate from pyrolysis.

With respect to the soil use of HTC-derived biochars, the mean residence time is usually in the range of several decades (Steinbeiss *et al.* 2009). Various process parameters have an effect on the soil stability of carbon. Reaction temperature has a particularly significant impact on the mean residence time (Gaji *et al.* 2012; Chapter 8). Further results show that straw-based digestates can be made seven times more stable by using HTC (Schulze 2013).

Bibliography

Adu-Gyamfi, N., Ravella, S. R. and Hobbs, P.J. (2012). Optimizing anaerobic digestion by selection of the immobilizing surface for enhanced methane production. *Bioresource Technology* 120: 248–255.

Ahiablame, L. M., Engel, B. and Chaubey, I. (2012). Effectiveness of low impact development practices: literature review and suggestions for future research. *Water Air Soil Pollut.* 223: 4253–4273.

Albelda, T.M., Frias, J.C., Garcia-Espana, E. and Schneider, H.-J. (2012). Supramolecular complexation for environmental control. *Chemical Society Reviews* 41, 3859-3877.

Altland, J.E. and Locke, J.C. (2012). Biochar affects macronutrient leaching from a soilless substrate. *HortScience* 47: 1136–1140.

Bacaoui, A., Dahbi, A., Yaacoubi, A., Bennouna, C., Maldonado-Hodar, F.J., Rivera-Utrilla, J., Carrasco-Marin, F. and Moreno-Castilla, C., (2002). Experimental design to optimize preparation of activated carbons for use in water treatment. *Environ. Sci. and Tech.* 36, 3844–3849.

Bach, M. and Frede, H.-G. (2012). Trend of herbicide loads in the river Rhine and its tributaries. *Integrated Environmental Assessment and Management* 8: 543–552.

Barth, K. and Zucker, H. (1955). Zur Verwendung von Holzkohle in der Geflügelfütterung unter besonderer Berücksichtigung einer möglichen Vitamin-Adsorption. *Zeitschrift Für Tierernährung Und Futtermittelkunde* 10(1–3): 300–307.

Beck, D.A., Johnson, G.R. and Spolek, G.A. (2011). Amending green roof soil with biochar to affect runoff water quantity and quality. *Environ. Pollution* 159(8–9): 2111–2118.

Beesley, L., Moreno-Jiménez, E., Gomez-Eyles, J.L., Harris, E., Robinson, B. and Sizmur, T. (2011). A review of biochars' potential role in the remediation, revegetation and restoration of contaminated soils. *Environ. Pollution* 159: 3269–3282.

Berndtsson, J.C. (2010). Green roof performance towards management of runoff water quantity and quality: a review. *Ecological Engineering* 36: 351–360.

Borja, R., Sánchez, E., Weiland, P. and Travieso, L. (1993). Effect of ionic exchanger addition on the anaerobic digestion of cow manure. *Environ. Technol.* 14: 891–896.

Borja, R., Sánchez, E. and Durán, M.M. (1996). Effect of the clay mineral zeolite on ammonia inhibition of anaerobic thermophilic reactors treating cattle manure. *J. Environ. Sci. Heal. Part Environ. Sci. Eng. Toxicol.* 31: 479–500.

Budai, A., Zimmerman, A. R., Cowie, A. L., Webber, J. B. W., Singh, B. P., Glaser, B., Masiello, C. A., Andersson, D., Shields, F., Lehmann, J., Camps Arbestain, M., Williams, M., Sohi, S. and Joseph, S. (2013). Biochar carbon stability test method: an assessment of methods to determine biochar carbon stability. Available at www.biochar-international.org/sites/default/files/IBI_Report_Biochar_Stability_Test_Method_Final.pdf.

Busch, D., Kammann, C., Grünhage, L. and Müller, C. (2012) Simple biotoxicity tests for evaluation of carbonaceous soil additives: Establishment and reproducibility of four test procedures. *Journal of Environmental Quality* 41: 1023–1032.

Cato, M.P. (1935) *On Agriculture*. Cambridge, MA: Harvard University Press.

Chen, B., Zhou, D. and Zhu, L. (2008) Transitional adsorption and partition of nonpolar and polar aromatic contaminants by biochars of pine needles with different pyrolytic temperatures. *Environmental Science and Technology* 42: 5137–5143.

Chen, B., Chen, Z. and Lv, S. (2011). A novel magnetic biochar efficiently sorbs organic pollutants and phosphate. *Bioresource Technology* 102(2): 716–723.

Chen, Y., Cheng, J.J. and Creamer, K.S. (2008) Inhibition of anaerobic digestion process: a review. *Bioresour. Technol.* 99: 4044–4064.

Chen, Y.-X., Huang, X.-D., Han, Z.-Y., Huang, X., Hu, B., Shi, D.-Z. and Wu, W.-X. (2010). Effects of bamboo charcoal and bamboo vinegar on nitrogen conservation and heavy metals immobility during pig manure composting. *Chemosphere* 78: 1177–1181.

Cheng, C.-H., Lehmann, J., Thies, J.E., Burton, A.J. and Engelhard, M. (2006). Oxidation of black carbon by biotic and abiotic processes. *Organic Geochemistry* 37: 1477–1488.

Commission Regulation (EC) 2000/60 on establishing a framework for Community action in the field of water policy. Available at http://ec.europa.eu/health/endocrine_disruptors/docs/wfd_200060ec_directive_en.pdf.

Commission Regulation (EC) 178/2002 on laying down the general principles and requirements of food law, establishing the European Food Safety Authority and laying down procedures in matters of food safety: Available at http://eur-lex.europa.eu/LexUriServ/LexUriServ.do?uri=OJ:L:2002:031:0001:0024:en:PDF.

Commission Regulation (EC) No 834/2007 on organic production and labelling of organic products and repealing Regulation (EEC) No 2092/91. Available at http://eur-lex.europa.eu/LexUriServ/LexUriServ.do?uri=OJ:L:2007:189:0001:0023:EN:PDF.

Commission Regulation (EC) No 2009/128 of the European Parliament and of the Council of 21 October 2009 establishing a framework for Community action to achieve the sustainable use of pesticides. Available at http://eur-lex.europa.eu/legal-content/EN/NOT/?uri=CELEX:02009L0128-20091125.

Cooney, D. O. and Struhsaker, T.T. (1997). Adsorptive capacity of charcoals eaten by Zanzibar red colobus monkeys: implications for reducing dietary toxins. *International Journal of Primatology* 18(2).

Cross, A. and Sohi, S.P. (2013). A method for screening the relative long-term stability of biochar. *GCB Bioenergy* 5(2): 215–220.

DOE (2012). Linking carbon trading systems. Available at www.climatechange.gov.au/australia-and-european-commission-linking-emissions-trading-systems.

Dumroese, R.K., Heiskanen, J., Englund, K. and Tervahauta, A. (2011). Pelleted biochar: chemical and physical properties show potential use as a substrate in container nurseries. *Biomass and Bioenergy* 35: 2018–2027.

EBC (2012). *European Biochar Certificate: Guidelines for a Sustainable Production of Biochar.* Arbaz: European Biochar Foundation. Available at www.european-biochar.org/biochar/media/doc/ebc-guidelines.pdf.

Elad, Y., David, D.R., Harel, Y.M., Borenshtein, M., Kalifa, H.B., Silber, A. and Graber, E.R. (2010). Induction of systemic resistance in plants by biochar, a soil-applied carbon sequestering agent. *Phytopathology* 100: 913–921.

Elmer, W.H. and Pignatello, J.J. (2011). Effect of biochar amendments on mycorrhizal associations and fusarium crown and root rot of asparagus in replant soils. *Plant Disease* 95: 960–966.

Friedman, E. A., Feinstein, E. I., Beyer, M. M., Galonsky, R. S. and Hirsch, S. R. (1978). Charcoal-induced lipid reduction in uremia. *Kidney International*. 8(suppl.): 170–176.

Funke, A., Mumme, J., Koon, M. and Diakité, M. (2013). Cascaded production of biogas and hydrochar from wheat straw: energetic potential and recovery of carbon and plant nutrients. *Biomass and Bioenergy* 58: 229–237.

Gaji , A., Ramke, H.-G., Hendricks, A. and Koch, H.-J. (2012). Microcosm study on the decomposability of hydrochars in a Cambisol. *Biomass and Bioenergy* 47: 250–259.

Geeta, G.S., Raghevendra, S. and Reddy, T.K.R. (1986). Increase in biogas production from bovine excreta by addition of various inert materials. *Agric. Wastes* 17: 153–156.

Gerlach, A. and Schmidt, H.P. (2012). The use of biochar in cattle farming. *Ithaka Journal* 2012(1): 80–84.

Graber, E.R., Tsechansky, L., Gerstl, Z. and Lew, B. (2012). High surface area biochar negatively impacts herbicide efficacy. *Plant and Soil* 353: 95–106.

Greensher, J., Mofenson, H.C., Picchioni, A.L. and Fallon, P. (1979). Activated charcoal updated. *Journal of the American College of Emergency Physicians* 8(7): 261–263.

Hanaki, K., Hirunmasuwan, S. and Matsuo, T. (1994). Protection of methanogenic bacteria from low pH and toxic materials by immobilization using polyvinyl alcohol. *Water Res.* 28: 877–885.

Hansen, K.H., Angelidaki, I. and Ahring, B.K. (1999). Improving thermophilic anaerobic digestion of swine manure. *Water Res.* 33: 1805–1810.

Hansen, H.H., Storm, I.M.L.D. and Sell, A.M. (2012). Effect of biochar on in vitro rumen methane production. *Acta Agric. Scand. Sect. A - Anim. Sci.* 62: 305–309.

Harel, Y.M., Elad, Y., Rav-David, D., Borenstein, M., Shulchani, R., Lew, B. and Graber E.R. (2012). Biochar mediates systemic response of strawberry to foliar fungal pathogens. *Plant and Soil* 357: 245-257.

Haring, F. (1937). Mitt. f. d. Landwirtschaft 52 S.308-309 Hrsg: Reichsnährstand

Hayden, J. W. and Comstock, E. G. (1975). Use of activated charcoal in acute poisoning. *Clinical Toxicology* 8(5): 515–533.

Hilber, I., Blum, F., Hale, S., Cornelissen, G., Schmidt, H-P., Bucheli, T. D. (2012). Total and bioavailable PAH concentrations in biochar – a future soil improver. *Proceedings EGU General Assembly 2012*, 22–27 April, 2012 Vienna, Austria.

Hina, K., Hedley, M., Camps-Arbestain, M. and Hanly, J. (2013). Comparison of pine bark, biochar and zeolite as sorbents for NH4$^+$-N removal from water. *CLEAN – Soil, Air, Water* 43(1): 86–91.

Ho, L. and Ho, G. (2012). Mitigating ammonia inhibition of thermophilic anaerobic treatment of digested piggery wastewater: use of pH reduction, zeolite, biomass and humic acid. *Water Res.* 46: 4339–4350.

Hua, L., Wu, W., Liu, Y., Mcbride, M. and Chen, Y. (2009). Reduction of nitrogen loss and Cu and Zn mobility during sludge composting with bamboo charcoal amendment. *Environ. Sci. and Poll. Res.* 16: 1–9.

Humborg, C., Morth, C.-M., Sundbom, M. and Wulff, F. (2007) Riverine transport of biogenic elements to the Baltic Sea – past and possible future perspectives. *Hydrology and Earth System Sciences* 11: 1593–1607.

IPCC (2001). *Third Assessment Report*. Cambridge: Cambridge University Press. Available at http://www1.ipcc.ch/ipccreports/tar/vol4/english/index.htm.

IPCC (2014). *Summary for Policy Makers*. AR5, Working Group 3. Available at http://report.mitigation2014.org/spm/ipcc_wg3_ar5_summary-for-policymakers_approved.pdf.

Ithaka Institute (2013). Barn protocol for biochar use in animal farming. Available at www.ithaka-institut.org/en/ct/56 (accessed 5 June 2015).

Kadam, P.C. and Boone, D.R. (1996). Influence of pH on ammonia accumulation and toxicity in halophilic, methylotrophic methanogens. *Appl. Environ. Microbiol.* 62: 4486–4492.

Kammann, C., Schmidt, H.-P., Messerschmidt, N., Linsel, S., Steffens, D., Müller, C., Koyro, H.-W., Conte, P., Joseph, S. (2015). Plant growth improvement mediated by nitrate capture in co-composted biochar. *Scientific Reports* 5: 11080 (doi: 10.1038/srep11080).

Kana, J.R., Teguia, A., Mungfu, B.M. and Tchoumboue, J. (2010). Growth performance and carcass characteristics of broiler chickens fed diets supplemented with graded levels of charcoal from maize cob or seed of *Canarium schweinfurthii* Engl. *Tropical Animal Health and Production* 43(1): 51–56.

Kumar, S., Jain, M.C. and Chhonkar, P.K. (1987). A note on stimulation of biogas production from cattle dung by addition of charcoal. *Biol. Wastes* 20: 209–215.

Kuzyakov, Y., Subbotina, I., Chen, H., Bogomolova, I. and Xu, X. (2009). Black carbon decomposition and incorporation into soil microbial biomass estimated by 14C labeling. *Soil Biology and Biochemistry* 41(2): 210–219.

Kyan, T., Shintani, M., Kanda, S., Sakurai, M., Ohashi, H., Fujisawa, A. and Pongdit, S. (1999). *Kyusei Nature Farming and the Technology of Effective Microorganisms: Guidelines for Practical Use*, revised edition. Bangkok: International Nature Farming Research Center.

Leng, R.A., Inthapanya, S. and Preston, T.R. (2012). Biochar lowers net methane production from rumen fluid in vitro. *Livestock Research for Rural Development* 24(6): 24103.

Leng, R.A., Preston, T.R. and Inthapanya, S. (2013). Biochar reduces enteric methane and improves growth and feed conversion in local 'Yellow' cattle fed cassava root chips and fresh cassava foliage. *Livestock Research for Rural Development* 24(11): 2–7.

Liu, J., Schulz, H., Brandl, S., Miehtke, H., Huwe, B. and Glaser, B. (2012) Short-term effect of biochar and compost on soil fertility and water status of a Dystric Cambisol in NE Germany under field conditions. *Journal of Plant Nutrition and Soil Science* 175: 698–707.

Luder, W. (1947). Adsorption durch Holzkohle. Dissertation, Universität Bern, Switzerland.

Mangold, E. (1936). Die Verdaulichkeit der Futtermittel in ihrer Abhängigkeit von verschiedenen Einflüssen. *Forschungsdienst – Reichsarbeitsgemeinschaften D. Landwirtschaftswissenschaft* 1: 862–867.

Mašek, O., Brownsort, P., Cross, A. and Sohi, S. (2011). Influence of production conditions on the yield and environmental stability of biochar. *Fuel* 103: 151–155.

Meier, U. and Dinkel, F. (2002). Ammoniakemissionen aus Gülle und deren Minderungs-massnahmen unter besonderer Berücksichtigung der Vergärung. Bundesamt für Energie. Available at www.bfe.admin.ch/php/.../streamfile.php?file.

Michaud, S., Bernet, N., Buffière, P., Roustan, M. and Moletta, R. (2002). Methane yield as a monitoring parameter for the start-up of anaerobic fixed film reactors. *Water Res.* 36: 1385–1391.

Michaud, S., Bernet, N., Buffière, P. and Delgenès, J.P. (2005). Use of the methane yield to indicate the metabolic behaviour of methanogenic biofilms. *Process Biochem.* 40: 2751–2755.

Mohan, S.V., and Karthikeyan, J. (1997). Removal of lignin and tannin colour from aqueous solution by adsorption onto activated charcoal. *Environmental Pollution* 97(1–2): 183–187.

Montalvo, S., Díaz, F., Guerrero, L., Sánchez, E. and Borja, R. (2005). Effect of particle size and doses of zeolite addition on anaerobic digestion processes of synthetic and piggery wastes. *Process Biochem.* 40: 1475–1481.

Mumme, J. (2015). *Verbundvorhaben Anaerkon: anaerobe Konversion von Biomasse zu hochwertigen Energieträgern und Kohlenstoffsenken.* Project report. Report number: 03SF0381A. Hanover: Technische Informationsbibliothek.

Mumme, J., Eckervogt, L., Pielert, J., Diakité, M., Rupp, F. and Kern, J. (2011). Hydro-thermal carbonization of anaerobically digested maize silage. *Bioresour. Technol.* 102: 9255–9260.

Mumme, J., Srocke, F., Heeg, K. and Werner, M. (2014): Use of biochars in anaerobic digestion. *Bioresource Technology* 164: 189–197.

Muralikrishna, G., Schwarz, S., Dobleit, G., Fuhrmann, H. and Krueger, M., (2011). Fermentation of feruloyl and non-feruloyl xylooligosaccharides by mixed fecal cultures of human and cow: a comparative study in vitro. *European Food Research and Technology* 232: 601–611.

Nau, C.A., Neal, J., and Stembridge, V.A. (1960). A study of the physiological effects of carbon black. *Archives of Environmental Health* 1(6): 512–533.

Oberndorfer, E., Lundholm, J., Bass, B., Coffman, R.R., Doshi, H., Dunnett, N., Gaffin, S., Köhler, M., Liu, K.K.Y. and Rowe, B. (2007) Green roofs as urban ecosystems: ecological structures, functions, and services. *BioScience* 57(10): 823–833.

Ondarts, M., Hort, C., Sochard, S., Platel, V., Moynault, L. and Seby, F. (2012). Evaluation of compost and a mixture of compost and activated carbon as biofilter media for the treatment of indoor air pollution. *Environmental Technology* 33: 273–284.

O'Toole, A., Knoth de Zarruk, K., Steffens, M. and Rasse, D.P. (2013). Characterization, stability, and plant effects of kiln-produced wheat straw biochar. *Journal of Environmental Quality* 42(2): 429–436.

Otto, E.R., Yokoyama, M., Hengemuehle, S., von Bermuth, R.D., van Kempen, T. and Trottier, N.L. (2003). Ammonia, volatile fatty acids, phenolics, and odor offensiveness in manure from growing pigs fed diets reduced in protein concentration. *Journal of Animal Science* 81: 1754–63.

Pataki, D.E., Carreiro, M.M., Cherrier, J., Grulke, N.E., Jennings, V., Pincetl, S., Pouyat, R.V., Whitlow, T.H. and Zipperer, W.C. (2011). Coupling biogeochemical cycles in urban environments: ecosystem services, green solutions, and misconceptions. *Frontiers in Ecology and the Environment* 9(1): 27–36.

Prost, K., Borchard, N., Siemens, J., Kautz, T., Sequaris, J-M., Moeller, A. and Amelung, W. (2013). Biochar affected by composting with farmyard manure. *Journal of Environmental Quality* 42: 164–172.

Quilliam, R.S., Rangecroft, S., Emmett, B.A., Deluca, T.H. and Jones, D.L., (2012). Is biochar a source or sink for polycyclic aromatic hydrocarbon (PAH) compounds in agricultural soils? *GCB Bioenergy* 5(2): 96–103.

Ravier, I., Haouisee, E., Clément, M., Seux, R. and Briand, O. (2005). Field experiments for the evaluation of pesticide spray-drift on arable crops. *Pest Manag Sci.* 61(8): 728–736.

Safarik, I., Horska, K., Pospiskova, K. and Safarikova, M. (2012). Magnetically responsive activated carbons for bio - and environmental applications. *International Review of Chemical Engineering – Rapid Communications* (4)3: 346–352.

Sarkhot, D.V., Berhe, A.A. and Ghezzehei, T.A. (2012). Impact of biochar enriched with dairy manure effluent on carbon and nitrogen dynamics. *Journal of Environmental Quality* 41: 1107–1114.

Schirrmann, U. (1984). Aktivkohle u. ihre Wirkung auf Bakterien. Dissertation, TU München, Munich.

Schmidt, H. P. (2014). The use of biochar as building material – cities as carbon sinks. *Ithaka Journal* 1. Available at www.ithaka-journal.net/pflanzenkohle-zum-hauser-bauen-stadte-als-kohlenstoffsenken?lang=en.

Schmidt, M.W.I., Torn, M.S., Abiven, S., Dittmar, T., Guggenberger, G., Janssens, I.A, Kleber, M., Lehmann, J., Manning, D., Nannipieri, P., Rasse, D.P., Weiner, S. and Trumbore, S. (2011). Persistence of soil organic matter as an ecosystem property. *Nature* 478(7367): 49–56.

Schulze, M. (2013). Hydrothermale Karbonisierung – Einfluss ausgesuchter Parameter bei der Produktion von Biokohlen auf ihre Stabilität im Boden [Hydrothermal carbonization – influences of selected production parameters on the biochars' stability in soil]. Diploma thesis, Universität Potsdam, Germany.

Sharpley, A. N., Kleinman, P.J.A, Jordan, P. Bergström, L. and Allen, A.L. (2009). Evaluating the success of phosphorus management from field to watershed. *J. Env.Qual.* 38: 1981–1988.

Smith, J.L., Collins, H.P. and Bailey, V.L. (2010). The effect of young biochar on soil respiration. *Soil Biology and Biochemistry*, 42(12): 2345–2347.

Spokas, K.A., Novak, J.M., Stewart, C.E., Cantrell, K.B., Uchimiya, M., DuSaire, M.G. and Ro, K.S. (2011). Qualitative analysis of volatile organic compounds on biochar. *Chemosphere* 85: 869–882.

Sprott, G.D. and Patel, G.B. (1986). Ammonia toxicity in pure cultures of methanogenic bacteria. *Syst. Appl. Microbiol.* 7: 358–363.

Starkenstein, E. (1915). *Medizinische Wochenschrift, Feldärztliche Beilage.* Munich: Verlag von J.F. Lehmann.

Steinbeiss, S., Gleixner, G. and Antonietti, M. (2009). Effect of biochar amendment on soil carbon balance and soil microbial activity. *Soil Biology and Biochemistry* 41: 1301–1310.

Steinberg, C.E.W. (2012). Whatever doesn't kill you might make you stronger: Hormesis. In C.E.W. Steinberg (ed.), *Stress Ecology: Environmental Stress as Ecological Driving Force and Key Player in Evolution.* Heidelberg: Springer.

Steiner, C., Melear, N., Harris, K. and Das K.C. (2011). Biochar as bulking agent for poultry litter composting. *Carbon Management* 2: 227–230.

Struhsaker, T.T., Cooney, D.O. and Siex, K.S. (1997). Charcoal consumption by Zanzibar red colobus monkeys: its function and its ecological and demographic consequences. *Int. J. Primatol.* 18: 61–72.

Sung, S.W. and Liu, T. (2003). Ammonia inhibition on thermophilic anaerobic digestion. *Chemosphere* 53: 43–52.

Swiss Biochar (2013). Carbon feed. Available at http://swiss-biochar.com/eng/carbonfeed.php (accessed 10 May 2013).

Tada, C.,Yang,Y., Hanaoka,T., Sonoda,A., Ooi, K. and Sawayama, S. (2005). Effect of natural zeolite on methane production for anaerobic digestion of ammonium rich organic sludge. *Bioresour. Technol.* 96: 459–464.

Taghizadeh-Toosi, A., Clough, T., Sherlock, R. and Condron, L. (2012). Biochar adsorbed ammonia is bioavailable. *Plant and Soil* 350: 57–69.

Tian,Y., Sun, X., Li, S.,Wang, H.,Wang, L., Cao, J. and Zhang, L. (2012). Biochar made from green waste as peat substitute in growth media for *Calathea rotundifola* cv. Fasciata. *Scientia Horticulturae* 143: 15–18.

Van, D.T.T., Mui, N.T. and Ledin, I. (2006). Effect of method of processing foliage of Acacia mangium and inclusion of bamboo charcoal in the diet on performance of growing goats. *Animal Feed Science and Technology* 130: 242–256.

Volkmann, A. (1935). *Behandlungsversuche der Kaninchen- bzw. Katzencoccidiose mit Viscojod und Carbo medicinalis.* Leipzig:Verlag Edelmann.

Wang, H., Lin, K., Hou, Z., Richardson, B. and Gan, J. (2010). Sorption of the herbicide terbuthylazine in two New Zealand fore st soils amended with biosolids and biochars. *Journal of Soil and Sediments* 10(2): 283–289.

Ward, A.J., Hobbs, P.J., Holliman, P.J. and Jones, D.L. (2008). Optimisation of the anaerobic digestion of agricultural resources. *Bioresour. Technol.* 99: 7928–7940.

Wirth, B. and Mumme, J. (2013). Anaerobic digestion of waste water from hydrothermal carbonization of corn silage. *Applied Bioenergy* 1: 1–10.

Yadvika, S., Sreekrishnan, T.R., Kohli, S. and Rana, V. (2004). Enhancement of biogas production from solid substrates using different techniques: a review. *Bioresour. Technol.* 95: 1–10.

Yao,Y., Gao, B., Inyang, M., Zimmerman, A. R., Cao, X., Pullammanappallil, P. and Yang, L. (2011). Removal of phosphate from aqueous solution by biochar derived from anaerobically digested sugar beet tailings. *Journal of Hazardous Materials* 190(1–3): 501–507.

Yatzidis, H. and Oreopoulos, D. (1976). Early clinical trials with sorbents. *Kidney International* 7(suppl.): S215–217.

Zwart, D.C. and Kim, S.-H. (2012). Biochar amendment increases resistance to stem lesions caused by *Phytophthora* spp. in tree seedlings. *HortScience* 47: 1736–1740.

Chapter 12

Biochar horizon 2025

Hans-Peter Schmidt and Simon Shackley

In the late 2000s, when the biochar euphoria was at its highest, the breakthrough from biochar research and development to large-scale implementation into farmers' practice seemed imminent. The story about improving soil fertility, ecosystem services and saving the climate all at once added up to a convincing win–win–win narrative. In principle, this hopeful ideal is still possible, but several factors have led biochar to be as yet only a niche product. On the one hand, evidence of positive results in European agriculture is rare, though results are more promising from other parts of the world where there are poorer soils and fewer chemical inputs. Furthermore, legislation of its use in European soil is complex and confusing, industrial production has not taken off due mostly to technical problems incurring high unit production costs and the existence of low product margins in a market which is largely yet to be built. On the other hand, scientific projects have multiplied and various results – not only those limited to agronomy – have shown that biochar may be positioned more than ever as a potential key material for the bio-based economy.

Whereas five years ago biochar was mostly used as mono-constituent soil amendment, its application and use has diversified since that time. Nowadays, no farmer would even think of spreading 10 or more tonnes of dusty biochar onto their soil. Rather, they would use biochar material parsimoniously in combination with organic or mineral fertiliser. It would be applied closer to the roots (Blackwell *et al.* 2010; Graves 2013), as opposed to top dressing it homogeneously over the entire field, as used to be the assumed best practice in the mid-to-late 2000s. Knowledge about the specific functions of biochar in soils has improved tremendously though there are still too many gaps to allow for the design of biochars for special soil ameliorating purposes and targeted at specific crops and/or crop management regimes.

As has happened over the past five years, there is still a lot of trial-and-error research trying this or that poorly characterised biochar in this or that (more or less well characterised) soil and climate – a design set-up where both positive or negative effects can typically only be explained hypothetically. The more biochar became fashionable, the more new research groups started biochar projects often repeating these same methodological errors. Thanks to a European Cooperation on

Science and Technology (COST) action, the fragmented European biochar research has recently become more coordinated or at least more collaborative. However, as with many nascent fields of research and development, the science, technology and practice does not always share the same concepts, language, methodology or 'rule of thumb' in how to do and evaluate things, which creates barriers for progress, communication and understanding.

Based on anecdotal evidence, European farmers have been using biochar quite successfully in animal farming, manure treatment and composting with subsequent use as a soil amendment mostly for horticulture (this being higher value-added than arable farming). Biochar containing soil substrates are sold commercially in garden centres in several European countries. Urban arboriculturalists use organic biochar substrates for urban tree care, notably in Stockholm, where this practice was generalised (Embren 2013). While those professionals have apparently found pathways to use biochar profitably, academic researchers can sometimes look dismissively on these real world examples as, in their opinion, they exhibit some of the following drawbacks:

- scientific proof is missing;
- precautionary principles are not always followed (e.g. a full risk assessment is usually not undertaken); and
- scientific research in highly complex multifactorial systems are difficult to design, implement and rarely publishable.

The reductionist paradigm still dominates vast swathes of science, requiring an experimental research design that is enormously challenging when dealing with something as complex and multi-variable as biochar, soils, interaction between soil amendments and underlying soils, the weather, crop management, how these dynamics change over time, and so on.

Farmers are often led by their own experience, intuition and understanding, as well as being sometimes prone to esoteric beliefs and influenced by inform-ation and opinions from other farmers and commercial advisors. This does not equip the farmer to deliver scientific results, of course, and that is clearly not their intention. Farmers' knowledge and understanding is not as highly regarded by the scientific profession as peer-reviewed science. However, the example of anthropogenic dark earths, not only in Amazonia but in many other countries across several continents, suggests that sometimes farmers discover things by accident, by intuition or by tried-and-tested experience (or more likely a combination of all three), which leads them to understand how to turn poor soil into highly fertile land. And as scientists still struggle to rediscover how ancient farmers made *terra preta*, it might be a good idea to overcome prejudice from both sides and for academics and practitioners to start to collaborate more closely with one another.

There are examples in this book of how such fruitful collaborations between science and practice can work for the improvement of biochar research. Following his chemical analyses of original *terra preta* soils, Bruno Glaser deduced that it was not the biochar *per se* that makes *terra preta* so valuable, but a complex ancient waste management system where biochar played an important role in holding and chemically fixing soil organic matter and nutrients (Glaser *et al.* 2001). On the basis of this conclusion, in 2009 Glaser initiated an international collaboration with a professional compost company led by Gerald Dunst in Austria (as discussed in more detail in Chapter 9). In large-scale composting trials, they demonstrated the benefits that addition of biochar bring to the composting process and succeeded in turning organic wastes into *terra preta*-like substrates (Schulz *et al.* 2013).

Contrary to the habitual micro-composting in malodourous lab buckets that can still be seen in many publications, Glaser and Dunst composted 20 tonnes of biomass for each treatment using professional equipment, producing sufficient substrates for multiple large-scale scientific and farmer trials. Gerald Dunst has become not only one of the leading European biochar producers but also one of the most important producers of biochar containing substrates, with his Austrian company Sonnenerde (see www.sonnenerde.at). His biochar products have been used on at least 100 farmers' fields and university trial plots and improvements to soil fertility were demonstrated even on fertile Middle European fields when combined appropriately with organic nutrients.

The hype surrounding biochar has been fuelled by the results from application to poor, acid, highly leached and eroded tropical and subtropical soils where crop yields are very low. Since in such poor soils most organic amendments and conservation farming practices tend to improve yields, the results of trials with biochar and biochar-compost may have led to some over confidence in its efficacy. Critics have stated that even if biochar-compost triples the maize yield in Ghana, Nepal or Bolivia, the yield is still three to four times lower than in intensive fields in the Northern corn belt and that such results are no proof at all that biochar really upgrades soil quality. The argument has some validity because if it were the case that the benefits of biochar were to be found only in very poor, highly eroded soils – where many alternative and sustainable methods of good farmers' practices would have more or less the same effect (such as applying compost, mulching, legume crop rotation, mixed cultivation, coordinated grassing, agroforestry, etc.) – then biochar is unlikely to have a strong future in mainstream agriculture.

Biochar's nutrient holding capacity (NHC) effect may be what is most important in understanding its benefits to subtropical and tropical farming and may well work in harmony with the above-mentioned conservation farming practices. When biochar is applied to animal manure, compost, urine or other nutrient-rich organic waste materials, it only takes a few minutes to reduce noxious odours indicating that nutrient losses from these substrates are at least partially inhibited. Biochar is known to be a strong adsorber of both polar and apolar organic substances. Its cation exchange capacity (CEC) can be important even though it is lower than for many composts or soil organic matter (SOM)-rich soils. It has an anion exchange capacity (AEC) but this is much too small to explain biochar's well-demonstrated ability to retain nitrate and phosphate – both negatively charged ions.

Jassal *et al.* (2015) showed that biochar may sorb up to 5 per cent N by mass and that the main N-capturing mechanism of nitrate, ammonium or urea is not through chemical but physical entrapment in the porous system of the biochar. Kammann *et al.* (2015) recently demonstrated that biochar literally soaks up nitrate, phosphate, dissolved organic carbon (DOC) and other easily soluble substances during aerobic composting. Pellegrino Conte has concomitantly shown that water shelled anions and cations can be trapped inside biochar's carbonaceous micropores by unconventional water bonding (Conte *et al.* 2014; Fang *et al.* 2014) and it seems that the also-trapped DOC becomes an important binding partner for water shelled ions inside the micro- and mesopores of the biochar (Pignatello *et al.* 2015). This might explain why biochar that has been placed in compost, manure, urine or other wet organic substances – and which is then picked-out again and analysed – contains large amounts of nitrate, ammonium, phosphate and other soluble plant nutrients. These molecules are not easily extractable by conventional potassium chloride (KCl) extraction methods, though they are apparently still available to plants and microbial populations. In fact, as Kammann *et al.* (2015) have shown, sequential extractions from biochar particles picked out of organic matter in which they have been co-composted continues to deliver nitrate, DOC and phosphate – which are not completely depleted even after 10 extraction cycles.

If these preliminary findings can be confirmed, the highest value of biochar in agriculture might lie in organic nutrient capture and recycling. In mixed-use farms, biochar could thus first be used in manure management to decrease nutrient losses, reduce odours and to improve, when applied to soil, organic fertiliser efficiency while reducing greenhouse gas (GHG) emissions during the whole cascade of use from manure storage, to composting and application. Some of the biochars might even be used as feed additives to initiate the nutrient capturing effect of the biochar already in the digestive system of animals while improving feed efficiency and animal fitness. Even though scientific proof of the latter is still scarce, more than 90 per cent of the biochar holding a European Biochar Certificate (EBC 2012; Schmidt *et al.* 2012) is first used in animal farming (partly as a feed supplement and mostly as bedding and manure treatment) and only after those initial uses is it then used as a soil amendment.

With further scientific progress to be expected on understanding the specific mechanisms related to biochar's nutrient capturing, nutrient holding and bio-availability, biochars might be optimised and specially designed for improved performance as slow release fertilisers. However, at current prices for industrially produced biochar, it is difficult to imagine a profitable scenario for larger-scale applications in manure treatment with subsequent use in soil – even when further improved performances of biochar might make it possible to use smaller quantities.

Very different would be the scenario when farmers could produce – on farm – their own biochar from residues such as straw, husks, shrubs, cuttings and prunings in small scale, low cost biochar kilns. Farmers could thus combine biochar production with on-farm waste management, biomass heat generation and on-site organic fertiliser production. On-farm biochar production would be a major step towards a more circular economy in agriculture. Whether such on-farm biochar production would be cheaper than large-scale production is at this stage still a debatable point. On the one hand, centralised, large-scale production benefits form economies of scale, meaning that biochar production costs can be lower than for small-scale production (Shackley *et al.* 2011). On the other hand, if small-scale, on-farm production can utilise equipment with a much lower capital cost per unit biochar produced – and can make the most of flexibility in the labour force – then biochar production could be potentially cheaper at this scale. Transportation costs would be lower and a convenient route for disposal of some on-farm wastes would open up. On-farm production and use would also avoid the margins of sellers and resellers, which usually double the cost per tonne of biochar. Utilisation of the heat generated during pyrolysis could also find an application in many farm settings. The cost per tonne of biochar could thus come down to below €200 compared with the current market price of €500 to €900 per tonne (see Chapter 9). Furthermore, as the amount of biochar produced per farm and year would be dependent on the biomass surplus of each farm, it would be in direct relation to the farm's carbon balance and potential of photosynthesis.

With the Kon-Tiki flame curtain kilns developed in 2014 in Switzerland (Schmidt and Taylor 2014; Schmidt *et al.* 2015), and rapidly spreading by open source technology transfer to many farmers (in more than 35 countries by summer 2015), a first step towards the democratisation of biochar production may already be taking place. A 1750 litre Kon-Tiki kiln can produce 500kg of biochar (dry matter basis) and close to 2 MWh of heat from shrubs, husks, prunings and other organic farm waste in 2 to 3 hours, though needing one worker. The cost per kiln varies with design and country but is within a range of €50 to €5000, a feasible investment for more or less every farmer.

With low-cost, on-farm biochar production, biochar could be used in various cascades from silage and animal feeding, to bedding and manure management, to waste water treatment and urine recycling, to grain store house building, as well as vegetable and fruit storage. The potential ubiquity of biochar in many farming processes could improve the ecological and carbon footprint, nutrient cycling and overall sustainability of farming. With each year the amount of biochar and soil organic matter in soil could be increased steadily, creating an expanding carbon sink in agricultural soils.

Another, and probably parallel, evolving scenario would be the industrial production of biochar from waste materials. As there are gate fees to be obtained for the disposal and treatment of waste materials such as sewage sludge, floating sea debris, waste wood and urban green waste, the cost of biochar could be lowered when these fees are factored into the overall cost benefit analysis (Shackley *et al.* 2011). However, in order to meet the requirement to be safe and legally acceptable, waste feedstocks and subsequent biochars will require considerable analysis and testing, to reduce contaminants and comply with the EU Waste Framework Directive (WFD) and its 'end-of-waste' criteria (Chapter 10; British Biochar Foundation 2014).

The development of industrial processes with advanced process control and targeted biomass and mineral blending could lead to the design of high performance biochar-mineral fertiliser complexes that may partially replace traditional mineral fertilisers. As Stephen Joseph and co-authors (2013) showed in their influential paper, 'Shifting paradigms: development of high-efficiency biochar fertilizers based on nano-structures and soluble components', the blending of biochar, minerals and fertiliser either during or after the pyrolysis process may enhance fertiliser efficiency. If the agro-chemical industry discovers any advantage in these processes, be it to reduce fertiliser manufacturing costs, improve plant growth responses or plant disease resistance, the blending of chemical fertilisers with biochar could spread rapidly on an industrial level. However, when (as at present) NPK prices are lower than biochar prices on a kg basis, the reduction in fertiliser need would have to be very consequential to become economically beneficial, assuming that N and P additions are not yet subject to regulatory limits to avoid excess run-off.

The European fertiliser regulations are currently undergoing a set of revisions and under discussion is whether to include recycled minerals as a new class of fertiliser, with separate regulations also covering organic and inorganic soil amendment and soil conditioning products (with biochar being explicitly listed as one such new product). Such regulations would be in line with political intentions to pave the way towards a more bio-based and ecologically circular economy. The development of the EBC has led to the first steps being taken in controlling the quality and sustainability of biochar and its production. More specific national-level regulatory frameworks are also being developed, given the particularity of national regulatory frameworks on additions to soil and the use of waste products (such as the Biochar Quality Mandate in the UK context (British Biochar Foundation 2014). Depending on the development of the biochar sector, on-farm biochar production and new pathways of biochar use, the EBC will need to be updated and revised over the next several years. Based on the current results of biochar research, the sustainability of its implementation into the farming and industrial practice does not seem to be a major obstacle if certification standards such as the EBC are respected. However, diligent quality control to prevent exceeding maximum permissible limits (MPLs) for toxicants which cause environmental damage, strict process-based controls and demanding requirements to demonstrate the sustainability of feedstocks, could increase the costs of biochar production in Europe. If those requirements become too excessive the industry might be forced to look for other uses for biochar outside of agriculture or, indeed, at other uses of the spare biomass.

Such other possible uses of biochar include: as a building material, in ceramics, in the paper industry, in electronics such as batteries and solar panels, in textiles and for 3D printing of consumer goods. The cost of meeting environmental regulations is lower in these applications, while margins are likely to be substantially higher. Even if biochar were to be used in much higher quantities as a constituent of industrial products such as these, the production of biochar would still be linked to agriculture. Biochar will remain a direct or indirect product of more or less intensive farming as no other process than photosynthesis, followed by the conversion of biomass to biochar via pyrolysis, is as effective at concentrating atmospheric carbon into processable solids. Moreover, most of the material and industrial products containing biochar may eventually be recycled and returned back into the soil where it will serve as a soil amendment or at least as a soil carbon sink. Biochar is part of the emergence of the bio-economy, whereby a range of industries are increasingly using biological processes and organic carbon to efficiently replace fossil carbon sources.

We are beginning to see an increase in use of biological processes and organic carbon whether that be in soils, in urban structures like houses, roads, tubes and tunnels or in myriad consumer products. However, the underlying condition – and constraint – on this is the global capacity of photosynthesis and how this might change in the future. Only if soils become more fertile, especially the globally increasing surfaces of degraded and deforested land, such that there is an increase in the capacity to nurture and support greater biomass growth, can the new circular bio-based economy be established. Biochar could play a key role in making this transition possible.

On the basis of these reflections, we finish the book by taking a non-scientific, and admittedly speculative, perspective on the future of biochar in five to ten years' time:

- 0.05 per cent of small-scale farmers globally will produce their own biochar and use it mainly to reduce nutrient losses from animal farming and decomposing farm residues (i.e. as part of the process of producing their own fertilisers and improving the on-farm circular economy).
- Coffee, banana, cotton, pineapple and cacao will probably lead the way in the implementation of biochar into farmers' practice as it may address over-fertilisation, salinisation, soil erosion, soil pathogens and soil acidity, which have already had severe impacts on the production of these cash crops.
- The agro-chemical industry will discover the potential of biochar blending with their products and launch first biochar-based fertilisers and then integrated biochar-fertiliser seed products.
- Municipal waste and sewage sludge pyrolysis will start to become a serious option.
- Urban and infrastructure developments will start to include biochar-based materials.
- Electric pyrolysis systems for household waste management and heating will enter the consumer market.

- Biochar paper will appear in hospitals, toilet papers, wallpapers, packaging materials and take-away coffee cups.
- Biochar-based energy storage systems will generate an increasing number of patents.
- 20 per cent of Europeans will know what biochar is.

Bibliography

Blackwell, P., Krull, E., Butler, G., Herbert, A. and Solaiman, Z. (2010). Effect of banded biochar on dryland wheat production and fertiliser use in south-western Australia: an agronomic and economic perspective. *Australian Journal of Soil Research* 48: 531–545.

British Biochar Foundation (2014). *Biochar Quality Mandate.* Available at www.british-biocharfoundation.org/wp-content/uploads/BQM-V1.0.pdf.

Conte, P., Hanke, U.M., Marsala, V., Cimo, G., Alonzo, G. and Glaser, B. (2014). Mechanisms of water interaction with pore systems of hydrochar and pyrochar from poplar forestry waste. *Journal of Agricultural and Food Chemistry* 5(2): 116–121.

EBC (2012). *European Biochar Certificate: Guidelines for a Sustainable Production of Biochar.* Arbaz: European Biochar Foundation. Available at www.european-biochar.org/biochar/media/doc/ebc-guidelines.pdf.

Embren, B. (2013). *The Stockholm Solution.* Available at www.trees.org.uk/aa/documents/amenitydocs/2013_documents/wed_05_Bjorn_Embren-The_Stockholm_Solution.pdf (accessed 10 April 2015).

Fang, Q., Chen, B., Lin, Y. and Guan, Y. (2014). Aromatic and hydrophobic surfaces of wood-derived biochar enhance perchlorate adsorption via hydrogen bonding to oxygen-containing organic groups. *Environ. Sci. Technol.* 48: 279–288.

Glaser, B., Haumaier, L., Guggenberger, G. and Zech, W. (2001). The 'Terra Preta' phenomenon: a model for sustainable agriculture in the humid tropics. *Naturwissenschaften* 88: 37–41.

Graves, D. (2013). A comparison of methods to apply biochar into temperate soils. In N. Ladygina and F. Rineua (eds), *Biochar and Soil Biota,* 202–260. London: CRC Press.

Jassal, R.S., Johnson, M.S., Molodovskaya, M., Black, T.A., Jollymore, A. and Sveinson, K. (2015). Nitrogen enrichment potential of biochar in relation to pyrolysis temperature and feedstock quality. *Journal of Environmental Management* 152: 140–144.

Joseph, S., Graber, E., Chia, C., Munroe, P., Donne, S., Thomas, T., Nielsen, S., Marjo, C., Rutlidge, H., Pan, G., Li, L., Taylor, P., Rawal, A. and Hook, J. (2013). Shifting paradigms: development of high-efficiency biochar fertilizers based on nano-structures and soluble components. *Carbon Manag.* 4: 323–343.

Kammann, C.I., Schmidt, H.P., Messerschmidt, N., Linsel, S., Steffens, D., Müller, C., Koyro, H.-W., Conte, P. and Joseph, S. (2015). Plant growth improvement mediated by nitrate capture in co-composted biochar. *Sci. Rep.* 5: article 11080.

Pignatello, J.J., Uchimiya, M., Abiven, S. and Schmidt, M.W.I. (2015). Evolution of biochar properties in soil. In J. Lehmann and S. Joseph (eds), *Biochar for Environmental Management: Science, Technology and Implementation,* 2nd edn, 195–233. Abingdon: Routledge.

Schmidt, H.P. and Taylor, P. (2014). Kon-Tiki: the democratization of biochar production. *Biochar Journal.* Available at www.biochar-journal.org/en/ct/39 (accessed 14 March 2015).

Schmidt, H-P., Bucheli, T., Kammann, C., Glaser, B., Abiven, S. and Leifeld, J. (2012). The European Biochar Certificate. Available at www.european-biochar.org/en.

Schmidt, H., Pandit, B., Martinsen, V., Cornelissen, G., Conte, P. and Kammann, C. (2015) Fourfold increase in pumpkin yield in response to low-dosage root zone application of urine-enhanced biochar to a fertile tropical soil. *Agriculture* 5: 723–741. Available at http://www.mdpi.com/2077-0472/5/3/723 (accessed 4 December 2015).

Schulz, H., Dunst, G. and Glaser, B. (2013). Positive effects of composted biochar on plant growth and soil fertility. *Agron. Sustain. Dev.* 33: 817–827.

Shackley, S., Hammond, J., Gaunt, J. and Ibarrola, R. (2011). The feasibility and costs of biochar deployment in the UK. *Carbon Manag.* 2: 335–356.

Index

Milton Keynes UK
Ingram Content Group UK Ltd.
UKHW021623071024
449327UK00020BA/1165